CRIMEAN-CONGO HEMORRHAGIC FEVER
A Global Perspective

Crimean-Congo Hemorrhagic Fever

A Global Perspective

edited by

Onder Ergonul

Marmara University, School of Medicine, Istanbul, Turkey

and

Chris A. Whitehouse

U.S. Army Medical Research Institute of Infectious Disease (USAMRIID), Fort Detrick, U.S.A.

Springer

A C.I.P. Catalogue record for this book is available from the Library of Congress.

ISBN 978-1-4020-6105-9 (HB)
ISBN 978-1-4020-6106-6 (e-book)

Published by Springer,
P.O. Box 17, 3300 AA Dordrecht, The Netherlands.

www.springer.com

Printed on acid-free paper

Onder Ergonul would like to dedicate this book to all those who have been affected by Crimean-Congo hemorrhagic fever, in particular, the many health-care workers who have lost their lives while caring for their patients.

Chris A. Whitehouse would like to dedicate this book to Arlone K. Whitehouse (1926–2005).

ACKNOWLEDGMENTS

We thank Marie Johnson and the editorial and production staff at Springer for their hard work in making this project a reality. We are grateful to Sebnem Eren, M.D. for contributing artwork used on the cover. We also thank our families for allowing us to steal precious time away from them to work on this book.

Onder Ergonul and Chris A. Whitehouse

CONTENTS

EDITOR BIOGRAPHIES

Onder Ergonul is an associate professor of Infectious Diseases and Clinical Microbiology at the Marmara University School of Medicine in Istanbul, Turkey. He graduated from Hacettepe University School of Medicine in 1989 in Ankara, Turkey. He completed his residency in 1996 in Infectious Diseases and Clinical Microbiology Department of Ankara University. Concentrating on quantitative methods, he received his Master of Public Health degree from Harvard University School of Public Health in 2003. Between 2000 and 2002, he was a research fellow in the Clinical Epidemiology division of Infectious Diseases Department at the University of Utah, School of Medicine, USA. Between 2003 and 2006, he worked in Ankara Numune Education and Research Hospital in Ankara, Turkey.

Dr. Ergonul is a member of the European Society of Clinical Microbiology and Infectious Diseases (ESCMID), Clinical Microbiology and Infectious Diseases Society of Turkey (KLIMIK). He has authored and coauthored numerous scientific publications about clinical and epidemiologic aspects of infectious diseases. His main research interest is infectious diseases epidemiology, particularly on emerging infections and hospital infections.

Chris A. Whitehouse is a microbiologist in the Diagnostic Systems Division at the US Army Medical Research Institute of Infectious Diseases at Fort Detrick, Maryland, and is an adjunct associate professor in the Department of Microbiology, Immunology, and Tropical Medicine at the George Washington University.

He received his undergraduate degree in Biology from Old Dominion University, Norfolk, Virginia. Dr. Whitehouse holds a Master's degree in Parasitology and a Ph.D. in Microbiology and Immunology, both from the University of Louisville, Kentucky. Between 1997 and 1999, he was a postdoctoral fellow in the Department of Microbiology and Immunology at the University of Kentucky, and from 1999 to 2000, he was a postdoctoral fellow at the Center for Vector-Borne Diseases, University of Rhode Island.

Dr. Whitehouse is a member of the American Society for Microbiology, the American Society of Tropical Medicine and Hygiene, and the American Committee on Arthropod-Borne Viruses. He serves on the Editorial Boards of the *Journal of Clinical Microbiology* and *Molecular and Cellular Probes*.

Dr. Whitehouse has authored and coauthored numerous scientific publications in the fields of virology, vector-borne diseases, and diagnostic microbiology. His main research interests are the ecology, epidemiology, and molecular diagnostics of arboviral and tick-borne diseases.

AUTHOR LIST

Editors:
Onder Ergonul, M.D., M.P.H.
Chris A. Whitehouse, Ph.D.

Authors (in alphabetical order):

Ahmet Aciduman, M.D., Ph.D.
Ankara Etlik İhtisas Hospital
Neurosurgery Clinic
Ankara, Turkey
ahmetaciduman@yahoo.com

Berna Arda, M.D., Ph.D.
Professor of Medical Ethics and History of Medicine
Ankara University, School of Medicine
Department of Medical History and Medical Ethics
Ankara, Turkey
Berna.Arda@medicine.ankara.edu.tr

Tatjana Avšič-Županc, Ph.D.
Professor of Microbiology
Institute of Microbiology and Immunology
Medical Faculty of Ljubljana
Zaloška 4
1000 Ljubljana, Slovenia
tatjana.avsic@mf.uni-lj.si

Zhenqiang Bi, Ph.D.
Biorisk Reduction for Dangerous Pathogens (BDP)
Department of Epidemic and Pandemic Alert Response (CDS/EPR)
World Health Organization
20 Avenue Appia
CH-1211 Geneva 27, Switzerland
biz@who.int

Mike Bray, M.D., M.P.H.
Medical Officer
Biodefense Clinical Research Branch
Division of Clinical Research
NIAID, National Institutes of Health
Room 5128, 6700A Rockledge Dr.
Bethesda, MD 20892
MBray@niaid.nih.gov

Felicity J. Burt, Ph.D.
Department of Virology
Faculty of Health Sciences
University of the Free State
Box 339
Bloemfontein 9300, South Africa
burtFJ.MD@mail.uovs.ac.za

A.M. Butenko, Ph.D.
Head, Department of Arboviruses
Ivanovsky Institute of Virology
Russian Academy of Medical Sciences
Moscow, Russia
arboelisa@mail.ru

Sadegh Chinikar, Ph.D.
Head of Laboratory of Arboviruses and Viral Haemorrhagic
Fevers National Center
Pasteur Institute of Iran
Tehran, Iran
sadeghchinikar@yahoo.com

Ben S. Cooper, Ph.D.
Statistics, Modelling and Bioinformatics Department
Centre for Infections
Communicable Disease Surveillance Centre

Health Protection Agency
London, UK
ben.cooper@hpa.org.uk

Dimiter S. Dimitrov, Ph.D, ScD
Senior Investigator
Protein Interactions Group, CCR Nanobiology Program
Center for Cancer Research
NCI-Frederick, NIH
Building 469, Room 105
P.O. Box B, Miller Drive
Frederick, MD 21702-1201
dimitrov@ncifcrf.gov

Onder Ergonul, M.D., M.P.H.
Associate Professor
Infectious Diseases and Clinical Microbiology Department
Marmara University, School of Medicine
Altunizade, Istanbul, Turkey
onderergonul@yahoo.com

Agustin Estrada-Pena, Ph.D.
Professor
Department of Parasitology
Veterinary Faculty
Miguel Servet 177
50013 Zaragoza, Spain
estrada.agustin@gmail.com

Ramon Flick, Ph.D.
BioProtection Systems Corporation
2901 South Loop Drive
Suite 3360, Bldg. 3
Ames, IA 50010-8646, USA
rflick@bpsys.net

Pierre Formenty, DVM, Ph.D.
Biorisk Reduction for Dangerous Pathogens Team (BDP)
Department of Epidemic and Pandemic Alert Response (CDS/EPR)
World Health Organization
20 Avenue Appia
CH-1211 Geneva 27, Switzerland
formentyp@who.int

Fernando Gonzalez-Martin, Ph.D.
International Health Regulations (IHR)
Department of Epidemic and Pandemic Alert and Response (CDC/EPR)
World Health Organization
20 Avenue Appia
CH-1211 Geneva 27, Switzerland
gonzalezmartininf@who.int

Roger Hewson, Ph.D.
Virus Research
Novel and Dangerous Pathogens
Centre for Emergency Preparedness and Response
Health protection Agency
Porton Down
Salisbury, SP4 0JG
England, UK
Roger.Hewson@hpa.org.uk

G.G. Karganova, Ph.D.
M.P. Chumakov Institute of Poliomyelitis and Viral Encephalitides
Russian Academy of Medical Sciences
Moscow, Russia

Ali Mirazimi, Ph.D.
Head of BSL-4 Research Program
Center for Microbiological Preparedness
Swedish Institute for Infectious Disease Control
Solna, Sweden
ali.mirazimi@smi.ki.se

Aysegul Nalca, M.D., Ph.D.
Center for Aerobiological Sciences
United States Army Institute of Infectious Diseases (USAMRIID)
Fort Detrick, Frederick, MD 21702-5011, USA
aysgeul.nalca@amedd.army.mil

Janusz T. Paweska, DVSc
Head of Special Pathogens Unit
National Institute for Communicable Diseases
Private Bag X4
Sandringham 2131, South Africa
januszp@nicd.ac.za

Sarah E. Randolph, Ph.D.
Professor of Parasite Ecology
Department of Zoology
University of Oxford, South Parks Road
Oxford OX1 3PS, UK
sarah.randolph@zoo.ox.ac.uk

David J. Rogers, Ph.D.
Department of Zoology
University of Oxford, South Parks Road
Oxford OX1 3PS, UK
david.rogers@zoo.ox.ac.uk

Masayuki Saijo, M.D., Ph.D.
Department of Virology 1
National Institute of Infectious Diseases
4-7-1 Gakuen, Musashimurayama
Tokyo 208-0011, Japan
msaijo@nih.go.jp

Glenn Schnepf, Ph.D.
Medical Officer
CDS/EPR/ARO/EDP
World Health Organization
20 Avenue Appia
CH-1211 Geneva 27, Switzerland
schnepfg@who.int

Robert Swanepoel, DTVM, Ph.D.
Special Pathogens Unit
National Institute for Communicable Diseases
Private Bag X4
Sandringham 2131, South Africa
bobs@nicd.ac.za

Arnaud Tarantola, M.D., M.S.
International and Tropical Department
Institut National de Veille Sanitaire, Paris, France
12 rue du Val d'Osne
94415 Saint-Maurice, Cedex, France
a.tarantola@invs.sante.fr

Irina Tarasevich, Ph.D.
Professor of Microbiology
Academy of Medical Science RF
Department of Epidemiology
Laboratory of Rickettsial Ecology of Gamaleya Institute
Director of WHO collaborating Center for Rickettial Reference and Research
Moscow, Russia
taras@postman.ru

Pierre Tattevin, M.D.
Infectious Diseases and Reanimation Clinic
Pontchaillou University Hospital
rue Le Guilloux, 35033 Rennes, Cedex, France
pierre.tattevin@chu-rennes.fr

Michael J. Turell, Ph.D.
Research Entomologist
Virology Division
USAMRIID
1425 Porter Street
Fort Detrick, MD 27702-5011, USA
Michael.Turell@amedd.army.mil

Ramazan Uzun, DVM, Ph.D.
Public Health Officer
Communicable Diseases Unit
Ministry of Health
Ankara, Turkey
ramazanuzun@yahoo.com

Zati Vatansever, Ph.D.
Associate Professor
Department of Parasitology
Faculty of Veterinary Medicine
Ankara University
Ankara, Turkey
zativet@gmail.com

Chris A. Whitehouse, Ph.D.
Microbiologist
Diagnostic Systems Division
USAMRIID
Fort Detrick, Frederick, MD 21702-5011, USA
chris.whitehouse@amedd.army.mil

John P. Woodall, Ph.D.
Director, Nucleus for Investigating Emerging Infectious Diseases
Institute of Medical Biochemistry
Center for Health Sciences
Federal University of Rio de Janeiro
Brazil
woodall@promedmail.org

Hervé Zeller, Ph.D.
Microbiologist
Unité de Biologie des Infections Virales Emergentes
Institut Pasteur
21 avenue Tony Garnier, 69365
Lyon, Cedex 07, France
zeller@cervi-lyon.inserm.fr

PREFACE

It has been over 63 years since the first recognized outbreak of a previously unknown human hemorrhagic disease erupted in war-torn Crimea. The disease, initially called Crimean hemorrhagic fever, is now referred to as Crimean-Congo hemorrhagic fever (CCHF) and the causative agent was eventually found to be a new tick-borne arbovirus, CCHF virus (CCHFV). Since that time, CCHF has become one of the most geographically widely distributed tick-borne diseases in the world; the disease, or the presence of the virus, has been reported from at least 31 countries in Africa, Asia, southeast Europe, and the Middle East.

In 1979, the eminent tick biologist, Harry Hoogstraal, published his encyclopedic review on the ecology and epidemiology of CCHF. Since then, there have been tremendous developments in the knowledge of CCHF, including improved detection, characterization, and identification tools for the causative virus. In addition, new information is beginning to come to light on the pathogenesis of disease and potentially new methods to treat the disease. Also, the disease is occurring in places where it had not previously occurred. For example, there have been over 1,000 human cases in Turkey since 2002, when the first cases in that country were reported.

Crimean-Congo Hemorrhagic Fever: A Global Perspective is the first book written which is specifically devoted to CCHF. Our aim in writing this book was to present updated information on several key aspects of the disease and the virus which causes it. CCHF is a global disease, and writing this book was truly an international effort. In total, there are 34 authors from at least 13 different countries who have contributed to this book. These authors are leading scientists in their fields and provide a global perspective on this global disease. There are six main sections in this book. Section I gives an introduction to CCHF and presents a historical background on the disease. Section II discusses the molecular biology and genetics of CCHFV. Section III is the largest section and discusses the epidemiology and ecology of CCHF, including several chapters focusing on CCHF in individual countries. Section IV describes what is currently known about the clinical features, pathogenesis, and treatment options for CCHF. Section V provides information on the prevention and control of CCHF. Section VI discusses potential areas for future research and includes frequently asked questions that may prove helpful to public health officials or others, who may not be experts on CCHF, but nevertheless may be required to provide information on the disease to the public. We hope that this book will shed light on this viral hemorrhagic disease and provide useful information for all the clinicians, virologists, entomologists, and public health officials who deal with this important disease.

Onder Ergonul, Istanbul, Turkey
Chris A. Whitehouse, Frederick, Maryland, USA

SECTION I

INTRODUCTION AND HISTORICAL PERSPECTIVES

CHAPTER 1

INTRODUCTION

ONDER ERGONUL, M.D., M.P.H.[1] AND CHRIS A. WHITEHOUSE, PH.D.[2]*

[1]*Marmara University, School of Medicine, Infectious Diseases and Clinical Microbiology Department, Istanbul, Turkey. E-mail: ondererergonul@yahoo.com*

[2]*Diagnostic Systems Division, United States Army Institute of Infectious Diseases (USAMRIID), Fort Detrick, Frederick, MD 21702-5011, USA. E-mail: chris.whitehouse@amedd.army.mil*

Crimean-Congo hemorrhagic fever (CCHF) is a fatal viral infection described from parts of Africa, Asia, Eastern Europe, and the Middle East [15, 20, 50, 52]. The CCHF virus (CCHFV) belongs to the genus *Nairovirus* in the family Bunyaviridae and causes a severe disease in humans, with a reported mortality rate of 3–30% [15]. The geographic range of CCHFV is the most extensive of the medically significant tick-borne viruses. Humans become infected through the bites of ticks, by contact with a patient with CCHF during the acute phase of infection, or by contact with blood or tissues from viremic livestock [52].

The widespread geographical distribution of CCHFV, its ability to produce severe human disease with high fatality rates, and fears about its intentional use as a bioterrorism agent [7] (Box 1-1) makes CCHFV a significant human pathogen. Moreover, ecological complexity of vector-borne diseases, therapeutic controversy, and human-to-human transmission of a zoonotic infection make it a highly challenging research topic.

There has been a significant increase in the number of published articles on CCHFV during the last 5 years. There have been over 400 publications on CCHF, with almost half being published after 2000. The majority of these are in English, but many significant studies were conducted and published in Russian by Soviet researchers since the 1940s.

The geographic range of CCHFV is known to be the most extensive of the tick-borne viruses important to human health, and the second most widespread of all medically important arboviruses, after dengue viruses [49]. The history of reported outbreaks or epidemics (Box 1-2) is presented in Table 1-1. Before 1970,

*The views, opinions, and findings contained herein are those of the authors and should not be construed as an official Department of the Army position, policy, or decision unless so designated by other documentation.

O. Ergonul and C. A. Whitehouse (eds.), Crimean-Congo Hemorrhagic Fever, 3–11.

Box 1-1. Bioterrorism and viral hemorrhagic fevers

According to the National Institute of Allergy and Infectious Disease of the National Institutes of Health in the USA, hemorrhagic fever viruses are categorized as "Category A" bioterrorism agents. These include the arenaviruses (Junin virus, Machupo virus, Guanarito virus, and Lassa fever virus), the bunyaviruses (Hantaviruses and Rift Valley fever), the flaviruses (Dengue), and filoviruses (Ebola and Marburg). However, CCHFV is currently classified as a Category C agent [7].

the majority of the cases had been reported from the former Soviet Union (Crimea, Astrakhan, Rostov, Uzbekistan, Kazakhstan, and Tadzhikistan), Bulgaria, as well as virus circulation in parts of Africa such as the Democratic Republic of Congo and Uganda [20, 41, 49, 54]. The first outbreak in China was reported in 1965 [55]. The awareness of hemorrhagic cases in Africa increased in the 1960s resulting in a series of in-depth studies from the Republic of South Africa [18, 44–46]. Outbreaks in Africa were reported from the Congo and Tanzania [45], Mauritania [35], Burkina Faso [34], and Senegal [8]. Since the late 1970s, outbreaks have been reported in several Middle Eastern countries including Iraq [1, 47], Pakistan [5], United Arab Emirates [4, 37, 43], Saudi Arabia [13], Sultanate of Oman [53], and in western parts of China [29]. Since 2000, new outbreaks have been reported from Pakistan [3, 39, 42], Iran [24], Senegal [26], Albania [28], Yugoslavia [10, 28], Bulgaria [30], Turkey [15], Kenya [12], and Mauritania [25]. In some of the countries, only serologic evidence of CCHFV infection exists. These include India [38], Egypt [9], Portugal [17], Hungary [21], France [20], and Benin [20]. In Greece, there is serological evidence of infection among humans [2], and a single viral isolate (AP92) made from *Rhipicephalus bursa* ticks collected from goats in and near Thessaloniki in 1975 [31]. Interestingly, this strain is the most divergent of all the CCHFV strains including those isolated in the neighboring Balkan countries of Albania, Kosovo, and Bulgaria [30]. Mountains approximately 1,500–2,500 m high separate these countries from Greece, and it has been suggested that this might have resulted in the genetic isolation of CCHFV in Greece [30]. However, whether the genetic differences between AP92 and other CCHFV strains is the result of different tick species, different geographic location, or another reason, remains to be elucidated.

Box 1-2. What is an epidemic?

An epidemic is defined as the occurrence in a community or region of cases of an illness, specific health-related behavior, or health-related events which clearly exceed the normal expectancy [23]. The area and the time in which the cases occur are precisely specified. An epidemic is relative to the usual frequency of the disease in the same area, among the specified population, at the same time of the year. A single case of a communicable disease long absent from a population, or first invasion by a disease not previously recognized may constitute an epidemic. Conversely, multiple cases of some communicable diseases may present only in the endemic level [51].

Table 1-1. Reported outbreaks of CCHF since 1945–2006 [1, 2, 5, 12, 13, 15, 20, 24, 27–30, 34, 37, 39, 42, 43, 47, 49, 53]

Geographical location	Years	No. of cases	Case fatality rate (%)	Occupation
Southeast Europe				
Crimea	1944–1945	200	10	Military members
Astrakhan	1953–1963	104	17	Agricultural workers
Rostov	1963–1969	323	15	Agricultural workers
Bulgaria	1953–1974	1105	17	Agricultural workers, HCWs
	1975–1996	279	11	Agricultural workers
	1997–2003	138	21	Agricultural workers
Albania	2001	7	Survived	Agricultural workers, HCWs
Kosovo	2001	18	33	Agricultural workers
Turkey	2002–2005	1103	5	Agricultural workers, HCWs
Asia				
China	1965–1994	260	21	Agricultural workers
	1997	26	24	Agricultural workers
Kazakhstan	1948–1968	75	50	Agricultural workers
Tadzhikistan	1943–1970	97	23	Agricultural and laboratory workers
Pakistan	1976	14	29	Shepherd, HCWs
	1994	3	Unknown	HCWs
	2000	9	55	Agricultural workers, HCWs
Middle East				
United Arab Emirates	1979	6	50	HCWs
	1994–1995	11	73	Agricultural workers
Sharjah	1980	1	Survived	Storekeeper
Iraq	1979–1980	55	64	Agricultural workers
Saudi Arabia	1990	7		Agricultural workers
Sultanate of Oman	1995–1996	4	Unknown	Agricultural workers
Iran	2003	81	18	Agricultural workers
Africa				
Zaire (Congo)	1956	2	Survived	Physician
Uganda	1958–1977	12	8	Laboratory workers
Mauritania	1983	1	Survived	Camel herd owner
	2004	38	29	Agricultural workers, HCWs
Burkina Faso	1983	1	Survived	Unknown
Republic of South Africa	1981–1986	32	31	Farmers, HCWs
Tanzania	1986	1	Survived	Student
Southwest Africa	1986	1	Survived	Unknown
Kenya	2000[48]	1	Died	Agricultural worker

HCWs: Health-care workers

CCHFV is a member of the genus *Nairovirus* of the family *Bunyaviridae*. Like other bunyaviruses, CCHFV is an enveloped particle with a single-stranded RNA genome of negative polarity [52]. Other genera within the family include *Orthobunyavirus*, *Hantavirus*, *Phlebovirus*, and *Tospovirus*. The genus *Nairovirus* includes 34 described viruses and is divided into seven serogroups [48]. The most important groups are the CCHF group, which includes CCHFV, Hazara virus, and the Nairobi sheep disease group, which includes Nairobi sheep disease and Dugbe viruses [52]. Only three members of *Nairovirus* genus are known to cause disease among humans: CCHFV, Dugbe virus, and Nairobi sheep disease virus.

Using serological methods, CCHFV from various geographic regions were originally thought to be very similar. However, the development of nucleic acid sequence analysis techniques has revealed extensive genetic diversity. The majority of nucleic acid sequence analyses are based on fragments of the S-segment RNA; however, recently whole genome analyses have been performed on a number of strains. Phylogenetic analysis of strains responsible for outbreaks has been very important to the understanding of transmission dynamics of the virus.

CCHFV circulates in the nature in an enzootic tick–vertebrate–tick cycle, and there is no evidence that the virus causes disease in animals other than humans and newborn mice. CCHFV infection has been demonstrated more commonly among smaller wildlife species such as hares and hedgehogs that act as hosts for the immature stages of the tick vectors [20, 49]. On the other hand, antibodies against CCHFV have been detected in the sera of horses, donkeys, goats, cattle, sheep, and pigs in various regions of Europe, Asia, and Africa [49]. It must be borne in mind that antibody studies, particularly if the prevalence is low, are not as meaningful as obtaining actual virus isolates. Although ground-feeding birds have not been shown to produce detectable viremia [56], birds may play a role in the transportation of CCHFV-infected ticks. The potential roles of migratory birds and the movement of livestock carrying ticks in the spread of the virus over distant geographical areas have been described [20, 22, 33]. However, there is no precise data on the role of birds and bird-parasitizing ticks in CCHF in the literature. This is a wide open area for future research.

CCHFV are transmitted by *Hyalomma* genus ticks, particularly by *Hyalomma marginatum marginatum*. These are two-host ticks with the larvae feeding on small mammals and ground-feeding birds, and the adults feeding on large mammals such as livestock and wild boar (Fig. 1-1). CCHFV was first isolated from adult *Hyalomma* ticks in the 1960s [20, 49]. Viral isolates were also obtained from field-collected eggs and unfed immature stages of *H. marginatum marginatum*, demonstrating evidence of transovarial (from infected mother through the eggs), and transstadial (i.e. from larvae to nymph to adult) transmissions [49]. The known occurrence of CCHF in Europe, Asia, and Africa coincides with the world distribution of ticks of the genus *Hyalomma* [20, 49]. *H. marginatum marginatum* is also known as the Mediterranean *Hyalomma*, and it may be the main vector of CCHFV in Europe. CCHFV has also been isolated from *Hyalomma anatolicum anatolicum* and other *Hyalomma* spp. Isolations from other tick genera, such as *Rhipicephalus*, *Ornithodoros*, *Boophilus*, *Dermacentor*, and *Ixodes* spp. [20, 49] may be locally significant.

Changes in climatic conditions have been suggested to be one of the factors that could facilitate the reproduction of ticks (as well as other arthropod

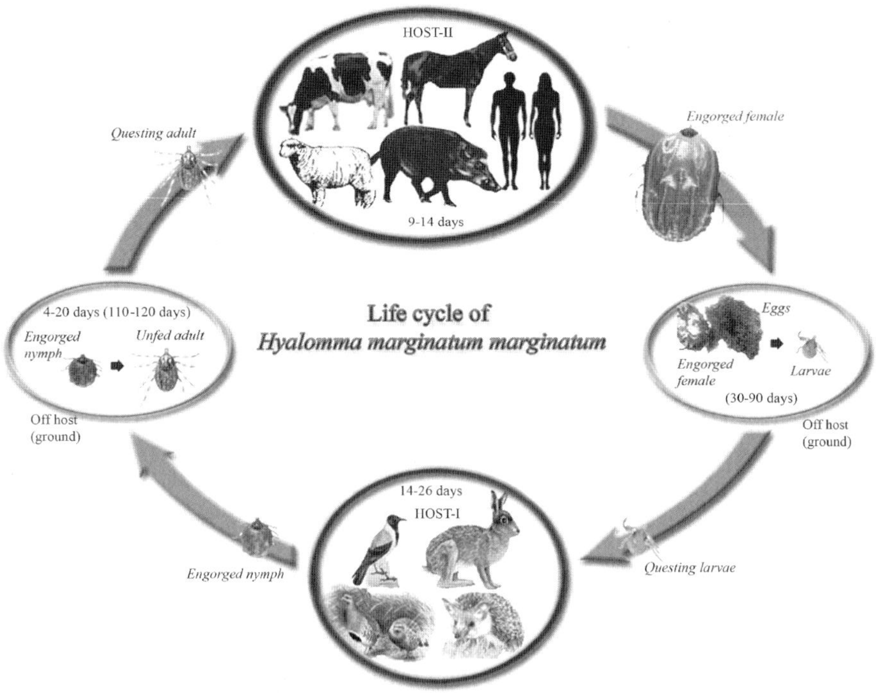

Fig. 1-1. Life cycle of *Hyalomma marginatum marginatum* ticks (Courtesy of Dr. Zati Vatansever). (*See Color Plates*)

vectors), and consequently, increase the overall incidence of tick-borne diseases [16]. In the northern hemisphere, *H. marginatum marginatum* is usually activated by the increasing temperature in the spring (usually beginning in April), and the immature stages are active in summer between May and September [20]. The CCHF outbreak in Turkey can be illustrative. For example, the number of days with the temperature ≥5°C in April, and the daily mean temperature in April, in the outbreak region of Turkey were reported to be increased in the years prior to the first human cases [14]. However, the climate change to date is not necessarily the cause of the marked increased incidence of a variety of tick-borne diseases in many parts of Europe over the last two decades [32]. In general, CCHF outbreaks have developed on a background of favorable climatic factors and environmental changes beneficial for survival of large numbers of *Hyalomma* ticks and of the hosts of both their immature and adult stages [20]. In the former Soviet Union, environmental changes included wartime neglect of agricultural lands, introduction of susceptible military personnel, or new settlers into infected area, wide-scale changes in agricultural practices, converting floodplains into farmland, and flood control [20]. During World War II, after the enemy occupation of the Crimea (1941–1944), normal agricultural activities were disrupted and the common sport of hunting European hares was abandoned. When Soviet troops reoccupied the hilly Crimean steppes in 1944, hares become excessively abundant and neglected pastures were overgrown by weeds, and the first CCHF outbreak in modern era was documented [20]. Interestingly, a similar explanation was suggested from Turkey. According to this scenario [15], prior to the start of the CCHF outbreak in 2002, the fields at the region had been abandoned from hunting and pasturing between 1995 and 2001, because of terrorist activities in the region. During this period, numbers of small mammals such as hares, and wild animals such as boars increased. After 2001, the fields became unrestricted for hunting and pasturing so that cattle and sheep were exposed to potentially infected tick populations [15].

Epidemiologically, CCHF cases are distributed mainly among those who work outdoors and exposed to ticks. The great majority of the affected cases were persons who work in agriculture and/or husbandry [15]. Health-care workers (HCWs) are the second most commonly affected group. HCW are under serious risk of infection, particular those who directly care for CCHF patients with frank hemorrhaging.

CCHF is characterized by a sudden onset of high fever, chills, severe headache, dizziness, and back and abdominal pains. Additional symptoms include nausea, vomiting, diarrhea, neuropsychiatric, and cardiovascular changes. In severe cases, hemorrhagic manifestations, such as petechiae and/or ecchymosis, can develop. The highest case fatality rate (CFR) of 80% was reported from China [55], and the lowest (5%) was reported from Turkey. The possible reasons for such diversity in CFR are discussed in Chapter 16 by Ergonul. The pathogenesis of CCHF is not well understood. However, a common pathogenic feature of viral hemorrhagic fever viruses is their ability to disable the host immune response by attacking and manipulating the cells that initiate the antiviral response [19].

Early diagnosis is critical for the patients and potential nosocomial infections, and for the prevention of the transmission in the community. Suspected cases should be evaluated to plan the management, which includes supportive care, particularly hematological support. The differential diagnosis will differ depending on the particular geographic region. Laboratory diagnosis includes the reverse transcription polymerase chain reaction (RT-PCR), which is the method of choice for the rapid detection of CCHFV RNA [11]. This method is highly specific, sensitive, and rapid [36]. Immunoglobulin M (IgM) and Immunoglobulin G (IgG) antibodies are detectable by enzyme-linked immunosorbent assay (ELISA) and immunofluorescent antibody (IFA) from about 7 days after the onset of disease [40]. ELISA methods are quite specific and much more sensitive than IFA and neutralization tests [6].

Supportive therapy is an essential part of the case management for CCHF patients. It includes the administration of thrombocytes, fresh frozen plasma, and erythrocyte preparations. The antiviral drug, ribavirin, has been found effective against the CCHFV in vitro [50]; although its exact mechanism of action against the virus is unclear. Although its clinical use in CCHF is controversial, it is the only antiviral drug currently available. The benefits of ribavirin treatment have not been examined under the strict conditions of a randomized clinical trial, and the drug is not approved for the treatment of CCHF by the US Food and Drug Administration. Emerging data on viral replication has a significant potential for the development of new drugs, and will be discussed in greater detail in Chapter 19.

Without an effective vaccine against CCHF and few treatment options, prevention against infection is key. Various preventive measures are detailed in Chapter 20 by Whitehouse. In addition, a list of frequently asked questions (FAQ) has been compiled and is listed at the end of the book.

REFERENCES

1. Al-Tikriti SK, Al-Ani F, Jurji FJ, Tantawi H, Al-Moslih M, Al-Janabi N, Mahmud MI, Al-Bana A, Habib H, Al-Munthri H, Al-Janabi S, K AL-J, Yonan M, Hassan F, Simpson DI (1981) Congo-Crimean haemorrhagic fever in Iraq. Bull World Health Organ 59:85–90
2. Antoniadis A, Casals J (1982) Serological evidence of human infection with Congo-Crimean hemorrhagic fever virus in Greece. Am J Trop Med Hyg 31:1066–1067
3. Athar MN, Baqai HZ, Ahmad M, Khalid MA, Bashir N, Ahmad AM, Balouch AH, Bashir K (2003) Short report: Crimean-Congo hemorrhagic fever outbreak in Rawalpindi, Pakistan, February 2002. Am J Trop Med Hyg 69:284–287
4. Baskerville A, Satti A, Murphy FA, Simpson DI (1981) Congo-Crimean haemorrhagic fever in Dubai: histopathological studies. J Clin Pathol 34:871–874
5. Burney MI, Ghafoor A, Saleen M, Webb PA, Casals J (1980) Nosocomial outbreak of viral hemorrhagic fever caused by Crimean hemorrhagic fever-Congo virus in Pakistan, January 1976. Am J Trop Med Hyg 29:941–947
6. Burt FJ, Leman PA, Abbott JC, Swanepoel R (1994) Serodiagnosis of Crimean-Congo haemorrhagic fever. Epidemiol Infect 113:551–562
7. CDC (2005) Bioterrorism http://wwwbtcdcgov/Agent/Agentlistasp
8. Chapman LE, Wilson ML, Hall DB, LeGuenno B, Dykstra EA, Ba K, Fisher-Hoch SP (1991) Risk factors for Crimean-Congo hemorrhagic fever in rural northern Senegal. J Infect Dis 164:686–692
9. Darwish MA, Imam IZ, Omar FM, Hoogstraal H (1978) Results of a preliminary seroepidemiological survey for Crimean-Congo hemorrhagic fever virus in Egypt. Acta Virol 22:77
10. Drosten C, Minnak D, Emmerich P, Schmitz H, Reinicke T (2002) Crimean-Congo hemorrhagic fever in Kosovo. J Clin Microbiol 40:1122–1123
11. Drosten C, Kummerer BM, Schmitz H, Gunther S (2003) Molecular diagnostics of viral hemorrhagic fevers. Antiviral Res 57:61–87
12. Dunster L, Dunster M, Ofula V, Beti D, Kazooba-Voskamp F, Burt F, Swanepoel R, DeCock KM (2002) First documentation of human Crimean-Congo hemorrhagic fever, Kenya. Emerg Infect Dis 8:1005–1006
13. el-Azazy OM, Scrimgeour EM (1997) Crimean-Congo haemorrhagic fever virus infection in the western province of Saudi Arabia. Trans R Soc Trop Med Hyg 91:275–278
14. Ergonul O, Akgunduz S, Kocaman I, Vatansever Z, Korten V (2005) Changes in temperature and the Crimean-Congo hemorrhagic fever outbreak in Turkey. In: 15th European Congress of Clinical Microbiology and Infectious Diseases, Copenhagen. Clin Microbiol Infect, p 360
15. Ergonul O (2006) Crimean-Congo haemorrhagic fever. Lancet Infect Dis 6:203–214
16. Estrada-Pena A (2001) Forecasting habitat suitability for ticks and prevention of tick-borne diseases. Vet Parasitol 98:111–132
17. Filipe AR, Calisher CH, Lazuick J (1985) Antibodies to Congo-Crimean haemorrhagic fever, Dhori, Thogoto and Bhanja viruses in southern Portugal. Acta Virol 29:324–328

18. Gear JH, Thomson PD, Hopp M, Andronikou S, Cohn RJ, Ledger J, Berkowitz FE (1982) Congo-Crimean haemorrhagic fever in South Africa. Report of a fatal case in the Transvaal. S Afr Med J 62:576–580
19. Geisbert TW, Jahrling PB (2004) Exotic emerging viral diseases: progress and challenges. Nat Med 10:S110–121
20. Hoogstraal H (1979) The epidemiology of tick-borne Crimean-Congo hemorrhagic fever in Asia, Europe, and Africa. J Med Entomol 15:307–417
21. Horvath LB (1976) Precipitating antibodies to Crimean haemorrhagic fever virus in human sera collected in Hungary. Acta Microbiol Acad Sci Hung 23:331–335
22. Khan AS, Maupin GO, Rollin PE (1997) An outbreak of Crimean-Congo hemorrhagic fever in the United Arab Emirates, 1994–1995. Am J Trop Med Hyg 57:519–525
23. Last JM (1995) A Dictionary of Epidemiology. Oxford University Press, New York
24. Mardani M, Jahromi MK, Naieni KH, Zeinali M (2003) The efficacy of oral ribavirin in the treatment of Crimean-Congo hemorrhagic fever in Iran. Clin Infect Dis 36:1613–1618
25. Nabeth P, Cheikh DO, Lo B, Faye O, Vall IO, Niang M, Wague B, Diop D, Diallo M, Diallo B, Diop OM, Simon F (2004) Crimean-Congo hemorrhagic fever, Mauritania. Emerg Infect Dis 10:2143–2149
26. Nabeth P, Thior M, Faye O, Simon F (2004) Human Crimean-Congo hemorrhagic fever, Senegal. Emerg Infect Dis 10:1881–1882
27. Papa A, Bino S, Llagami A, Brahimaj B, Papadimitriou E, Pavlidou V, Velo E, Cahani G, Hajdini M, Pilaca A, Harxhi A, Antoniadis A (2002) Crimean-Congo hemorrhagic fever in Albania, 2001. Eur J Clin Microbiol Infect Dis 21:603–606
28. Papa A, Bozovi B, Pavlidou V, Papadimitriou E, Pelemis M, Antoniadis A (2002) Genetic detection and isolation of Crimean-Congo hemorrhagic fever virus, Kosovo, Yugoslavia. Emerg Infect Dis 8:852–854
29. Papa A, Ma B, Kouidou S, Tang Q, Hang C, Antoniadis A (2002) Genetic characterization of the M RNA segment of Crimean-Congo hemorrhagic fever virus strains, China. Emerg Infect Dis 8:50–53
30. Papa A, Christova I, Papadimitriou E, Antoniadis A (2004) Crimean-Congo hemorrhagic fever in Bulgaria. Emerg Infect Dis 10:1465–1467
31. Papadopoulos O, Koptopoulos G (1980) Crimean-Congo hemorrhagic fever (CCHF) in Greece: isolation of the virus from *Rhipicephalus bursa* ticks and a preliminary serological survey. Zentbl Bakteriol Hyg 1:189–193
32. Randolph SE (2004) Evidence that climate change has caused 'emergence' of tick-borne diseases in Europe? Int J Med Microbiol 293 (Suppl 37):5–15
33. Rodriguez LL, Maupin GO, Ksiazek TG, Rollin PE, Khan AS, Schwarz TF, Lofts RS, Smith JF, Noor AM, Peters CJ, Nichol ST (1997) Molecular investigation of a multisource outbreak of Crimean-Congo hemorrhagic fever in the United Arab Emirates. Am J Trop Med Hyg 57:512–518
34. Saluzzo JF, Digoutte JP, Cornet M, Baudon D, Roux J, Robert V (1984) Isolation of Crimean-Congo haemorrhagic fever and Rift Valley fever viruses in Upper Volta. Lancet 1:1179
35. Saluzzo JF, Digoutte JP, Camicas JL, Chauvancy G (1985) Crimean-Congo haemorrhagic fever and Rift Valley fever in south-eastern Mauritania. Lancet 1:116
36. Schwarz TF, Nsanze H, Longson M, Nitschko H, Gilch S, Shurie H, Ameen A, Zahir AR, Acharya UG, Jager G (1996) Polymerase chain reaction for diagnosis and identification of distinct variants of Crimean-Congo hemorrhagic fever virus in the United Arab Emirates. Am J Trop Med Hyg 55:190–196
37. Schwarz TF, Nsanze H, Ameen AM (1997) Clinical features of Crimean-Congo haemorrhagic fever in the United Arab Emirates. Infection 25:364–367
38. Shanmugam J, Smirnova SE, Chumakov MP (1976) Presence of antibody to arboviruses of the Crimean Haemorrhagic Fever-Congo (CHF-Congo) group in human beings and domestic animals in India. Indian J Med Res 64:1403–1413

39. Sheikh AS, Sheikh AA, Sheikh NS, Rafi US, Asif M, Afridi F, Malik MT (2005) Bi-annual surge of Crimean-Congo haemorrhagic fever (CCHF): a five-year experience. Int J Infect Dis 9:37–42
40. Shepherd AJ, Swanepoel R, Leman PA (1989) Antibody response in Crimean-Congo hemorrhagic fever. Rev Infect Dis 11 (Suppl 4):S801–806
41. Simpson DIH, Knight EM, Courtois G, Williams MC, Weinbern MP, Kibukamusoke JW (1967) Congo virus: a hitherto undescribed virus occurring in Africa, human isolations and clinical notes. East Afr Med J 44:86–92
42. Smego RA, Jr., Sarwari AR, Siddiqui AR (2004) Crimean-Congo hemorrhagic fever: prevention and control limitations in a resource-poor country. Clin Infect Dis 38:1731–1735
43. Suleiman MN, Muscat-Baron JM, Harries JR, Satti AG, Platt GS, Bowen ET, Simpson DI (1980) Congo/Crimean haemorrhagic fever in Dubai. An outbreak at the Rashid Hospital. Lancet 2:939–941
44. Swanepoel R, Struthers JK, Shepherd AJ, McGillivray GM, Nel MJ, Jupp PG (1983) Crimean-Congo hemorrhagic fever in South Africa. Am J Trop Med Hyg 32:1407–1415
45. Swanepoel R, Shepherd AJ, Leman PA, Shepherd SP, McGillivray GM, Erasmus MJ, Searle LA, Gill DE (1987) Epidemiologic and clinical features of Crimean-Congo hemorrhagic fever in southern Africa. Am J Trop Med Hyg 36:120–132
46. Swanepoel R, Gill DE, Shepherd AJ, Leman PA, Mynhardt JH, Harvey S (1989) The clinical pathology of Crimean-Congo hemorrhagic fever. Rev Infect Dis 11 (Suppl 4): S794–800
47. Tantawi HH, Al-Moslih MI, Al-Janabi NY, Al-Bana AS, Mahmud MI, Jurji F, Yonan MS, Al-Ani F, Al-Tikriti SK (1980) Crimean-Congo haemorrhagic fever virus in Iraq: isolation, identification and electron microscopy. Acta Virol 24:464–467
48. van Regenmortel MHV, Fauquet CM, Bishop DML, Carstens EB, Estes MK, Lemon SM, Maniloff J, Mago MA, McGeoch DJ, Pringle CR, Wicknen RB (2000) 7th report of the International Committee of Taxonomy of Viruses. Virus Taxonomy, pp 599–621
49. Watts DM, Ksiasek TG, Linthicum KJ, Hoogstraal H (1988) Crimean-Congo hemorrhagic fever. In: Monath TP (ed.) The Arboviruses: Epidemiology and Ecology. CRC Press, Boca Raton, FL
50. Watts DM, Ussery MA, Nash D, Peters CJ (1989) Inhibition of Crimean-Congo hemorrhagic fever viral infectivity yields in vitro by ribavirin. Am J Trop Med Hyg 41:581–585
51. Weber DJ, Menajovsky B, Wenzel R (2001) Investigations of Outbreaks. In: Thomas JC, Weber DJ (eds) Epidemiologic Methods for the Study of Infectious Diseases. Oxford University Press, New York, pp 291–310
52. Whitehouse CA (2004) Crimean-Congo hemorrhagic fever. Antiviral Res 64:145–160
53. Williams RJ, Al-Busaidy S, Mehta FR, Maupin GO, Wagoner KD, Al-Awaidy S, Suleiman AJ, Khan AS, Peters CJ, Ksiazek TG (2000) Crimean-Congo haemorrhagic fever: a seroepidemiological and tick survey in the Sultanate of Oman. Trop Med Int Health 5:99–106
54. Woodall JP, Williams MC, Simpson DI (1967) Congo virus: a hitherto undescribed virus occurring in Africa. II. Identification studies. East Afr Med J 44:93–98
55. Yen YC, Kong LX, Lee L, Zhang YQ, Li F, Cai BJ, Gao SY (1985) Characteristics of Crimean-Congo hemorrhagic fever virus (Xinjiang strain) in China. Am J Trop Med Hyg 34:1179–1182
56. Zeller HG, Cornet JP, Camicas JL (1994) Experimental transmission of Crimean-Congo hemorrhagic fever virus by West African wild ground-feeding birds to *Hyalomma marginatum rufipes* ticks. Am J Trop Med Hyg 50:676–681

CHAPTER 2

A HISTORICAL PERSPECTIVE OF INFECTIOUS DISEASES WITH REFERENCE TO CRIMEAN-CONGO HEMORRHAGIC FEVER

BERNA ARDA, M.D., PH.D.[1] AND AHMET ACIDUMAN, M.D., PH.D.[2]

[1] *Professor of Medical Ethics and History of Medicine, Ankara University, School of Medicine, Ankara, Turkey. Tel.: +90 312310 30 10 – 361; Fax: +90 312 310 63 70; E-mail: berna.arda@medicine.ankara.edu.tr*

[2] *Ankara Etlik İhtisas Hospital, Neurosurgery Clinic, Ankara, Turkey. Tel.: +90 312 223 98 17; Fax: +90 312 312 37 24; E-mail: ahmetaciduman@yahoo.com*

2.1. HISTORICAL PROCESS OF THE DISEASE CONCEPT

In studying the evolution of the concept of disease, Kraupl Taylor argues that the term "disease" emphasizes more the pathological side of the disease itself, while the term "illness" corresponds to clinical signs [13]. From the perspective of the history of medicine, the term "disease" should be used in its widest meaning, because if the label of "disease" is used in the sense of its definitions made by contemporary medical scientists, many disease definitions in the medical history will have to remain unmentioned. However, despite their great differences from their definitions today, the smallpox disease defined by Galen in the 3rd century is a "disease" as is the smallpox defined by Rhazes (Fig. 2-1) in the 10th century. Similarly, in the 1700s, the smallpox disease defined by Jenner is also a "disease". Remarkably, way back in history, the signs of a disease, considered only a symptom today, were regarded as a disease on their own. Hence, symptoms such as abdominal pain, hemorrhage, diarrhea, vomiting, and fever were cited as diseases in the Hippocratic era, Galen, and/or Razi and Avicenna, the distinguished representatives of the Eastern world [8]. This approach has cultural and in fact, mythological bases, which can be embodied by the example of Febris, the goddess of fever and malaria. She represents fever and febrile diseases. In antiquity, unlike in our times, not much was known about the mechanism of fever in disease and therefore was considered a disease on its own. In periods of medicine with mystical explanations, supernatural forces were blamed. Consequently, it is not surprising that the goddess Febris was held responsible for fever.

O. Ergonul and C. A. Whitehouse (eds.), Crimean-Congo Hemorrhagic Fever, 13–22.

Fig. 2-1. Abū Bakr al-Rāzī (854–932), "Rhazes", eminent physician of the eastern medieval world.

In the medical paradigm of our times, the term "disease" ("antite morbide") is used in its most specific meaning [2]. To label a clinical picture with this term, a through knowledge of all its characteristics is the primary requirement. A clinical picture described under the heading of "X disease" in any textbook is explained in subtitles covering its nosological and symptomatological characteristics, and progression and process. Its etiology and pathogenesis as well as "anatomopathology" and "histopathology" are also provided. Similarly, the treatment is defined under the same heading. In this book, the same scheme has been used for Crimean-Congo hemorrhagic fever (CCHF). If there is a lack of information associated with the disease, particularly with etiology and pathogenesis, the clinical picture will often be labeled as a "syndrome."

There is in fact a historical dimension to the labeling of "syndrome." We are all aware of the importance of scrutinizing a patient's symptoms for the development of a differential diagnosis of diseases. In times when symptoms were considered a disease in their own right, the signs that accompanied each other were termed "symptom," which literally means "a friend for the road walking the same path." Thus, it can be said that the term "syndrome" was a concept used in the medicine of antiquity. Hippocratic paradigm is also compatible: careful observation of the symptoms, follow-up of the patient, and disease as a natural result of natural reasons. In this respect, CCHF is a syndrome progressing with fever and hemorrhages. In 1937, Behçet was defined as a syndrome characterized with iris infection, ulcerative lesions of the mouth, and genital ulceration. When AIDS was defined as a new syndrome in 1981, skin eruptions, diarrhea, and accompanying symptoms associated with respiratory system directed physicians to the same point: "All of these constitute a different disease from the other diseases you have known so far."

The definition and explanation of the changes inflicted by the disease agent on the healthy organism forms the theories of pathology, which have developed parallel to various thought patterns throughout the ages. Whatever their type, all theories of disease are the products of intellectual endeavors to "understand and account for" the concept of disease. In fact, when naming and interpreting the concept of disease, the human mind has always followed this path. This notion is important in the definition and explanation of the disease entity, particularly with respect to infectious diseases. The history of medicine encompasses various theories and explanations on how a human becomes ill and the resulting effect.

2.2. CAUSE–EFFECT RELATIONSHIP AND SIMPLE EPIDEMIOLOGICAL AND CLINICAL OBSERVATIONS

Since ancient times, fighting disease and the protection of health has been one of the primary objectives of humankind. The behaviors of this kind that had initially helped form the "healing power of nature" (*vis medicatrix naturae*) with their mechanisms of protection made up of a chain of natural reactions or reflexes created "instinctive medical behaviors." Over time, certain applications that were proven to be useful through the trial and error method were regarded as acquired medical behaviors and the experiences of thousands of years constituted the content of empirical medical practice. The fact that there are social and psychological grounds to diseases, as well as a biological basis, is a result that was attained through observations of empirical medicine. It is known that civilization started at various locations on the earth including India, China, Mesopotamia, Egypt, Anatolia, and Aegean coasts of the Mediterranean. Thus, medical knowledge is a product of empirical medicine, but is also a part of our inheritance from many earlier civilizations.

The importance of agricultural reform and domestication of animals have been widely emphasized by medical historians. It is clear that ability to recreate nature and observe biological processes has improved human thought. Invention of agriculture was at the same time "cultivation" of diseases [9], of which parasites were the primary mediators. The pathogens in domesticated animals found favorable living conditions in the human organism and, thereafter, adapted themselves accordingly. Mice, rats, mosquitoes, fleas, lice, and ticks have been known to act as disease carriers since antiquity. For example, a louse was found among the hair of a mummy dated around 3000 BC, which belongs to first royalty period of Egypt. However, it is not known whether a clinical picture similar to CCHF was ever among this group of diseases. The agents of certain infectious diseases, such as smallpox and measles, have been well adapted to humans. They were transmitted from person to person because of their highly contagious nature.

As humans began to transform nature to a living area, they also began to become more heavily infected, i.e. parasitized, by their own microorganisms. Thus, humans started to develop immunity and live with these microorganisms. This way a compromise between the assailant and human was established. Immunity is a key concept for the history of infections. Clearly, similar conditions occur genetically in some diseases. In falciparum malaria, for example, persons with sickle cell anemia are more resistant to infection by *Plasmodium falciparum*. These types of examples show that when exposed to a certain pathogen for a given amount of time, we learn the ways to overcome the disease provoked by this pathogen.

After the era when mystical explanations were given for healing, humanity developed a long-standing disease theory: "four elements or humoral pathology theory." The basis of this theory rests on the beliefs rooted in ancient times. Based on the civilizations of the Far East, humans have been thought to be a small model of the universe. Accordingly, the factors constituting the universe (macrocosmos) and human (microcosmos) are similar. The four elements in nature "fire," "air," "water," and "soil" correspond to "yellow bile," "blood," "phlegm," and "black bile," respectively. Based on the humoral pathology theory of the Hippocratic School, disease is the disturbance of the balance of these four elements. To regain health, the balance had to be reestablished. It is possible to see the initial traces of homeostasis theory of 20th century based on simple observation in this theory of pathology. Humoral pathology observed by Avicenna in the Eastern world has been effective for ages. However, it was replaced by "cellular pathology" by the work of Rudolph Virchow in 19th century [6, 12].

Definition of microorganisms, description of their transmission mechanisms, and prevention of contamination have contributed significantly in gaining the desired medical outcomes. The dissolution of "self reproduction" (spontaneous generation) theory and reproduction of bacilli under laboratory conditions paved the way for the development of vaccines and antimicrobial treatments, a turning point in the history of infectious diseases. Thus, many new developments in microbiology and infectious diseases were seen during the 20th century; however, this is also an age when threats of bioterrorism are realized [10].

2.3. RESEARCH INTO THE HISTORY WITH THE DEFINITIONS OF TODAY

In our times, CCHF is a medically and scientifically defined tick-borne arboviral disease [14]. This definition means that the etiology, pathogenesis, and symptomology are largely known. Starting from the case definition and symptoms of

CCHF, we studied the written documents of some of the civilizations of the past that had settled in the geography known to be endemic for CCHF.

Of course, it is very difficult to determine when a disease, as is defined today, first appeared in terms of the evolutionary development of medicine. Thus, in retrospective evaluations of diseases, it is necessary to be cautious and be aware that the name of a particular disease may also refer to many other diseases with similar symptoms. Medical historians, while researching the history of certain infectious diseases, agree that it is not possible to accurately go beyond 3000 BC.

It might be predicted that the smallpox virus was transmitted by cattle. The statues of Gods erected in the name of this disease in the temples of Indian civilizations are concrete evidence for the prevalence of this virus in Southern Asia [9]. Identification of plague in many different periods, in fact encompassing a wide array of diseases under the same name, is an interesting, if not inevitable, incident. Certain documents show that plague first appeared in Egypt and caused a widespread epidemic in 1347; thus, the entire world learned about this disease. However, it is not surprising that many diseases such as typhoid, paratyphoid, cholera, and typhus were described under the name of "plague." It was not until the late 19th century, when the specific causative agents of many of these diseases were identified, that they were no longer simply labeled as "plague." Consequently, "plague" has been a label representing lethality and destruction. It was referred to as "ta'un" in the Ottoman language. Similarly, research into today's terminology corresponding with the names of many fevers used by Thomas Sydenham in his book, "The Method of Treating Fevers" in 1666 is interesting as to the classification and shifts in the meanings of these terms throughout medical history. For example, a few of the terms used in this book in order to refer to various diseases are "goal fever," "hospital fever," "yellow fever," "malaria fever," "scarlet fever," "puerperal fever," "Malta fever," and "typhoid fever" [3]. Likewise, as for CCHF, a descriptive term taking the most dominant symptoms and its geographical location into consideration has been used, as is often the case for various viral diseases.

Having established a glorious civilization in Anatolia, the Hittites were dominant around 1650–1200 BC. Hittite culture was based on tolerance rather than violence and "an eye for an eye" notion. They believed diseases to be associated with supernatural forces. Accordingly, diseases were the result of the negligence of humans towards Gods, crimes against Gods, sins, disturbance of the dead, black magic, and/or the breaking of oaths or agreements. Ishara was the God of disease and Kamrusepas was the God of health. Yarris, on the other hand, was the God that protected the people and the kings from communicable diseases in the war arenas. The tablets with cuneiform scripts found among royal archive in Çorum Boğazkale (Hattusas), the capital of the Hittite Empire, provide information on the lifestyle of this great civilization. They indicate that the Hittites respected female physicians with the names of Makiya, Mammitum-um-mi, and Azzari. The Hittites are also known for their hygiene. They used to associate contamination with diseases. Punishment of those polluting the public water sources, isolation of the patients in case of epidemics, thus, starting a quarantine practice are among the medical practices attributed to this civilization.

In Hittite period Anatolia, certain communicable diseases named "henkan" were observed. In the 1300s BC, during the wars when the Hittites conquered the land of Egypt in Syria, several epidemics were recorded as evidenced by the prayer of "disease" by King Mursilis II (Fig. 2-2) [5].

It is possible that henkan describes an extremely wide range of diseases with skin eruptions, high fever, and diarrhea. During the time of this civilization, which lived in middle and southeast Anatolia, CCHF cases too, if ever existed, were probably considered under the name "henkan." However, in order to prove this hypothesis, a reference made to a case in the tablets that will not leave any doubts, accompanied by a definition of the disease, is needed. Furthermore, the disease should also be traced in the following civilizations located within the same geographical area [4].

Fig. 2-2. The praying verses of Mursilis II for defeating plague on a tablet (1321–1295 or 1340–1310 BC). Text belongs to the Hittite Catalogue CTH 378/I version A: Ankara Anatolian Civilization Museum, Tablet Archive, No. Bo 2801.

(*Continued*)

Fig. 2-2. cont'd. This text has approximately ten broken fragments of the tablet which obtained different excavation periods and after that philological studies all of these are joined. The all text has been published by René Lebrun; Hymnes et Prières Hittites, *Homo Religiosus* 4, Louvain-La-Neuve (1980) 193–203. Tablets have taken photographed from the museum with special permission by Professor Cem Karasu, the chairperson of Hittitology Department in Ankara University.

2.4. HISTORICAL DEFINITIONS TO DATE

Abū Bakr al-Rāzī (854–932), known as Rhazes (Fig. 2-1) in the Western world, a distinguished physician of the Eastern world, was born in the city of Rayy. He learned medicine in a relatively late period of his life. Rhazes, who valued patient follow-up, is believed to have synthesized Hippocratic and Galen medicine. Among his books on medicine, philosophy, and mathematics, his most important work was *Kitāb al-Hāwī* (*Continens*), related to the treatment of diseases. Avoiding prescribing numerous drugs to his patients, one of his greatest contributions to medicine was the differential diagnosis of smallpox and measles. Strikingly, three cases defined by Rhazes bear similarities to CCHF in today's definition [1]; thus, they provide important information as to the earliest definitions of the disease:

Case 1:

A woman had had fever and diarrhea called hulife; later, something like black blood had started to leak from her. The master asked if that black thing boils on earth. They said it did not. The master said if it did, it would be considered black bile because the one not boils is considered burned blood trapped in the liver. Now, couple of bowls are made red-hot on fire and then dropped in ayran (yoghurt drink) of cattle and extinguished; later the patient drank the ayran. With the permission of Allah, that blood was stopped, and the patient achieved full recovery.

Case 2:

A young person had headache, high fever, and redness in the eyes, and his urine color was red too. The master ordered him to use violet lozenges and then barley extract (juice), to put şiyaf-i şakika (headache drops) to the ear on his painful side, vinegar and rose extract (juice), and rose oil on his hand, and ordered him to eat tafşil (a meal cooked with meat, husked lentils, and vinegar) and hallü zeyt (vinegar and olive oil).

Case 3:

One person had fever and bleeding from below, also palpitation. The master ordered him to use sumac extract (juice) and copal lozenges and commended him to eat meals cooked with sumac.

Various references made to Rhazes as "the master" and reference made to the other books of Rhazes, such as the *Kitāb al-Mansūrī* (*Liber Almansoris*) while talking about the treatment [1], confirms that the cases were quoted from *al-Hāwī* (*Continens*). The detailed description of the first case is suggestive of a CCHF-like disease (Fig. 2-3). Likewise, complete recovery within a few days is suggestive of a "viral" disease.

Zayn al-Din Sayyed Isma'il ibn al-Husayn al-Jorjani, who lived in 12th century and died in 1136 in Merv, was a successor of Avicenna. His best-known work, *Zakhirah-i Khvarazm'Shahi* had become a widely used handbook and sustained this quality in the following centuries. It was also reported that Ibn (al-) Baytar or Baitar benefited from this book [7]. Interestingly, Hoogstraal had suggested that a disease defined in this encyclopedic book written in Persian might be CCHF. This case from the 12th century is from the Tajikistan region. The signs of the disease were blood in urine, vomit, and phlegm. It was stated that arthropods were the cause of the disease, and normally, the disease lives as a parasite in a black bird. Applying sandalwood essence and "bodzkhar" to the area of bite and feeding the patient with fresh goat milk were the treatment methods used. For centuries, similar pictures of the disease had been reported in the folk culture of Uzbekistan [7]. The relationship mentioned by Hoogstraal is one of the first simple observational examples of cause–effect relationships in

Fig. 2-3. Detailed description of CCHF-like cases. These pages belong to Razi's Hamidiye manuscript. This manuscript registered with number 1013 in İstanbul Süleymaniye Manuscript Library. It has 151 pages totally and written at the beginning of 18th century.

the 12th century. However, these determinations are highly speculative with respect to the methodology of history.

2.5. CONCLUSIONS

History is a discipline of process evaluation with retrospective data collection beginning at the present as one of its methods. Undoubtedly, tracing any disease in any society at any given time requires a comprehensive and multidisciplinary approach. Evaluation of the effects of civilizations on health and disease provides significant insight into the history of medicine, and further, the history of infectious diseases supplies many pertinent examples. While civilizations aimed to keep diseases under control, they themselves were often vectors of disease. The African continent, initially referred to as "white man's grave" because of several tropical fevers, became a site of colonization. The detection of the malaria agent and the use of quinine as an effective control also propelled imperialism. Thus, it is not surprising that Africa turned into "black man's grave" at the end of this process. Colonial medicine usually aimed at protecting the health of the colonists. Therefore, despite limited efforts of missioners, much of modern medicine of the time could not reach the local people [9, 11]. In the light of this information, in evaluating CCHF the reader should recognize that comprehension of only the technical and medical aspects of the disease will be insufficient to understand the history associated with the disease. As was detailed above, it is possible to find many traces of the history of infectious diseases. Various microorganisms can be detected through examination of historical remains, such as mummies. This facilitates the inquiries of medical historians. However, throughout biological evolution, for an infectious agent that first appeared hundreds or perhaps thousands of years ago, our words from the perspective of medical history remain highly limited, and our claims have to be modest.

As the story of CCHF in the 20th century unfolded in the Crimean peninsula, it was only the initial stage of the act. This was a stage where humans would get to know their foe and fight against it with great vigor. Do we know in which century the story actually started? Is the disease mentioned by Jorjani in the 12th century the same as CCHF defined in the textbooks today? Or can we claim that the cases defined by Rhazes in the late 9th century and early 10th century, long before Jorjani, were CCHF? Certainly, we will continue to produce hypotheses and questions regarding our past. As we proceed forward, it is certain that history will shed light on the missing pieces of infectious diseases and the role they play in our lives.

Acknowledgments

We would like to thank Professor Cem Karasu, chairperson of Hittitology Department, Ankara University Faculty of Letters, for his kind support.

REFERENCES

1. Al-Razi (~1732–1733) Hamidiye 1013 manuscript. In: Süleymaniye Manuscript Library, İstanbul, folios 1b, 2a, 124b, 125a, 128a, 139a
2. Arda B (1997) Disease Concept of Western Medieval Age (Batı Ortaçağı'nda hastalık kavramı). Güneş Kitabevi, Ankara (in Turkish)
3. Arda B (2001) FMF: some considerations about its historical and demographic features. Türkiye Klinikleri Tip Tarihi 1:106–110
4. Arda B, Aksu M (2004) What the Hittites' tablets tell us? A short historical view of deafness on the basic of genetics. Turk J Med Sci 34:357–358
5. Bayat AH (2003) History of Medicine. Sade Matbaası, İzmir, pp 60–61, 65 (in Turkish)
6. Castiglioni A (1958) A History of Medicine, 2nd edn. Alfred A Knopf, New York, pp 18, 50
7. Hoogstraal H (1979) The epidemiology of tick-borne Crimean-Congo hemorrhagic fever in Asia, Europe and Africa. J Med Entomol 15:307–417
8. Kahya E, Demirhan Erdemir A (2000) Medicine and health foundations since Ottoman to Republic of Turkey in the light of scientific studies (Bilimsel çalışmalar ışığında Osmanlı'dan Cumhuriyet'e tıp ve sağlık kurumları). Türkiye Diyanet Vakfı Yayınları, Ankara, p 71 (in Turkish)
9. Kiple K (1996) The History of Disease (in History Medicine). Cambridge University Press, Cambridge, pp 16–30
10. Klietmann WF, Ruoff KL (2001) Bioterrorism: implications for the clinical microbiologist. Clin Microbiol Rev 14:364–381
11. Porter R (1996) Illustrated History Medicine. Cambridge University Press, Cambridge, pp 6–15
12. Singer C (1944) A Short History of Medicine. Oxford University Press, Oxford, pp 61–82
13. Taylor FK (1979) The Concepts of Illness, Disease and Morbus. Cambridge University Press, Cambridge, pp 18–117
14. Whitehouse CA (2004) Crimean-Congo hemorrhagic fever. Antiviral Res 64:145–160

CHAPTER 3

PERSONAL REFLECTIONS

JOHN P. WOODALL, PH.D.

Director, Nucleus for Investigating Emerging Infectious Diseases, Institute of Medical Biochemistry, Center for Health Sciences, Federal University of Rio de Janeiro, Brazil.
E-mail: woodall@promedmail.org

The saga of the first isolations of the virus that was eventually named, erroneously, Crimean-Congo hemorrhagic fever virus (CCHFV), is intriguing. In 1956, what is now the Democratic Republic of the Congo, capital Kinshasa, was still firmly under Belgian colonial rule as the Belgian Congo, capital Leopoldville, in honor of the Belgian monarch. The major city of the interior was known as Stanleyville, in honor of the Welsh-born American explorer Sir Henry Morton Stanley (1841–1904), who went to Africa to find David Livingstone, a quest that culminated in his immortal greeting "Dr. Livingstone, I presume?" Subsequently, Stanley accepted the invitation of Leopold II of Belgium to head another expedition, which resulted in the creation of the Congo Free State in 1884, renamed the Belgian Congo in 1908 (Fig. 3-1). After independence in 1960, the country was named Zaire, and Stanleyville was renamed Kisangani. Following a rebellion and change of government in 1997, the country's name was changed yet again, to the Democratic Republic of the Congo.

The Belgians provided a network of health services, and at the time our story begins, in 1956, Dr. Ghislaine Courtois was head of the Provincial Medical Laboratory in Stanleyville. On 6 March 1956 he saw, in his clinic, a 13-year-old boy who had fever, headache, nausea, vomiting, backache, generalized joint pains, and photophobia. He took a blood specimen by venepuncture. That same day he inoculated the blood into 3-day-old mice by both the intracerebral and intraperitoneal routes. One mouse became sick on postinoculation (p.i.) day 12, indicating that it had received only a small dose of virus, and its brain was inoculated into four 2-day-old mice, of which one became paralyzed on p.i. day 11. A passage from the brain of this mouse was made into four more 2-day-olds, and all were sick or dead by p.i. day 8. Thus, with difficulty, the strain was mouse-adapted. The serial number Dr. Courtois assigned to this isolate was V3011.

O. Ergonul and C. A. Whitehouse (eds.), Crimean-Congo Hemorrhagic Fever, 23–32.

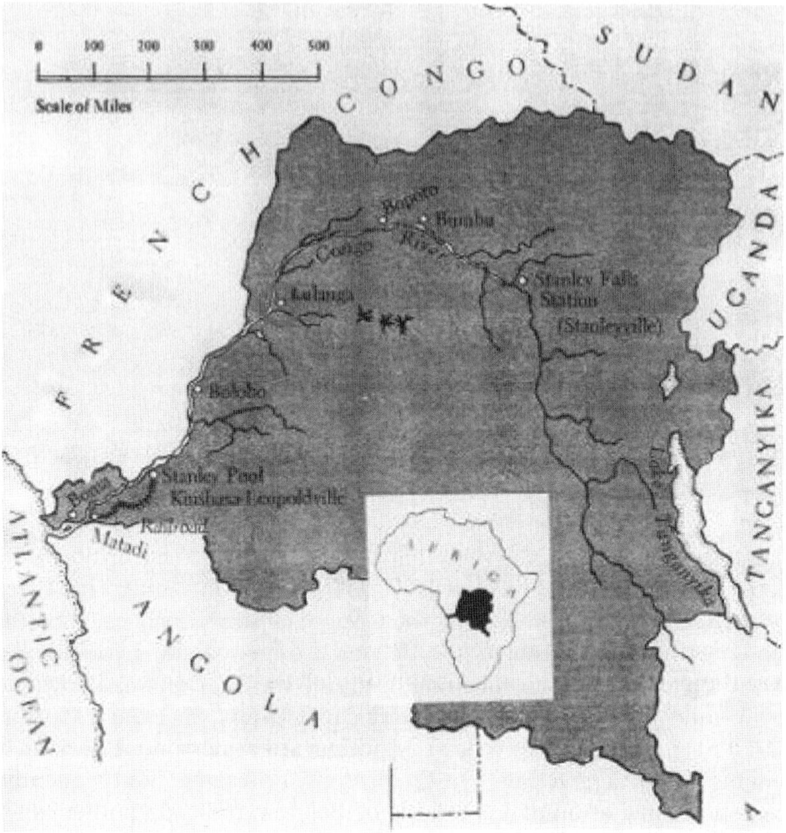

Fig. 3-1. Map of the Belgium Congo.

Dr. Courtois himself subsequently became ill with fever, nausea, and vomiting, and on 14 April 1956, his blood was inoculated into infant mice, all of which were sick by p.i. day 6, indicating that they had received a heavy dose of virus. The agent passed through a bacteria-tight Seitz filter (this was in the days before Millipore), establishing it as a virus. The serial number assigned to this isolate was V3010. The original strain was finally mouse-adapted by around 6 April; I do not know why the good doctor designated his own isolate, which was only adapted on 20 April, with an earlier number.

Dry ice was not available in large quantities at the time, and he did not have a –70°C freezer, so Dr. Courtois maintained the virus for months by brain passages in infant mice. He was unable to identify these viruses at Stanleyville with the reagents he had available, so in 1957 he sent infant mouse brains from the 35th passage of V3011 and 46th passage of V3010 in glycerol, which preserved them during transit at ambient temperature, to the East African Virus

Research Institute (EAVRI) in Entebbe, Uganda, which was the regional reference center. The two isolates were routinely tested by laboratory technician P.J. Mason for their host range in laboratory animals, and found to grow in adult mice but not in embryonated eggs. They were then relegated to the deep freeze.

3.1. INITIATION OF VIRAL STUDIES IN EAST AFRICA

The EAVRI began life in 1936 as the Yellow Fever Research Institute of the International Health Division of The Rockefeller Foundation (RF) (Fig. 3-2). The RF had set out to define the geographical limits of yellow fever in Africa and Latin America by establishing a chain of virus laboratories to perform virological and seroepidemiological studies in the field. In Africa, these were a laboratory in Yaba, Nigeria, West Africa, and the institute in Uganda, East Africa, both still British colonies. In the course of processing material from humans, wild vertebrates, and mosquitoes, these laboratories opened a Pandora's box of previously unknown viruses, many of them arthropod-borne and some causing significant human disease, and thus built up a regional reference collection. In 1949, the RF handed over the operation of their Institute to the East African High Commission, and it was renamed the EAVRI.

In 1959, I arrived at the EAVRI with my freshly minted Ph.D. from the Entomology Department of the London School of Hygiene and Tropical Medicine. My thesis was on the transmission of viruses by the mosquito *Aedes aegypti*. The two viruses I had studied, at the instigation of my professor, Dougie Bertram, were Semliki Forest and West Nile viruses, both first isolated at the EAVRI. I soon discovered that the EAVRI was top-heavy with entomologists but short of virologists, so began on-the-spot training as an arbovirologist

Fig. 3-2. Rockefeller Foundation/British Colonial Office yellow fever laboratory in Entebbe, Uganda, East Africa, circa 1950. (Courtesy of the Rockefeller Archive Center.)

Fig. 3-3. Staff of the East African Virus Research Institute (formerly yellow fever laboratory), Entebbe, Uganda, East Africa, circa 1962, during a celebration. Some of the authors of the original Congo virus papers may be seen in the back row: far left, D.I. Simpson (behind A.J. Haddow, Director); far right, J.P. Woodall (behind M.C. Williams).

under the tutelage of first, M. Paul Weinbren, a South African physician, and after he left the Institute, Miles C. Williams, a British physician and diplomate in tropical medicine who kindly mentored me to my lasting benefit (Fig. 3-3). After 2 years, in 1961, the director of the Institute, Dr. Alexander (Sandy) J. Haddow, M.D., D.Sc., told me of a treasure trove buried in a Revco –70° freezer – a lode of viruses sent in to the EAVRI for identification and untouched for lack of a virologist with enough time to study them. Sandy asked me to see what I could do with them. Among those I dug out, after 4 years in limbo, were V3010 and V3011, and six others which eventually turned out to be related to them, from humans from different parts of Uganda and from a cow from Kenya [7]. The fact that V3010 and V3011 turned out to be identical strongly suggests that Dr. Courtois suffered a laboratory infection, and was lucky to survive, since the case fatality rate of the virus is high. Serum taken from him 7 years later neutralized his virus.

The methods used for virus identification in that "steam age" of arbovirology were derived from methods for the study of influenza viruses, developed by Delphine Clarke of The Rockefeller Foundation Virus Laboratory (RFVL) in New York City. They were hemagglutination-inhibition (HI), complement fixation (CF), and the neutralization test. HI required sucrose-acetone- or fluorocarbon-extracted mouse brain as antigen, and fresh goose erythrocytes for agglutination. CF required fresh complement, obtained from guinea pig serum. We prepared our own, keeping for the purpose a flock of geese, that roamed the Institute grounds cackling loudly and defecating freely, and a breeding colony of guinea pigs. In those days you could always tell an arbovirus lab by the loud-mouthed flock of geese on the premises. Unable to prepare an HI antigen by either extraction method, I turned to CF, but found no relationship to 35 different arboviruses known to occur in Africa, or to herpes simplex virus. To take the identification further, the answer was to send the strains to the RFVL, New York, which was then the world reference center for arboviruses, being the depository for strains from its globe-girdling chain of RFVLs and collaborating institutes. That laboratory subsequently moved to New Haven, Connecticut, as the Yale Arbovirus Research Unit, and more recently to the University of Texas Medical Branch, Galveston, Texas.

Dr. Jordi Casals (Fig. 3-4), a grand old man of arbovirology at the RFVL, tested V3010 by CF against reference strains there, and found it was identical to another strain from Uganda, but to no other named virus at that time. NT and agar gel precipitation tests confirmed the relationship [1].

Since V3010 and 3011 were the earliest strains to be isolated, Jordi courteously invited me to name the new virus, and by extension all the related, subsequently isolated strains. In turn, I asked Dr. Courtois what he would like to

Fig. 3-4. Dr. Jordi Casals, circa 1966. (Courtesy of Dr. Martine Jozan.)

call it. Arboviruses are usually named after either the symptoms or the locality of isolation, e.g. yellow fever, West Nile, but the locality is preferably a small part of a country rather than the whole. However, when Dr. Courtois asked that it be named Congo virus, I complied and registered it in the first edition of the Catalogue of Arthropod-borne Viruses. Since V3010, in spite of its serial number, had been inoculated into mice after V3011, I took V3011 as the prototype.

Over the following years through 1965, we isolated five more strains of Congo virus from patients in Uganda, with long intervals between most of them which precluded any possibility of laboratory contamination between specimens. There were, however, two cases in EAVRI workers.

One was an animal technician who fell ill over the new year 1963–1964. He was hospitalized and appeared to improve, but suddenly developed serious hematemesis and died on the 7th day of illness. His work in the animal breeding quarters should not have brought him into contact with any virus, and since other cases had shown that the virus was endemic in the Entebbe area – and some years later it was shown to be tick-borne – it was impossible to determine whether his was a laboratory or a natural infection. Sequencing had not yet been discovered. Another lab technician, who had actually handled Congo virus, became ill 2 weeks after the death of his coworker, but did not develop any hemorrhagic symptoms and recovered completely. It should be noted that at the time we were still working at an open bench without gloves, and mouth pipetting, but none of the expatriate staff had an overt infection with any of these African viruses – we did not check if we had developed antibodies from an inapparent infection, because there were just too many different viruses to check. (Well, I did develop a beautiful case of Bunyamwera virus infection – the prototype of the supergroup to which CCHF virus belongs – with a striking rash on my trunk, but again it was a mosquito-borne virus endemic to the Entebbe area, so it could well have been a natural infection.)

So in 1967, 2 years after I had left the EAVRI to head the arbovirus laboratory at the New York State Health Department, the original papers entitled "Congo virus: a hitherto undescribed virus occurring in Africa", appeared in the East Africa Medical Journal, establishing, we thought, precedence [6, 8].

In retrospect it was a pity this report did not appear in a more international publication, but then neither did Chumakov et al.'s [3] paper on the isolation of the virus of Crimean hemorrhagic fever, published in Russian in 1968 in *Vopr Virusol* The Soviets, of course, had recognized Crimean hemorrhagic fever as a distinct disease since 1944, and by passaging it through human volunteers, established that it was a virus [2], but were unable to isolate it in the laboratory. They did not say how many of the volunteers died. But their standard practice was to use 21-day-old mice for isolation attempts, and they had never attempted it in infant mice before the late 1960s. When they did, they isolated the virus, but their registration in the Catalogue of Arthropod-borne Viruses was of a strain from a fatal case from Samarkand – that of the Golden Road – in Central Asia rather than the Crimea, whose serum was collected on 14 June 1967 and inoculated into infant mice at the Institute of Poliomyelitis and Viral Encephalitis in Moscow on 22 June 1967 [3], 4 months after our prototype Congo isolate from 1956 was published in February 1967.

3.2. NAMING THE VIRUS

The Moscow Institute's director, M.P. Chumakov (Fig. 3-5), or Mikhail Petrovich as he was known to all, was a much-revered figure in Soviet virology, having worked on polio since the 1950s. He was a heavyset bear of a man with a withered arm, supposedly from a bad attack of either polio or some other nasty virus. After sending this new virus to Jordi Casals at the RFVL for identification, he was horrified to learn that it was identical to Congo virus. Of course it was unthinkable that the virus from a disease that did not even have a name before 1957 could be associated with Crimean hemorrhagic fever, recognized for a decade earlier. It was still the time of the Cold War, and although the International Committee for the Taxonomy of Viruses tried for the compromise name *Congo*-Crimean hemorrhagic fever virus, the Soviets got their way, and around 1973 it was renamed CCHFV, against all the principles of scientific nomenclature based on priority in publication.

However, I can take heart from the following: its acronym, CCHF, can be read either way; in 1984 the Russians Tsar'kova VA and Nikolaev VP called it *Congo-Crimean* hemorrhagic fever in a paper in Vopr. Virusol.; as late as 1995 the CDC's *Morbidity and Mortality Weekly Report* published an item that stated

Fig. 3-5. M.P. Chumakov, 1966. (Courtesy of Dr. Martine Jozan.)

"This update applies to four viruses that cause syndromes of VHF: Lassa, Marburg, Ebola, and *Congo-Crimean* hemorrhagic fever viruses", and as recently as 2002 the respected journal Euro Surveillance, in a report on viral hemorrhagic fevers in Europe, referred to ". . . *Congo Crimean* Haemorrhagic Fever (CCHF) in Kosovo and Pakistan in 2001." There are, in fact, 23 publications from 1980 through 2002 listed by the National Library of Medicine's PubMed that contain the name *Congo-Crimean* hemorrhagic fever virus, placing Congo first, and 11 references to simply "Congo virus" (see bibliography). The problem is that anyone doing a literature search using the Soviet version of the name will miss all of them. These papers contain interesting records of the occurrence of the virus, or antibodies to it, in Greece, Portugal, the Republic of South Africa, Madagascar (the first isolation from there), the Maghreb, Dubai, Saudi Arabia, Kuwait, and Iraq. Intriguingly, the papers on Congo-Crimean haemorrhagic fever (CHF) in Iraq were authored by S.K. Al-Tikriti, clearly one of Saddam's clan (Saddam's full name is Saddam Hussein Al-Tikriti), whose other listed publications curiously cover work on methyl-mercury poisoning, causing one to wonder about his ultimate research objectives.

So there we have it – one of the first instances of a virus losing its real name due to politics, just as the prototype American hantavirus originally published as Four Corners virus [5] ended up being called Sin Nombre virus, i.e. virus with no name [4].

REFERENCES

1. Casals J (1969) Antigenic similarity between the virus causing Crimean hemorrhagic fever and Congo virus. Proc Soc Exp Biol Med 131(1):233–236
2. Chumakov MP (1947) [A new virus disease – Crimean hemorrhagic fever] Nov. Med 4:9–11 (Russian)
3. Chumakov MP, Butenko AM, Shalunova NV, Mart'ianova LI, Smirnova SE, Bashkirtsev IuN, Zavodova TI, Rubin SG, Tkachenko EA, Karmysheva VIa, Reingol'd VN, Popov GV, Savinov AP (1968) [New data on the viral agent of Crimean hemorrhagic fever] Vopr Virusol 13(3):377 (Russian)
4. Goldsmith CS, Elliott LH, Peters CJ, Zaki SR (1995) Ultrastructural characteristics of Sin Nombre virus, causative agent of hantavirus pulmonary syndrome. Arch Virol 140(12): 2107–2122
5. Hjelle B, Jenison S, Torrez-Martinez N, Yamada T, Nolte K, Zumwalt R, MacInnes K, Myers G (1994) A novel hantavirus associated with an outbreak of fatal respiratory disease in the southwestern United States: evolutionary relationships to known hantaviruses. J Virol 68(2):592–596
6. Simpson DI, Knight EM, Courtois G, Williams MC, Weinbren MP, Kibukamusoke JW (1967) Congo virus: a hitherto undescribed virus occurring in Africa. I. Human isolations – clinical notes. East Afr Med J 44(2):86–92
7. Woodall JP, Williams MC, Santos DF, Ellice JM (1962) The 3010 group. East Afr Vir Res Inst Rep July 1961–June 1962:221–222
8. Woodall JP, Williams MC, Simpson DI (1967) Congo virus: a hitherto undescribed virus occurring in Africa. II. Identification studies. East Afr Med J 44(2):93–98

BIBLIOGRAPHY

Congo virus (Congo alone)

Casals J (1969) Antigenic similarity between the virus causing Crimean hemorrhagic fever and Congo virus. Proc Soc Exp Biol Med 131(1):233–236

Causey OR, Kemp GE, Madbouly MH, David-West TS (1969) Congo virus from domestic livestock, African hedgehog, and arthropods in Nigeria. Am J Trop Med Hyg 19(5):846–850

David-West TS, Osunkoya BO (1973) The effect of anti-lymphocytic serum of infection by congo virus (Crimean haemorrhagic fever virus related) and Mokola virus (rabies virus related). Br J Exp Pathol 54(3):274–278

David-West TS, Cooke AR, David-West AS (1974) Seroepidemiology of Congo virus (related to the virus of Crimean haemorrhagic fever) in Nigeria. Bull WHO 51(5):543–546

David-West TS, Osunkoya BO, Sagoe AS (1970) Protective effect of anti-lymphocytic serum on congo virus infection. Brief report. Arch Gesamte Virusforsch 38(2):267–270

Fagbami AH, Tomori O, Fabiyi A, Isoun TT (1975) Experimental [sic] Congo virus (Ib – AN 7620) infection in primates. Virologie 26(1):33–37

Johnson BK, Ocheng D, Gitau LG, Gichogo A, Tukei PM, Ngindu A, Langatt A, Smith DH, Johnson KM, Kiley MP, Swanepoel R, Isaacson M (1983) Viral haemorrhagic fever surveillance in Kenya, 1980–1981. Trop Geogr Med 35(1):43-47

Meunier DM, Johnson ED, Gonzalez JP, Georges-Courbot MC, Madelon MC, Georges AJ (1987) [Current serologic data on viral hemorrhagic fevers in the Central African Republic] Bull Soc Pathol Exot Filiales 80(1):51–61 (French)

Okorie TG (1991) Comparative studies on the vector capacity of the different stages of *Amblyomma variegatum* Fabricius and *Hyalomma rufipes* Koch for Congo virus, after intracoelomic inoculation. Vet Parasitol 38(2–3):215–223

Congo–CHF (Congo first) virus

Al-Nakib W, Lloyd G, El-Mekki A, Platt G, Beeson A, Southee T (1984) Preliminary report on arbovirus-antibody prevalence among patients in Kuwait: evidence of Congo/Crimean virus infection. Trans R Soc Trop Med Hyg 78(4):474–476

Al-Tikriti SK, Al-Ani F, Jurji FJ, Tantawi H, Al-Moslih M, Al-Janabi N, Mahmud MI, Al-Bana A, Habib H, Al-Munthri H, Al-Janabi S, AL-Jawahry K, Yonan M, Hassan F, Simpson DI (1981) Congo/Crimean haemorrhagic fever in Iraq. Bull WHO 59(1):85–90

Antoniadis A, Casals J (1982) Serological evidence of human infection with Congo-Crimean hemorrhagic fever virus in Greece. Am J Trop Med Hyg 31(5):1066–1067

Baskerville A, Satti A, Murphy FA, Simpson DI (1981) Congo-Crimean haemorrhagic fever in Dubai: histopathological studies. J Clin Pathol 34(8):871–874

CDC (1984) Congo-Crimean hemorrhagic fever – Republic of South Africa. MMWR 33(38): 535–536, 541, 548

CDC (1995) Update: management of patients with suspected viral hemorrhagic fever – United States. MMWR 30;44(25):475–479

Chastel C, Bailly-Choumara H, Bach-Hamba D, Le Lay G, Legrand MC, Le Goff F, Vermeil C (1995) [Tick-transmitted arbovirus in Maghreb] Bull Soc Pathol Exot 88(3):81–85 (French)

Crowcroft NS, Morgan D, Brown D (2002) Viral haemorrhagic fevers in Europe – effective control requires a co-ordinated response. Euro Surveill 7(3):31–32

Davies FG (1997) Tick virus diseases of sheep and goats. Parassitologia 39(2):91–94

Digoutte JP, Saluzzo JF, Adam F (1985) [Recent data on hemorrhagic fevers in West Africa]. Bull Soc Pathol Exot Filiales 78(5 Pt 2):874–878 (French)

Eley SM, Delic JI, Henstridge RM, Bruce LG, Humphreys CR, Moore NF (1989) Bunyaviridae. Serological relationships. Microbiologica 12(4):351–367

Ellis DS, Smith MD, Lloyd G, Bowen ET, Simpson DI (1984) A method for increasing the surface details of resin embedded viruses using Congo/Crimean haemorrhagic fever virus as a model. J Virol Methods 8(1–2):57–61

Ellis DS, Southee T, Lloyd G, Platt GS, Jones N, Stamford S, Bowen ET, Simpson DI (1981) Congo/Crimean haemorrhagic fever virus from Iraq 1979: I. Morphology in BHK21 cells. Arch Virol 70(3):189–198

Filipe AR, Calisher CH, Lazuick J (1985) Antibodies to Congo-Crimean haemorrhagic fever, Dhori, Thogoto and Bhanja viruses in southern Portugal. Acta Virol 29(4):324–328

Gear JH, Thomson PD, Hopp M, Andronikou S, Cohn RJ, Ledger J, Berkowitz FE (1982) Congo-Crimean haemorrhagic fever in South Africa. Report of a fatal case in the Transvaal. S Afr Med J 62(16):576–580

Mathiot CC, Fontenille D, Digoutte JP, Coulanges P (1988) First isolation of Congo-Crimean haemorrhagic fever virus in Madagascar. Ann Inst Pasteur Virol 139(2):239–241

Mathiot CC, Fontenille D, Georges AJ, Coulanges P (1989) Antibodies to haemorrhagic fever viruses in Madagascar populations. Trans R Soc Trop Med Hyg 83(3):407–409

Mayers DL (1999) Exotic virus infections of military significance. Hemorrhagic fever viruses and pox virus infections. Dermatol Clin 17:29–40

[No authors listed] (1984) Leads from the MMWR. Congo-Crimean hemorrhagic fever – Republic of South Africa. JAMA 252(18):2533, 2537

Saluzzo JF, Leguenno B, Van der Groen G (1988) Use of heat inactivated viral haemorrhagic fever antigens in serological assays. J Virol Methods 22:165–172

Suleiman MN, Muscat-Baron JM, Harries JR, Satti AG, Platt GS, Bowen ET, Simpson DI (1980) Congo/Crimean haemorrhagic fever in Dubai. An outbreak at the Rashid Hospital. Lancet 2:939–941

Tikriti SK, Hassan FK, Moslih IM, Jurji F, Mahmud MI, Tantawi HH (1981) Congo/Crimean haemorrhagic fever in Iraq: a seroepidemiological survey. J Trop Med Hyg 84:117–120

Tsar'kova VA, Nikolaev VP (1984) [Detection of the antigens of the virus of Congo-Crimean hemorrhagic fever by solid-phase immunoenzyme analysis and by the indirect hemagglutination reaction]. Vopr Virusol 29:724–726 (Russian)

SECTION II

ETIOLOGIC AGENT

CHAPTER 4

MOLECULAR BIOLOGY OF THE CRIMEAN-CONGO HEMORRHAGIC FEVER VIRUS

RAMON FLICK, PH.D.

BioProtection Systems Corporation, 2901 South Loop Drive, Suite 3360, Bldg. 3, Ames, IA 50010-8646, USA. Tel.: (515) 296-3944; Fax: (515) 296-3945; E-mail: rflick@bpsys.net

4.1. CLASSIFICATION

Crimean-Congo hemorrhagic fever virus (CCHFV) is a member of the family Bunyaviridae. This family comprises more than 300 virus species grouped into five distinct genera: *Orthobunyavirus*, *Hantavirus*, *Phlebovirus*, *Nairovirus*, and *Tospovirus* [15]. The genus *Nairovirus* consists of seven different serogroups, but only two of them are human pathogens [9]. The CCHF group contains CCHFV and the nonhuman pathogenic Hazara virus; the Nairobi sheep disease group includes the pathogenic Nairobi sheep disease and Dugbe viruses [54, 57]. All members of the genus *Nairovirus* are transmitted by ticks (argasids and ixodids); CCHFV most efficiently by members of the genus *Hyalomma*, followed by *Rhipicephalus* and *Dermacentor* [29, 35]. The natural cycle of CCHFV includes transovarial and transstadial transmission among ticks and a tick–vertebrate host cycle involving wild (e.g. hares, hedgehogs) and domestic animals (e.g. ostriches, cattle) [29, 53, 58].

4.2. STRUCTURE AND GENOME ORGANIZATION

4.2.1. Coding regions

CCHF virions are spherical, approximately 90–100 nm in diameter. They are enveloped particles with a tripartite, single-stranded RNA genome of negative polarity [13, 48, 49] (Fig. 4-1A). All three genome segments contain one open reading frame (ORF) flanked by noncoding regions (Fig. 4-1B) [12, 16, 19]. Four structural proteins are encoded: the RNA-dependent RNA polymerase (L protein) expressed by the large (L) segment, the mature glycoproteins G_N and G_C are encoded by the medium (M) segment and the nucleoprotein (N) by the small (S) genome segment (Fig. 4-1B) [23]. As for other negative-sense RNA

O. Ergonul and C. A. Whitehouse (eds.), Crimean-Congo Hemorrhagic Fever, 35–44.

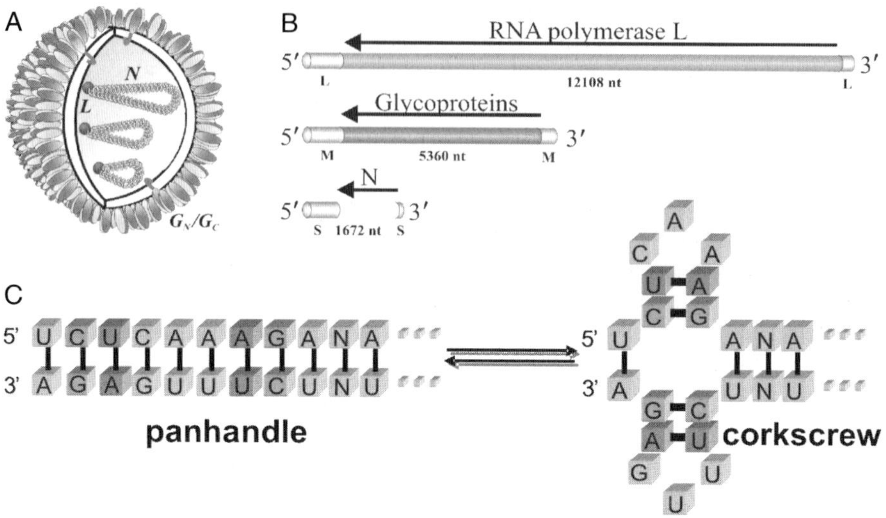

Fig. 4-1. CCHF virion structure and genome organization. (A) Schematic cross section of a CCHF virion. The three genome segments are encapsidated by the nucleoprotein N and associated with the viral polymerase L. In the host cell-derived lipid bilayered envelope the two mature viral glycoproteins G_N and G_C are incorporated. (B) Schematic representation of the three genome segments (L, M, and S) containing an ORF flanked by noncoding regions. (C) The highly conserved nucleotides at the genome segment ends can form via long-range or short-range intrastrand base pairs different predicted secondary structures (panhandle or corkscrew model) important for the binding of the viral polymerase.

viruses, the three genome segments are encapsidated by the nucleoprotein N and associated with the viral RNA (vRNA) polymerase L to form ribonucleoprotein particles (RNPs) [48, 49, 53] (Fig. 4-1A).

Together with hantaviruses, nairoviruses seem to have the simplest genome expression strategy among Bunyaviridae [23]. There is no convincing evidence for the expression of a nonstructural protein (e.g. NS_S, NS_M) by an overlapping ORF (e.g. genus *Orthobunyavirus*) or for an ambisense coding strategy as found in other bunyaviruses (e.g. S segment: genus *Phlebovirus* and *Tospovirus*) [14, 48]. However, recently a novel, apparently nonstructural protein generated by cleavage of the CCHFV glycoprotein polyprotein was discovered (see below) [46]. A unique feature of nairoviruses is the length of their L genome segment, which is almost twice as long as those of other bunyaviruses (with the exception of tospoviruses) [27, 31, 36, 39]. Moreover, the M segment of nairoviruses is 30–50% larger than the corresponding segments of members of other genera in the Bunyaviridae family [43, 47].

The polyprotein encoded by the CCHFV M segment is cotranslationally cleaved into two precursor molecules, $PreG_N$ and $PreG_C$, which are subsequently posttranslationally processed into the two mature glycoproteins G_N and G_C [46, 47, 55]. The glycoprotein processing of nairoviruses, particularly CCHFV, appears to be unique in that precursor molecules are generated first, which through additional posttranslational cleavage steps yield the mature glycoproteins. The mature G_N and G_C are inserted into the lipid envelope as spikelike structures [48]. They are responsible for virion attachment to so far unknown receptors on susceptible host cells and for induction of neutralizing antibodies [23].

Recently, the expression strategy and biosynthesis of the CCHFV glycoproteins have been studied in more detail including the identification of precursor cleavage sites and the determination of the exact N termini of the two major cleavage products, G_N and G_C. The novel protease SKI-1 [50], also responsible for the proteolytic processing of the arenavirus Lassa [34] and lymphocytic choriomeningitis (LCMV) glycoprotein precursors [7], has been identified as the cellular protease responsible for processing the N terminus of the mature G_N [55]. A yet unidentified protease is required for G_C processing and the exact C-terminus of G_N has not yet been determined. Two cleavage sites have been predicted for this processing step with the sequence RKLL and KKRKK, favoring the cellular proteases SKI-1 and furin, respectively, as the responsible proteases. More recently, it was shown that furin-like proteases indeed contribute to the posttranslational cleavage of the CCHFV glycoprotein precursor. An additional nonstructural protein (GP38) was discovered to be encoded within the M segment polyprotein by an ORF-located upstream of the G_N coding sequence and downstream from a variable mucin-like region [46]. The N terminus of the novel 38 kDa protein is generated by a furin or PC-mediated cleavage, whereas the C terminus is processed by SKI-1. Based on the lack of predicted transmembrane domains and its absence in virus particles, GP38 is likely a soluble protein that could play an important role in viral pathogenesis [46] as shown for other negative strand RNA viruses [30].

Bunyaviruses are known to mature by budding through the endoplasmic reticulum into cytoplasmic vesicles from Golgi membranes. The budding site seems to be defined by the retention of the glycoproteins G_N and G_C at that particular site [48, 49]. From a number of bunyaviral studies which have addressed the mechanisms of Golgi targeting and retention, one can conclude that G_N appears to carry the appropriate signal(s) [1, 2, 10, 11, 24, 33, 37, 38, 44, 51, 52]. That was recently confirmed for CCHFV using recombinantly expressed CCHFV glycoproteins [6, 25]. The cellular localization of these proteins was compared to authentic glycoproteins expressed during viral infection using indirect immunofluorescence assays (IFA), subcellular fractionation or western blot assays and confocal microscopy [25]. To further elucidate potential intracellular targeting or retention signals of the two glycoproteins G_N and G_C, GFP fusion proteins containing different parts of the CCHFV glycoprotein were analyzed for their intracellular targeting. The N-terminal glycoprotein G_N localized to the Golgi complex, a process mediated by retention or targeting signal(s) in the cytoplasmic domain and ectodomain [6, 25]. In contrast, the C-terminal glycoprotein G_C remained in the endoplasmic reticulum but could be rescued into the Golgi complex by coexpression of G_N [6, 25]. These data are consistent with the intracellular targeting of most bunyavirus glycoproteins and supports the general model for assembly and budding of bunyavirus particles in the Golgi compartment [14, 48].

Recently, two independent research groups published the complete nucleotide sequence of the CCHFV L segment [27, 31]. They observed 60% homology on nucleotide and amino acid level to the Dugbe virus L segment, the only other fully sequenced nairovirus [36]. Sequence alignments revealed that the most conserved areas were located within the polymerase module, containing important catalytic domains of vRNA-dependent RNA polymerases [28, 31, 39]. Interestingly, additional regions containing putative ovarian tumor (OTU)-like cysteine protease and helicase domains (zinc-finger motif, leucine zipper motif) were identified in the L segments of CCHF [27, 31] and Dugbe viruses, suggesting an autoproteolytic cleavage process for nairovirus L proteins as well as an helicase activity.

The CCHFV nucleoprotein was also the topic of recent nairovirus research on molecular biological level. It was demonstrated that CCHFV N is located in the perinuclear region in the absence of RNA genomic segments [3]. Based on the intracellular localization in the absence of other viral proteins or vRNA one has to assume that CCHFV N has an autonomous directionality sequence that is sufficient for its subcellular targeting. However, the targeting is dependent on actin filaments [3].

4.2.2. Noncoding regions

The ORF-flanking noncoding regions contain variable parts and at the extreme genome ends highly conserved nucleotides [16, 48] (Figs. 4-1B, C). These terminal sequences (8–13 nucleotides) are conserved between each RNA segment (L, M, and S) and also among members within the genus *Nairovirus* [19]. Because of their partial complementarity, intrastrand base-pair interactions between these terminal nucleotides can lead to noncovalently closed, circular RNAs [14, 23] (Fig. 4-1C). Such structures have been observed by electron microscopy of extracted vRNA from the phlebovirus Uukuniemi [26]. It is thought that these structures harbor the promoter elements necessary for initiation of both transcription and replication, the encapsidation signal for specific binding of the N protein, as well as the genome packaging signal. However, the role of individual nucleotides as well as the predicted secondary structure (panhandle versus corkscrew model [19]) within these potentially base-paired terminal genome regions (Fig. 4-1C) required for the specific interaction with the viral polymerase L and N, as demonstrated recently for other bunyaviruses [4, 5, 18, 19, 32, 42, 45], has yet to be determined for CCHFV or any other nairovirus.

4.3. REPLICATION CYCLE

Upon entry of the CCHFV into cells, the three virion ribonucleoprotein (RNP) species are released into the cytoplasm where the RNP-associated viral polymerase L catalyzes primary transcription resulting in the synthesis of the individual viral mRNAs [14, 48]. The vRNA segments also serve as templates for the synthesis of full-length complementary RNAs (cRNA), which in turn serve as templates for the synthesis of more vRNAs. This replicative cycle performs the amplification of viral genomic RNAs for subsequent packaging into progeny viruses and is catalyzed by the vRNA polymerase L (Fig. 4-2).

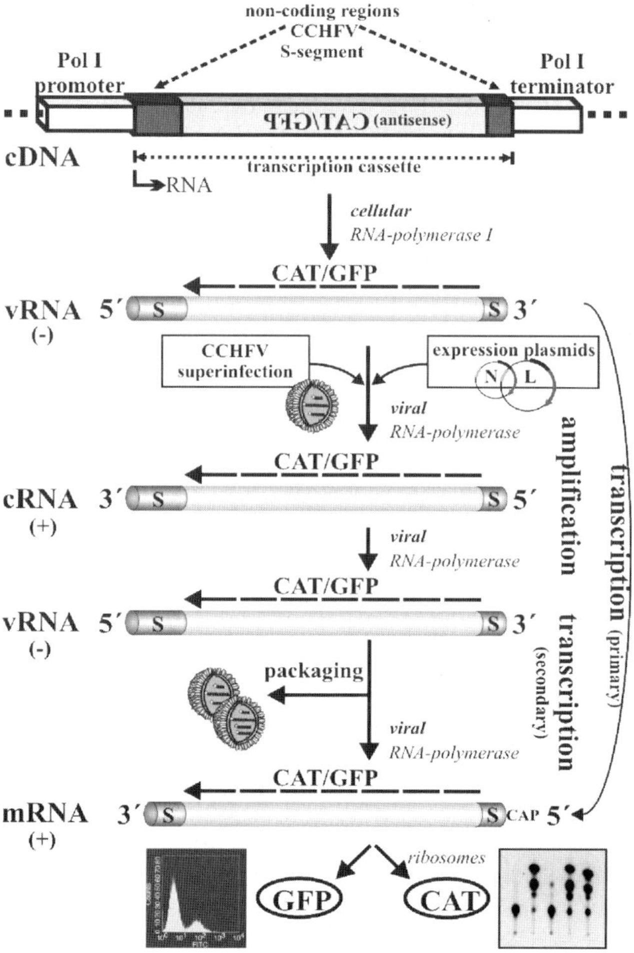

Fig. 4-2. RNA polymerase I-based reverse genetics system. Between the pol I promoter region and terminator sequence a CCHFV S segment–cDNA construct (transcription cassette) is exactly interspersed in antisense orientation. In addition, the ORF is replaced by a reporter gene encoding chloramphenicol acetyl transferase (CAT) or green fluorescent protein (GFP). After transfection into eukaryotic cell lines the cellular RNA pol I generates an artificial CCHF viral RNA (minigenome). After CCHF helper virus superinfection or transfection with viral expression plasmids, providing the polymerase L and nucleoprotein N from CCHF virus, transcription and replication of the artificial vRNA segment takes place. The CAT and GFP reporter genes facilitate the efficient detection of these processes. vRNA: viral RNA, cRNA: complementary RNA, mRNA: messenger RNA.

4.4. REVERSE GENETICS

The lack of an efficient reverse genetics system for *Nairovirus* members has so far precluded the detailed mutational dissection of *cis*-acting elements in the promoter region of the three RNA genome segments as well as the identification of *trans*-acting factors required for efficient genome encapsidation, transcription, replication, and packaging. However, recently the first report appeared describing attempts to develop such a system [20]. It allows driving replication and transcription of a reporter cDNA cassette (chloramphenicol acetyltransferase [CAT] or green fluorescent protein [GFP]) flanked by the 5′ and 3′ noncoding sequences from the S-RNA segment of CCHFV (Fig. 4-2). The transcription of the reporter cassette, flanked by the human polymerase I (pol I) promoter and murine pol I terminator [41, 59], takes place in the nucleus by the cellular RNA pol I. This results in an uncapped, nonpolyadenylated, and nonspliced RNA transcript with the correct viral 5′ and 3′ terminal sequences

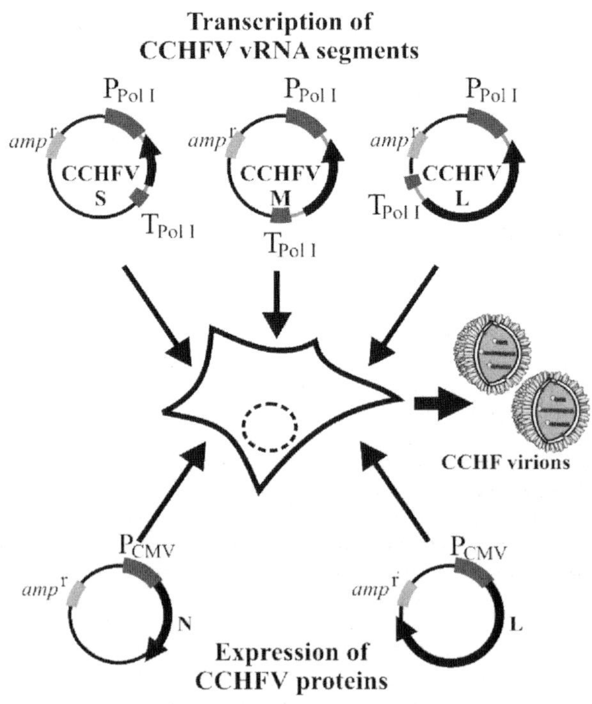

Fig. 4-3. Generation of recombinant CCHFV – infectious clone system. After transfection of pol I-driven vRNA transcription plasmids (producing L-, M-, and S-vRNA segments) and cotransfection of Cytomegalovirus (CMV)-driven viral protein expression plasmids (producing polymerase L and nucleoprotein N) transcription and replication of the vRNA segments takes place. Subsequently, the supernatant can be examined for CCHFV particles by passaging aliquots of the supernatant to fresh cell monolayer and analyse the virus titer by plaque assay.

[21, 59]. Following transport to the cytoplasm, the reporter RNA (minigenome) is replicated and transcribed by the viral L and N proteins supplied by superinfection with CCHF helper virus [20] (Fig. 4-2). Such a helper virus-driven minigenomes rescue system is a useful tool for the identification and dissection of *cis*-acting genome elements and allows the efficient study of the molecular biology of CCHFV. A similar system based on the pol I was initially used for Influenza A viruses [41, 59] and subsequently modified for the use with other bunyaviruses, i.e. Uukuniemi, Hantaan, La Crosse, and Rift Valley fever virus [8, 17–19, 22].

Using this CCHFV minigenome rescue system it could be confirmed that the viral ORF-flanking noncoding regions contain all information necessary for efficient genome segment encapsidation, transcription, and replication. Moreover, using the established CCHFV helper virus-driven system (Fig. 4-2), the successful transfer of reporter gene activity to fresh cells demonstrated the generation of recombinant CCHFV, thereby confirming minigenome packaging into progeny viruses [20]. Confirming the functionality of CCHFV nucleoprotein and viral polymerase for minigenome encapsidation, transcription, replication, and packaging is the basis for the subsequent development of an infectious clone system with the overall goal to produce recombinant viruses solely from cDNAs, without the use of any helper viruses [40, 56] (Fig. 4-3). Such a system would offer a unique opportunity to study the biology of nairoviruses and to develop therapeutic and prophylactic measures against CCHFV infections.

REFERENCES

1. Andersson AM, Melin L, Bean A, Pettersson RF (1997) A retention signal necessary and sufficient for Golgi localization maps to the cytoplasmic tail of a Bunyaviridae (Uukuniemi virus) membrane glycoprotein. J Virol 71:4717–4727
2. Andersson AM, Pettersson, RF (1998) Targeting of a short peptide derived from the cytoplasmic tail of the G1 membrane glycoprotein of Uukuniemi virus (Bunyaviridae) to the Golgi complex. J Virol 72:9585–9596
3. Andersson I, Simon M, Lundkvist A, Nilsson M, Holmstrom A, Elgh F, Mirazimi A (2004) Role of actin filaments in targeting of Crimean-Congo hemorrhagic fever virus nucleocapsid protein to perinuclear regions of mammalian cells. J Med Virol 72:83–93
4. Barr JN, Elliott RM, Dunn EF, Wertz GW (2003) Segment-specific terminal sequences of Bunyamwera bunyavirus regulate genome replication. Virology 311:326–338
5. Barr JN, Wertz GW (2004) Bunyamwera bunyavirus RNA synthesis requires cooperation of 3′- and 5′-terminal sequences. J Virol 78:1129–1138
6. Bertolotti-Ciarlet A, Smith J, Strecker K, Paragas J, Altamura LA, McFalls JM, Frias-Staheli N, Garcia-Sastre A, Schmaljohn CS, Doms RW (2005) Cellular localization and antigenic characterization of crimean-congo hemorrhagic fever virus glycoproteins. J Virol 79:6152–6161
7. Beyer WR, Popplau D, Garten W, von Laer D, Lenz O (2003) Endoproteolytic processing of the lymphocytic choriomeningitis virus glycoprotein by the subtilase SKI-1/S1P. J Virol 77:2866–2872
8. Blakqori G, Weber F (2005) Efficient cDNA-based rescue of La Crosse bunyaviruses expressing or lacking the nonstructural protein NSs. J Virol 79:10420–10428
9. Burt FJ, Spencer DC, Leman PA, Patterson B, Swanepoel R (1996) Investigation of tick-borne viruses as pathogens of humans in South Africa and evidence of Dugbe virus infection in a patient with prolonged thrombocytopenia. Epidemiol Infect 116:353–361

10. Chen SY, Compans RW (1991) Oligomerization, transport, and Golgi retention of Punta Toro virus glycoproteins. J Virol 65:5902–5909
11. Chen SY, Matsuoka Y, Compans RW (1991). Golgi complex localization of the Punta Toro virus G2 protein requires its association with the G1 protein. Virology 183:351–365
12. Clerex-Van Haaster CM, Clerex JP, Ushijima H, Akashi H, Fuller F, Bishop DH (1982) The 3′ terminal RNA sequences of bunyaviruses and nairoviruses (Bunyaviridae): evidence of end sequence generic differences within the virus family. J Gen Virol 61 (Pt 2):289–292
13. Clerx JP, Casals J, Bishop DH (1981) Structural characteristics of nairoviruses (genus *Nairovirus*, Bunyaviridae). J Gen Virol 55:165–178
14. Elliott RM (1996) The Bunyaviridae. Plenum Press, New York/London
15. Elliott RM, Bouloy M, Calisher CH, Goldbach R, Moyer JT, Nichol ST, Pettersson RF, Plyusnin A, Schmaljohn CS (2000) Family Bunyaviridae. In: van Regenmortel MHV, Fauquet CM, Bishop DHL, Cartens EB, Estes MK, Lemon S, Maniloff J, Mayo MA, McGeogch D, Pringle CR, Wickner RB (eds) Virus taxonomy-classification and nomenclature of viruses. Seventh report of the international committee on taxonomy of viruses. Academic Press, San Diego, pp 599–621
16. Elliott RM, Schmaljohn CS, Collett MS (1991) Bunyaviridae genome structure and gene expression. Curr Top Microbiol Immunol 169:91–141
17. Flick K, Hooper JW, Schmaljohn CS, Pettersson RF, Feldmann H, Flick R (2003) Rescue of Hantaan virus minigenomes. Virology 306:219–224
18. Flick K, Katz A, Overby A, Feldmann H, Pettersson RF, Flick R (2004) Functional analysis of the noncoding regions of the Uukuniemi virus (Bunyaviridae) RNA segments. J Virol 78:11726–11738
19. Flick R, Elgh F, Pettersson RF (2002) Mutational analysis of the Uukuniemi virus (Bunyaviridae family) promoter reveals two elements of functional importance. J Virol 76: 10849–10860
20. Flick R, Flick K, Feldmann H, Elgh F (2003) Reverse genetics for Crimean-Congo hemorrhagic fever virus. J Virol 77:5997–6006
21. Flick R, Hobom G (1999) Transient bicistronic vRNA segments for indirect selection of recombinant influenza viruses. Virology 262:93–103
22. Flick R, Pettersson RF (2001) Reverse genetics system for Uukuniemi virus (Bunyaviridae): RNA polymerase I-catalyzed expression of chimeric viral RNAs. J Virol 75:1643–1655
23. Flick R, Whitehouse CA (2005) Crimean-Congo hemorrhagic fever virus. Curr Mol Med 5:753–760
24. Gerrard SR, Nichol ST (2002) Characterization of the Golgi retention motif of Rift Valley fever virus G(N) glycoprotein. J Virol 76:12200–12210
25. Haferkamp S, Fernando L, Schwarz TF, Feldmann H, Flick R (2005) Intracellular localization of Crimean-Congo hemorrhagic fever (CCHF) virus glycoproteins. Virol J 2:42–56
26. Hewlett MJ, Pettersson RF, Baltimore D (1977) Circular forms of Uukuniemi virion RNA: an electron microscopic study. J Virol 21:1085–1093
27. Honig JE, Osborne JC, Nichol ST (2004) Crimean-Congo hemorrhagic fever virus genome L RNA segment and encoded protein. Virology 321:29–35
28. Honig JE, Osborne JC, Nichol ST (2004) The high genetic variation of viruses of the genus *Nairovirus* reflects the diversity of their predominant tick hosts. Virology 318:10–16
29. Hoogstraal H (1979) The epidemiology of tick-borne Crimean-Congo hemorrhagic fever in Asia, Europe, and Africa. J Med Entomol 15:307–417
30. Kawaoka Y (2004) Biology of negative strand RNA viruses: power of reverse genetics. Curr Top Microbiol Immunol 283
31. Kinsella E, Martin SG, Grolla A, Czub M, Feldmann H, Flick R (2004) Sequence determination of the Crimean-Congo hemorrhagic fever virus L segment. Virology 321:23–28
32. Kohl A, Bridgen A, Dunn E, Barr JN, Elliott RM (2003) Effects of a point mutation in the 3′ end of the S genome segment of naturally occurring and engineered Bunyamwera viruses. J Gen Virol 84:789–793

33. Lappin DF, Nakitare GW, Palfreyman JW, Elliott RM (1994) Localization of Bunyamwera bunyavirus G1 glycoprotein to the Golgi requires association with G2 but not with NSm. J Gen Virol 75 (Pt 12):3441–3451
34. Lenz O, ter Meulen J, Klenk HD, Seidah NG, Garten W (2001) The Lassa virus glycoprotein precursor GP-C is proteolytically processed by subtilase SKI-1/S1P. Proc Natl Acad Sci USA 98:12701–12705
35. Logan TM, Linthicum KJ, Bailey CL, Watts DM, Moulton JR (1989) Experimental transmission of Crimean-Congo hemorrhagic fever virus by *Hyalomma truncatum* Koch. Am J Trop Med Hyg 40:207–212
36. Marriott AC, Nuttall PA (1996) Large RNA segment of Dugbe nairovirus encodes the putative RNA polymerase. J Gen Virol 77 (Pt 8):1775–1780
37. Matsuoka Y, Chen SY, Compans RW (1994) A signal for Golgi retention in the bunyavirus G1 glycoprotein. J Biol Chem 269:22565–22573
38. Matsuoka Y, Chen SY, Holland CE, Compans RW (1996) Molecular determinants of Golgi retention in the Punta Toro virus G1 protein. Arch Biochem Biophys 336:184–189
39. Meissner JD, Seregin SS, Seregin SV, Vyshemirskii OI, Samokhvalov EI, Lvov DK, Netesov SV, Petrov VS (2006) A variable region in the Crimean-Congo hemorrhagic fever virus L segment distinguishes between strains isolated from different geographic regions. J Med Virol 78:223–228
40. Neumann G, Kawaoka Y (2004) Reverse genetics systems for the generation of segmented negative-sense RNA viruses entirely from cloned cDNA. Curr Top Microbiol Immunol 283:43–60
41. Neumann G, Zobel A, Hobom G (1994) RNA polymerase I-mediated expression of influenza viral RNA molecules. Virology 202:477–479
42. Osborne JC, Elliott RM (2000) RNA binding properties of bunyamwera virus nucleocapsid protein and selective binding to an element in the 5′ terminus of the negative-sense S segment. J Virol 74:9946–9952
43. Papa A, Ma B, Kouidou S, Tang Q, Hang C, Antoniadis A (2002) Genetic characterization of the m RNA segment of Crimean-Congo hemorrhagic fever virus strains, China. Emerg Infect Dis 8:50–53
44. Pettersson RF, Andersson A, Melin L (1995) Mapping a retention signal for Golgi localization of a viral spike protein complex. Cold Spring Harb Symp Quant Biol 60:147–155
45. Prehaud C, Lopez N, Blok MJ, Obry V, Bouloy M (1997) Analysis of the 3′ terminal sequence recognized by the Rift Valley fever virus transcription complex in its ambisense S segment. Virology 227:189–197
46. Sanchez AJ, Vincent MJ, Erickson BR, Nichol ST (2006) Crimean-Congo hemorrhagic fever virus glycoprotein precursor is cleaved by Furin-like and SKI-1 proteases to generate a novel 38-kilodalton glycoprotein. J Virol 80:514–525
47. Sanchez AJ, Vincent MJ, Nichol ST (2002) Characterization of the glycoproteins of Crimean-Congo hemorrhagic fever virus. J Virol 76:7263–7275
48. Schmaljohn CS, Hooper JW (2001) Bunyaviridae: The viruses and their replication. In: Knipe DM, Howley PM (eds) Fields Virology, vol 2, 4th edn. Lippincott Williams and Wilkins, Philadelphia, pp 1581–1602
49. Schmaljohn CS, Le Duc JW (1998) Bunyaviridae, 9th edn. Edward Arnold, London
50. Seidah NG, Mowla SJ, Hamelin J, Mamarbachi AM, Benjannet S, Toure BB, Basak A, Munzer JS, Marcinkiewicz J, Zhong M, Barale JC, Lazure C, Murphy RA, Chretien M, Marcinkiewicz M (1999) Mammalian subtilisin/kexin isozyme SKI-1: a widely expressed proprotein convertase with a unique cleavage specificity and cellular localization. Proc Natl Acad Sci USA 96:1321–1326
51. Shi X, Elliott RM (2004) Analysis of N-linked glycosylation of hantaan virus glycoproteins and the role of oligosaccharide side chains in protein folding and intracellular trafficking. J Virol 78:5414–5422
52. Spiropoulou CF, Goldsmith CS, Shoemaker TR, Peters CJ, Compans RW (2003) Sin Nombre virus glycoprotein trafficking. Virology 308:48–63
53. Swanepoel R (1995) Nairovirus Infections, Chapman & Hall, London, pp 285–293

54. van Regenmortel MHV, Fauquet CM, Bishop DHL, Carstens EB, Estes MK, Lemon SM, Maniloff J, Mayo MA, McGeoch DJ, Pringle CR, Wickner RB (2000) Seventh Report of the International Committee of Taxonomy of Viruses, Academic Press, San Diego, CA
55. Vincent MJ, Sanchez AJ, Erickson BR, Basak A, Chretien M, Seidah NG, Nichol ST (2003) Crimean-Congo hemorrhagic fever virus glycoprotein proteolytic processing by subtilase SKI-1. J Virol 77:8640–8649
56. Walpita P, Flick R (2005) Reverse genetics of negative-stranded RNA viruses: a global perspective. FEMS Microbiol Lett 244:9–18
57. Watts DM, Ksiazek TG, Linthicum KJ, Hoogstraal H (1988) Crimean-Congo hemorrhagic fever. In: Monath TP (ed.), The Arboviruses: Epidemiology and Ecology, vol II. CRC Press, Boca Raton, FL, pp 177–222
58. Zeller HG, Cornet JP, Camicas JL (1994) Experimental transmission of Crimean-Congo hemorrhagic fever virus by West African wild ground-feeding birds to *Hyalomma marginatum rufipes* ticks. Am J Trop Med Hyg 50:676–681
59. Zobel A, Neumann G, Hobom G (1993) RNA polymerase I catalysed transcription of insert viral cDNA. Nucleic Acids Res 21:3607–3614

CHAPTER 5

MOLECULAR EPIDEMIOLOGY, GENOMICS, AND PHYLOGENY OF CRIMEAN-CONGO HEMORRHAGIC FEVER VIRUS

ROGER HEWSON, PH.D.
Virus Research, Novel and Dangerous Pathogens, Centre for Emergency Preparedness and Response, Health Protection Agency, Porton Down, Salisbury, SP4 0JG, England, UK.
Tel.: +44 (0)1980 612390; Fax: +44 (0)1980 610848; E-mail: Roger.Hewson@hpa.org.uk

5.1. INTRODUCTION

Crimean-Congo hemorrhagic fever virus (CCHFV) constitutes a group of viruses of the genus *Nairovirus* (family *Bunyaviridae*). Like all members of the Bunyaviridae, the genome of CCHFV is composed of tripartite single-stranded RNA. These segments, designated small (S), medium (M), and large (L), minimally encode the nucleocapsid (N), envelope glycoproteins (Gn and Gc), and RNA-dependent RNA polymerase (RdRp), respectively [38].

Published descriptions of major epidemics, outbreaks, and the ecology of CCHFV have been reviewed extensively [18, 43, 45]. Interestingly a common theme is illustrated by the very wide distribution of the virus, which stretches over much of Asia, extending from the Xinjiang region of China to the Middle East and southern Russia, and to focal endemic areas over much of Africa and parts of southeastern Europe. Thus, CCHFV is the most widely distributed agent of severe haemorrhagic fever known.

5.2. MOLECULAR EPIDEMIOLOGY

Classic serological methods have been important in determining CCHF distribution; however, these assays do not readily differentiate between alternative strains of CCHFV. In order to characterize viral strains in more detail and facilitate a global epidemiological study, molecular methods based on partial and complete sequence data of the S segment have been used to identify certain S segment genotypes [9, 13, 36]. These genotypes show a strong relationship to the geographical area of parent virus isolation, leading to the terminology Asia 1,

O. Ergonul and C. A. Whitehouse (eds.), Crimean-Congo Hemorrhagic Fever, 45–55.

Asia 2, Europe 1, etc., which has been employed as a simple description of genotype (Fig. 5-1). Furthermore, these studies also show that similar genotypes are found in distant geographical locations (Fig. 5-2), supporting the idea that virus or infected ticks may be carried over long distances during bird migration [10]. Anthropogenic factors, such as the trade in livestock, may have also played a role in the dispersal of CCHFV. Thus, molecular epidemiological observations support a global and dynamic reservoir of CCHF virus.

Sequence information on L segments has lagged behind those of both S and M segments primarily due to the technical difficulties in working with these very long molecules. Nevertheless, several data from strains is available and while the number of alternative strains is on a different scale to those of S segments, there is evidence that the S and L segments from the same strains have similar evolutionary history (Fig. 5-3). For M segments however, the situation is different and it enables an insight into the ways CCHFV have evolved.

5.3. GENETIC VARIATION AND EVOLUTION

The driving force for evolution is provided by genetic change and variation in genomes. These lead to phenotypes which are molded by selective forces, thus genomes gradually change with their changing environments. RNA viruses, with their large population sizes, swift, and mutation-prone replication rates are generally considered capable of rapid evolution [16]. Additional evolutionary processes of (i) recombination, and for viruses with segmented genomes (ii) reassortment, also offer potentially important routes of generating genetic diversity. The genomes of arthropod-borne RNA viruses however, need to function and maintain high fitness in both arthropod and vertebrate host cells. This maintenance on two fronts is frequently thought to constrain the evolutionary processes acting on arbovirus genomes [44]. Thus, low levels of genetic diversity are frequently observed for arboviruses. The genome of CCHFV is interesting since, as well as showing features of high genetic stability [13], it also shows features of high flexibility [8]. CCHFV is often described as an emerging virus [22, 47]. Studies of its genetic fine structure aimed at developing a better understanding of the ways it can change and evolve are helping to illuminate its nature as an emerging pathogen. Complete genome entries of several CCHFV are now available in GenBank, and analysis of these sequences are enabling evolutionary hypothesis to be inferred and tested.

5.3.1. Recombination

Genetic homologous recombination – the formation of chimeric RNA molecules from sequences previously separated on different molecules – is an important means of variation open to RNA genomes. Indeed, it is clear that homologous recombination has been an important process that has shaped the evolution of RNA viruses per se [46]. However, the contribution of its effects

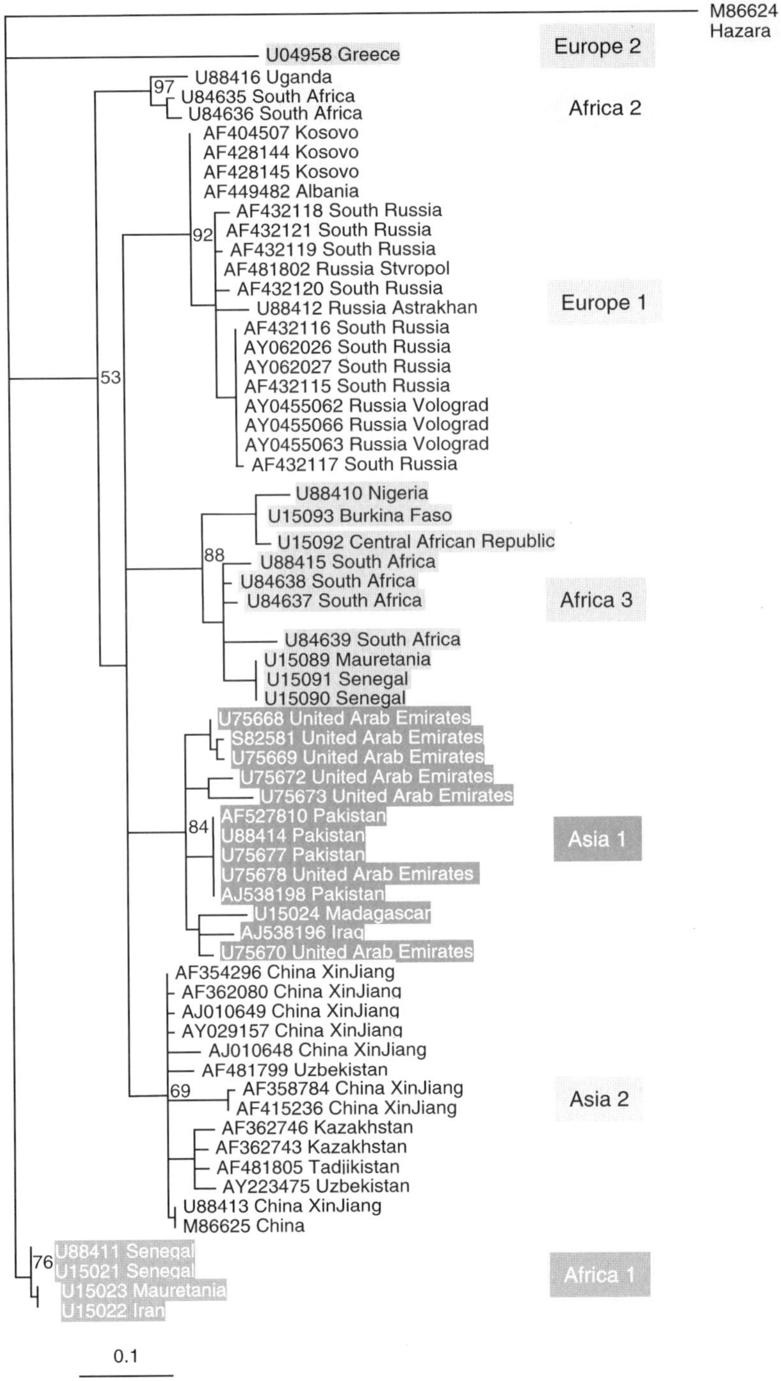

Fig. 5-1. Variation within CCHFV S segments and geographical correlation of genotypes. Maximum likelihood phylogenetic tree of CCHFV S RNA segments made from nucleotide alignments constructed from nucleotides 322–562 (Baghdad) enabling the incorporation of a maximum number of strains. Seven distinct lineages of S segment are extant.

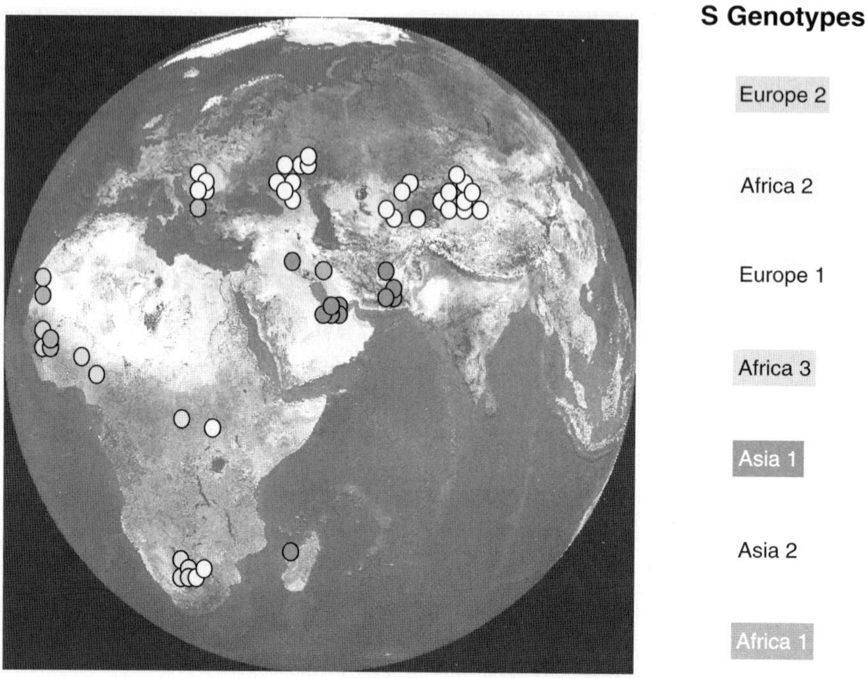

Fig. 5-2. Geographical correlation of genotypes. When superimposed onto the globe, the phylogenetic grouping of S RNA subtypes illustrates that the pattern of genetic diversity observed is largely related to the geographical distribution of the viruses. On some occasions, however, similar subtypes are sometimes found in distant geographical locations. It is possible that trade in livestock and perhaps long-distance carriage of virus or infected ticks during bird migration may have brought about links between such locations. (*See Color Plates*)

and the rate at which it occurs vary for different virus families. For example, it is known to be frequent in retroviruses [19], less common but periodic for positive-strand RNA viruses [24], but relatively infrequent in negative-strand RNA viruses [4, 32]. Yet, cases of recombination in the latter group do occur and evidence of it in the Bunyaviridae [39] and Arenaviridae [1] is well documented. Such reports have encouraged the search for recombination in CCHF viruses. Noteworthy evidence, including the demonstration of phylogenetic incongruence, often regarded as the best support for recombination [34], has been illustrated for the CCHF S segment [26]. Similar evidence for recombination in either of the M or L segments was not detected. A very recent study [8] also supported this latter observation in the majority of M and L segments. In addition, however, an analysis employing similarity plots, bootscanning and the informative sites tests, highlighted the possibility of recombination events within L segments of the Asian groups [8]. Interestingly, the cases of recombination are phylogenetically ancient and there is evidence that the sequences in question have diverged considerably after recombination. This suggests that

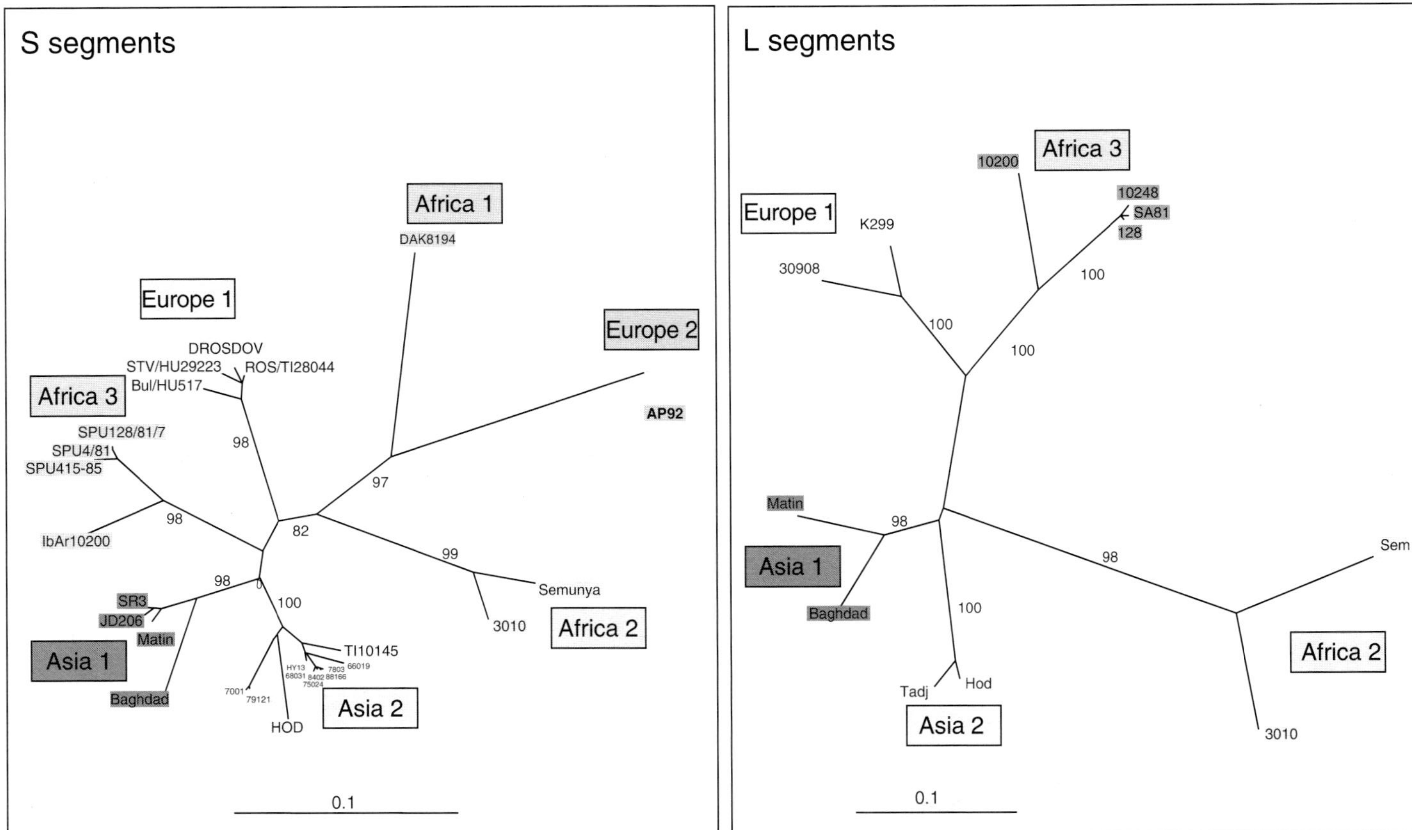

Fig. 5-3. Unrooted maximum-likelihood trees of full-length CCHFV S and L segment sequences showing phylogenetic relationships and correlation to geographic location. For the L segment, there are five different lineages or genotypes that, like S segments, have grouped according to their geographical location of isolation. From these data, it appears likely that the L segments also conform to the same grouping pattern as observed for S segments, although there are fewer L segment sequences. Additional sequence data provided by very recent work has enabled more comprehensive analysis and shows some exceptions to this idea. Nevertheless, while the more recent tree topologies of L and S segments are not analogous, they remain very similar.

recombination in CCHFV is a rare event and while it is difficult to estimate precise recombination rates, it is apparent that such rates are lower than those of point mutation. Nevertheless, an important consideration borne out of such work is that inferences about recombination events should only be entertained when molecular analysis have been constructed from complete segment sequence data. Additional consideration should also be given to the quality of published sequence data. A noteworthy example is provided by strains; (i) STV/HU29223 from European Russia (Stavropol) and (ii) Uzbek/TI10145 from Uzbekistan, which present some the of best evidence of genetic recombination in CCHFV as observed by phylogenetic incongruence [12]. However, this conclusion should be treated with caution as there is also evidence that the observed recombination may be an artifact [29].

5.3.2. Reassortment

RNA viruses with segmented genomes have the capacity to reassort their genomic segments into new genetically distinct viruses if the target cells are subject to dual infection. Indeed, this ability is believed to play a key role in the evolution, pathogenesis, and epidemiology of important pathogens such as influenza viruses, rotaviruses, and arthropod-borne orbiviruses [20, 25, 30]. Within the Bunyaviridae family as a whole, reassortment has been demonstrated for members of the genera *Orthobunyavirus* [2, 33, 42], *Phlebovirus* [40], *Hantavirus* [15, 23, 37], and *Tospovirus* [35], accordingly it is not surprising that segment reassortment in the *Nairovirus* genus has also been demonstrated [8, 14]. Here, evidence of reassortment in CCHFV is illustrated by a phylogenetic analysis of each strain or segment (Fig. 5-4). The phylogenetic groupings of S and L segments are consistent and show a correlation with the geography of parent strain isolation; however, the phylogenetic groupings of M segments are different. Distinct groups that were formed in S and L segments by Asia 1 and Asia 2 genotypes, for example, are not matched in the M segment phylogeny (Fig. 5-3). Although full-length sequence data is limited it is possible to ascertain that reassortment has taken place in the biogenesis of certain strains of CCHFV. For currently available data, the best evidence of reassortment is provided by the Matin strain isolated from Pakistan. If we consider groups for which there is full-length sequence data available on each segment (so that recombination events can be ruled out), then there appear to be strains with five types of S and L segment (Europe 1, Africa 2, Africa 3, Asia 1 [Middle East], and Asia 2 [Far East]) and five types of M segment (designated M1, M2, M3, M4, and M5). Even from the limited number of full-length sequences and the geographical location of virus isolations, it is possible to conceive that viruses of the Europe 1 lineage are composed of [S-Europe 1/L-Europe 1/M-4]; viruses of the Africa 2 lineage are [S-Africa 2/L-Africa 2/M-5]; viruses of the Africa 3 lineage are [S-Africa 3/L-Africa 3/M-2]; the majority of circulating viruses in the Middle East are composed of [S-Asia 1/L-Asia 1/M-2]; while in the Far East viruses contain the combinations [S-Asia 2/L-Asia 2/M-2], [S-Asia 2/L-Asia 2/M-1], and [S-Asia 2/L Asia 2/M-3]. From the available information it is possible to infer that strain Matin [S-Asia 1/L-Asia 1/M-1] is the result of reassortment between a Middle Eastern virus [S-Asia 1/L-Asia 1/M-2], and a Far-Eastern virus [S-Asia 2/L Asia 2/M-1]. It is likely that other strains have also arisen by segment reassortment. Indeed, very recent work has provided more complete sequence data from a broader range of strains [8] exposing many more examples of segment reassortment. It is clear that the majority of these events involve M segment reassortment, however, L segment reassortment viruses are also observed, albeit at a lower frequency. The reassortment events involving strains from widely separated geographical locations, illustrates that coreplication enabled by the movement and mixing of viruses is quite common. It follows that there may be a global reservoir of CCHFV, with local subreservoirs supporting high levels of virus circulation and permitting frequent coinfection (in which migratory birds play a significant role in virus dispersion).

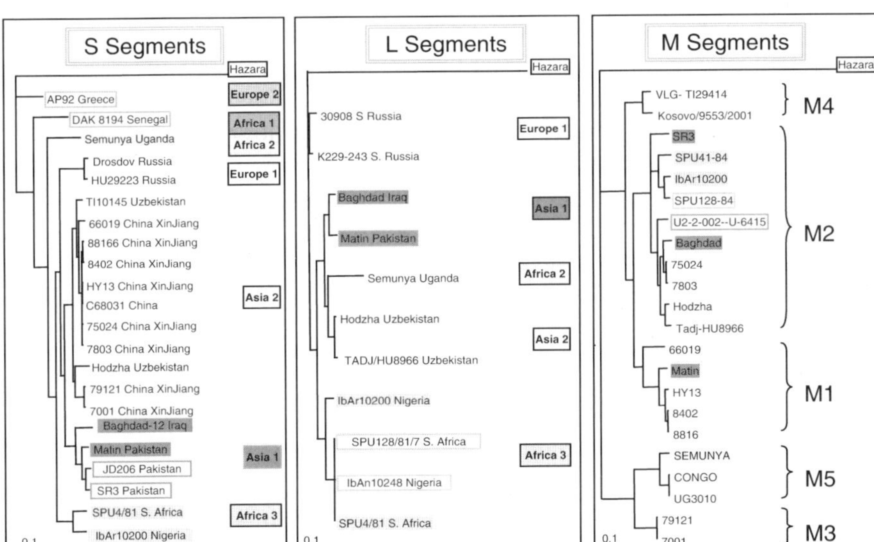

Fig. 5-4. Phylogenetic trees based on complete sequence show evidence of segment reassortment. Maximum likelihood phylogenetic groupings based on full-length sequence data and rooted against corresponding segments from Hazara virus as out groups. Filled boxes represent strains which are represented across all trees, colors correspond to grouping pattern described earlier. Comparison of trees shows that distinct geographical patterns formed by S and L segments are not maintained by M segments of the same strains. The best example of reassortment here is provided by the Matin strain. This was the first evidence of reassortment in CCHF viruses; more recent work has built on this notion and provided additional verification of RNA segment reassortment.

5.4. CONCLUSIONS

There is evidence that both recombination and reassortment are able to play roles in the evolution of CCHFV, in addition to general genetic drift. Obviously such genetic exchange requires coreplication of two or more strains within the same cell. The most likely coinfection environment where segment reassortment occurs is within ticks, where lasting virus infections persists for extended periods and superinfection with a second strain, during the strict requirement for blood meals, is very likely [28]. Given the currently available data on the low rate of recombination in CCHFV, and particularly the fact that the rate of recombination seems lower than general genetic drift, it appears that reassortment plays the most contributory role to the variability and flexibility of the CCHFV genome. Indeed, the low rate of recombination in negative-strand RNA viruses generally has led to suggestions that genome segmentation and reassortment have evolved to increase their fitness for survival [7, 31]. Specifically, while the high mutation rates of RNA viruses provide the raw material for evolutionary processes [21], mutations also introduce fitness compromising deleterious effects [6]. Genetic exchange through recombination or reassortment are recognized as adaptive methods of purging such effects [5, 27], thus in the practical absence of recombination, reassortment is able to take up the reins. In addition, reassortment enables alternative virus genotypes to be selected from a pool of functional segments.

The current evidence of reassortment in CCHFV [3, 8, 14] points principally to the exchange of M segments between viruses in mixed infections. In addition, the majority of data on L and S sequences show that in many cases these segments have evolved together as partners. Thus, in mixed virus infections where reassortment is a possibility, partner L and S segments have a propensity to end up in the same virus particle (due to the ostensibly strong interrelationships between the nuclear protein and RdRp) in order to constitute a viable new virus [3]. Some exceptions to this idea have been exposed by the availability of more sequences [8], and while it is clear that L and S segments trees are not analogous, they remain highly similar. M segments on the other hand seem to be more autonomous and could result in new virus phenotypes. Thus, as CCHFV are dispersed and introduced into new areas in which they are already endemic, the emergence of new CCHFV would principally be the result of M segment reassortment. Glycoprotein spikes encoded by M segments are well known for their ability to influence host range and cellular tropism [11, 41], furthermore, they are often associated with altered pathogenicity. These mechanisms, together with the likely contact and infection of new hosts, provide a foundation for the appearance of new CCHF disease and the emergence of new viruses [17].

These genomic studies highlight the importance of molecular surveillance to monitor and track the natural fluxes of virus and CCHF disease. A number of key questions can be asked in this context: For example, are certain viral genotypes more associated with severe disease? If so, are certain combinations of segments (or mutations) involved in the production of virulent strains? If there is a strain basis to disease, is viral genetic diversity increasing so that new strains with novel biological properties (such increased virulence or transmission potential) might appear? A practical conclusion of the evolutionary opportunities open to this virus is that CCHF diagnostic approaches and potential vaccine research strategies should be tested against isolates from all parts of the world, regardless of the intended location of use.

REFERENCES

1. Archer AM, Rico-Hesse R (1992) High genetic divergence and recombination in arenaviruses from the Americas. Virology 304:274–281
2. Chandler LJ, Hogge G, Endres M, Jacoby DR, Nathanson N, Beaty BJ (1991) Reassortment of La Crosse and Tahyna bunyaviruses in *Aedes triseriatus* mosquitoes. Virus Res 20:181–191
3. Chamberlain J, Cook N, Lloyd G, Mioulet V, Tolley H, Hewson R (2005) Co-evolutionary patterns of variation in small and large RNA segments of Crimean-Congo hemorrhagic fever virus. J Gen Virol 86:3337–3341
4. Chare ER, Gould EA, Holmes EC (2003) Phylogenetic analysis reveals a low rate of homologous recombination in negative-sense RNA viruses. J Gen Virol 84:2691–2703
5. Chao L (1994) Evolution of genetic exchange in RNA viruses. In: Morse SS (ed.) The Evolutionary Biology of Viruses. Raven Press Pub, New York, pp 233–250
6. Chao L (1990) Fitness of RNA virus decreased by Muller's ratchet. Nature 348:454–455
7. Chao L (1988) Evolution of sex in RNA viruses. J Theor Biol 133:99–122
8. Deyde VM, Khristova ML, Rollin PE, Ksiazek TG, Nichol ST (2006) Crimean-Congo hemorrhagic fever virus genomics and global diversity. J Virol 80:8834–8842
9. Drosten C, Minnak D, Emmerich P, Schmitz H, Reinicke T (2002) Crimean-Congo hemorrhagic fever in Kosovo. J Clin Microbiol 40:1122–1123
10. Gonzalez-Scarano F, Nathanson N (1996) Bunyaviridae. In: Fields BN, Knipe DM, Howley PM (eds) Virology, 4th edn. Lippincott-Raven Pub, Philadelphia, pp 1473–1504
11. Govorkova EA, Rehg JE, Krauss S, Yen HL, Guan Y, Peiris M, Nguyen TD, Hanh TH, Puthavathana P, Long HT, Buranathai C, Lim W, Webster RG, Hoffmann E (2005) Lethality to ferrets of H5N1 influenza viruses isolated from humans and poultry in 2004. J Virol 79:2191–2198
12. Hewson R, Chamberlain J (2003) (unpublished)
13. Hewson R, Chamberlain J, Clegg C, Jamil B, Hasan R, Gmyl A, Gmyl L, Smirnova SE, Lukashev A, Karganova G (2004) Crimean-Congo haemorrhagic fever virus: sequence analysis of the small RNA segments from a collection of viruses world wide. Virus Res 102:185–189
14. Hewson R, Gmyl A, Gmyl L, Smirnova SE, Karganova G, Jamil B, Hasan R, Chamberlain J, Clegg C (2004) Evidence of segment reassortment in Crimean-Congo haemorrhagic fever virus. J Gen Virol 85:3059–3070
15. Henderson WW, Monroe MC, St Jeor SC, Thayer WP, Rowe JE, Peters CJ, Nichol ST (1995) Naturally occurring Sin Nombre virus genetic reassortants. Virology 214:602–610
16. Holland J, Spindler K, Horodyski F, Grabau E, Nichol S, VandePol S (1982) Rapid evolution in RNA genomes. Science 215:1577–1585
17. Holmes EC (2004) The phylogeography of human viruses. Mol Ecol 13:745–756
18. Hoogstraal H (1979) The epidemiology of tick borne Crimean-Congo hemorrhagic fever in Asia, Europe and Africa. J Med Entomol 15:307–417
19. Hu WS, Temin HM (1990) Retroviral recombination and reverse transcription. Science 250:1227–1233
20. Iturriza-Gomara M, Isherwood B, Desselberger U, Gray J (2001) Reassortment in vivo: driving force for diversity of human rotavirus strains isolated in the United Kingdom between 1995 and 1999. J Virol 75:3696–3705

21. Jenkins GM, Rambaut A, Pybus OG, Holmes EC (2002) Rates of molecular evolution in RNA viruses: a quantitative phylogenetic analysis. J Mol Evol 54:152–161
22. Johnson KM, Morse S (1993) Emerging viruses in context: an overview of viral hemorrhagic fevers. In: Morse S. (ed.) Emerging Viruses. Oxford University Press, Oxford, pp 46–57
23. Klempa B, Schmidt HA, Ulrich R, Kaluz S, Labuda M, Meisel H, Hjelle B, Kruger DH (2003) Genetic interaction between distinct Dobrava hantavirus subtypes in *Apodemus agrarius* and *A. flavicollis* in nature. J Virol 77:804–809
24. Lai MMC (1992) RNA recombination in animal and plant viruses. Microbiol Rev 56:61–79
25. Li KS, Xu KM, Peiris JS, Poon LL, Yu KZ, Yuen KY, Shortridge KF, Webster RG, Guan Y (2003) Characterisation of H9 subtype influenza viruses from the ducks of southern China: a candidate for the next influenza pandemic in humans? J Virol 77:6988–6994
26. Lukashev A (2005) Evidence for recombination in Crimean-Congo hemorrhagic fever virus. J Gen Virol 86:2333–2338
27. Muller HJ (1964) The relation of recombination to mutational advance. Mutant Res 1:2–9
28. Nuttall PA, Jones LD, Davies CR (1991) Advances in disease vectors research: the role of arthropod vectors in arbovirus evolution. Adv Dis Vector Res 8:16–45
29. Petrova ID, Seregin SV, Petrov VS, Veshemirskii OI, Kuzina II, Lvov DK, Smokhvalov EI, Tyunnikov GI, Gutorov VV, Yashina LN, Netesov SV (2003) Genetic characteristics of the S segments of RNA from two strains of the Crimean-Congo hemorrhagic fever virus isolate in the south of Russia and in Uzbekistan (in Russian) Vopr Virusol 48:8–11
30. Pierce CM, Balasuriya UB, MacLachlan NJ (1998) Phylogenetic analysis of the S10 gene of field and laboratory strains of bluetongue virus from the United States. Virus Res 55:15–27
31. Pressing J, Reanney DC (1984) Divided genomes and intrinsic noise. J Mol Evol 20:135–146
32. Pringle CR, Parry JE (1982) Measurement of surface antigen by specific bacterial adherence and scanning electron microscopy (SABA/SEM) in cells infected by vesiculovirus ts mutants. J Gen Virol 59:207–211
33. Pringle CR, Lees JF, Clark W, Elliot R (1984) Genome subunit reassortment among bunyaviruses analysed by dot hybridization using molecularly cloned complementary DNA probes. Virology 135:244–256
34. Posada D, Crandall KA, Holmes EC (2002) Recombination in evolutionary genomics. Ann Rev Genet 36:75–97
35. Qiu WP, Geske SM, Hickey CM, Moyer JW (1998) Tomato spotted wilt Tospovirus genome reassortment and genome segment-specific adaptation. Virology 244:186–194
36. Rodriguez LL, Maupin GO, Ksiazek TG, Rollin PE, Khan AS, Schwarz TF, Lofts RS, Smith JF, Noor AM, Peters CJ, Nichol ST (1997) Molecular investigation of a multisource outbreak of Crimean-Congo hemorrhagic fever in the United Arab Emirates. Am J Trop Med Hyg 57:512–518
37. Rodriguez LL, Owens JH, Peters CJ, Nichol ST (1998) Genetic reassortment among viruses causing hantavirus pulmonary syndrome. Virology 242:99–106
38. Schmaljohn CS, Hooper JW (2001) Bunyaviridae. In: Fields BN, Knipe DM, Howley PM (eds) Virology, 4th edn. Lippincott-Raven Pub, Philadelphia, pp 1581–1602
39. Sibold C, Meisel H, Kruger DH, Labuda M, Lysy J, Kozuch O, Percoch M, Vaheri A, Plyusnin A (1999) Recombination in Tula hantavirus evolution: analysis of genetic lineages from Slovakia. J Virol 73:667–675
40. Turell MJ, Saluzzo JF, Tammariello RF, Smith JF (1990) Generation and transmission of Rift Valley fever viral reassortants by the mosquito *Culex pipiens*. J Gen Virol 71:2307–2312
41. To KF, Tong JH, Chan PK, Au FW, Chim SS, Chan KC, Cheung JL, Liu EY, Tse GM, Lo AW, Lo YM, Ng HK (2004) Tissue and cellular tropism of the coronavirus associated with severe acute respiratory syndrome: an in-situ hybridization study of fatal cases. J Pathol 202:157–163
42. Uruidi V, Bishop DH (1992) Non-random reassortment between the tripartite RNA genomes of La Crosse and snowshoe hare viruses. J Gen Virol 73:2255–2265
43. Watts DM, Kziasek TG, Linthicum KJ, Hoogstraal H (1988) Crimean-Congo hemorrhagic fever. In: Monath TP (ed.) The Arboviruses: Epidemiology and Ecology. vol 2. CRC Press, Boca Raton, FL, pp 177–260

44. Weaver SC (2006) Evolutionary influences in arboviral disease. Curr Top Microbiol Immunol 299:285–314
45. Whitehouse CA (2004) Crimean-Congo hemorrhagic fever. Antiviral Res 64:145–160
46. Worobey M, Holmes EC (1999) Evolutionary aspects of recombination in RNA viruses. J Gen Virol 80:2535–2543
47. WHO (2004) Report of the WHO/FAO/OIE consultation on emerging zoonotic diseases. http://whqlibdoc.who.int/hq/2004/WHO_CDS_CPE_ZFK_2004.9.pdf

SECTION III

EPIDEMIOLOGY AND ECOLOGY

CHAPTER 6

CRIMEAN-CONGO HEMORRHAGIC FEVER IN TURKEY

ZATI VATANSEVER, PH.D., RAMAZAN UZUN, DVM, PH.D., AGUSTIN ESTRADA-PENA, PH.D., AND ONDER ERGONUL, MD., M.P.H.*

**Corresponding author: Marmara University, School of Medicine, Infectious Diseases and Clinical Microbiology Department, Altunizade, Istanbul, Turkey. E-mail: ondergonul@yahoo.com*

6.1. HISTORY OF CCHF IN TURKEY

The first published seroepidemiologic study on Crimean-Congo hemorrhagic fever (CCHF) in Turkey was performed in the Agean region of Turkey in the 1970s [30]. According to this study, Crimean-Congo hemorrhagic fever virus (CCHFV) antibodies were detected in 96 out of 1,074 (9.2%) human serum samples, by hemaglutination inhibition test. Likewise, neutralizing antibodies against the virus were detected in 13 out of 96 (13.5%) samples. However, prior to 2002, no clinical cases of CCHF or virus detections in ticks were reported from Turkey.

In 2002 and 2003, febrile hemorrhagic patients were being admitted to various hospitals in Eastern Anatolia, mainly in Tokat and Sivas provinces. In addition, a significant number of the patients were referred to the tertiary hospitals of Ankara (the capital of Turkey). Because of such an unexpected clinical syndrome in the region, the Ministry of Health (MOH) of Turkey launched the first epidemiologic investigational study in July 2003. According to the study's report, the common epidemiological features of the patients included working in animal husbandry and history of tick bite. Clinically, all the patients had thrombocytopenia and most had leukopenia, elevated transaminases, especially aspartate transaminase (AST) and lactate dehydrogenase (LDH), fever, myalgia, nausea, and headache [33].

In Turkey, cases of viral hemorrhagic fever (VHF) had not been previously reported. Therefore, initially, endemic etiologic agents other than VHF were considered. Thus, sera of the hemorrhagic patients were tested for *Rickettsia*, *Ehrlichia*, *Leptospira*, and *Coxiella*; seven were reported as acute Q fever and treated accordingly [13]. Other than these bacteriologic causes, chemical or radioactive toxications were also considered. A scientific ad hoc committee of

O. Ergonul and C. A. Whitehouse (eds.), Crimean-Congo Hemorrhagic Fever, 59–74.

MOH defined the problem cases and described the diagnostic, therapeutic, and preventive measures in the summer of 2003. The etiologic agent was not identified at that time, but the cases were classified as mild, moderate, and severe according to their thrombocyte count. Those with a thrombocyte count of 100–150,000/mL were defined as mild, 70–100,000/mL were moderate, and <70,000 were severe cases. The case definition as defined by the MOH was based on epidemiologic, clinical, and laboratory characteristics, and included:

- Working with animal husbandry or history of tick bite
- Individuals, who had fever, myalgia, malaise, diarrhea
- Patients who had leukopenia, thrombocytopenia, elevated AST, alanine aminotransferase (ALT), and LDH levels

The case management was performed under the referral system of the MOH. All the hospital expenses of the patients were covered by the MOH. Sera were collected at the national level in Refik Saydam Hygiene Center of MOH in Ankara. Doxycycline was suggested for all the suspected cases in case of bacteriologic etiology. In addition to the medical measures, preventive studies against ticks and education of medical and veterinarian personnel were initiated.

Sera of the patients were sent to the Institut Pasteur in Lyon, France for further studies. In August 2003, the serologic and molecular investigations at the Institut Pasteur revealed that the etiologic agent was CCHFV. This was the first VHF syndrome recognized in Turkey. All the measures previously taken were reevaluated and updated accordingly. By 2004, the MOH of Turkey collaborated with the Centers for Disease Control and Prevention, USA, and an extended ad hoc scientific committee has been working in accordance with the MOH since that time.

6.2. CHARACTERISTICS OF THE CCHF OUTBREAK OF 2002–2006

According to the current protocol established by the MOH, sera of suspected CCHF patients were sent to the Central Virology Laboratory of the Refik Saydam Hygiene Center. The management of suspected CCHF cases in Turkey is described in Fig. 6-1.

By the end of 2006, there were 1,103 confirmed cases and 59 deaths [1] (Fig. 6-2) from 716 rural villages, mainly located in the transition zone between the Central Anatolian plateau and northern-most mountains (Fig. 6-3). Most of the areas share similar geographical characteristics, such as small mountains bisected by streams, which form valley systems. The villages where CCHF occurred are surrounded by oak-dominated scrub forests, which are inhabited by dense populations of wildlife, especially hare and wild boar. Crop fields and pastures, which border the forests, are the main areas where human tick bites occurred.

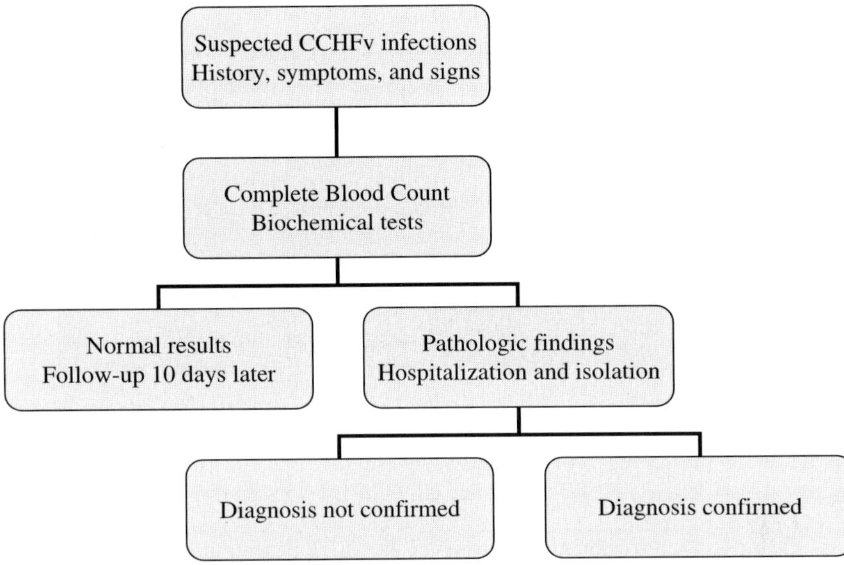

Fig. 6-1. Flow diagram showing the management of suspected CCHF cases in Turkey.

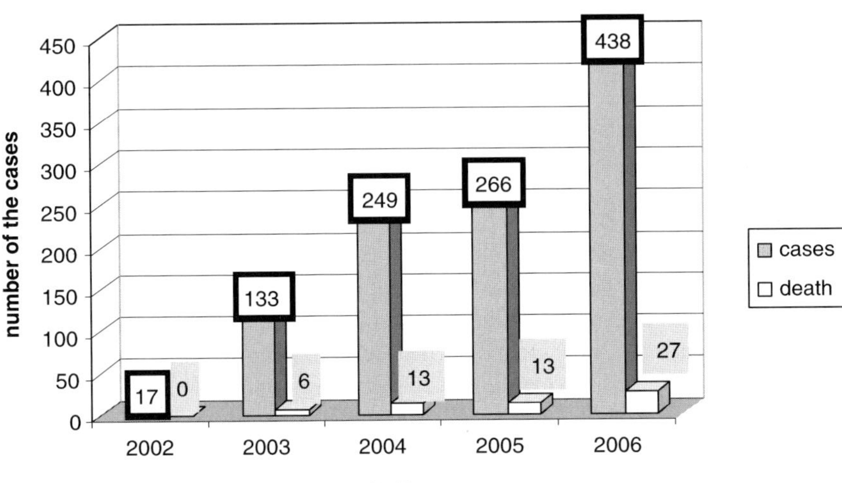

Fig. 6-2. Graph showing the numbers of CCHF cases and deaths in Turkey from 2002 to 2006.

Demographic and clinical characteristics of the CCHF patients from Turkey have been detailed in various publications [6, 8, 11, 17, 21]. This information provides a good description of the CCHF cases in Turkey (Tables 6-1 and 6-2). In general, the gender distribution is equal; however, the rate of females with CCHF was higher in one study from Black Sea region [17]; however, that study

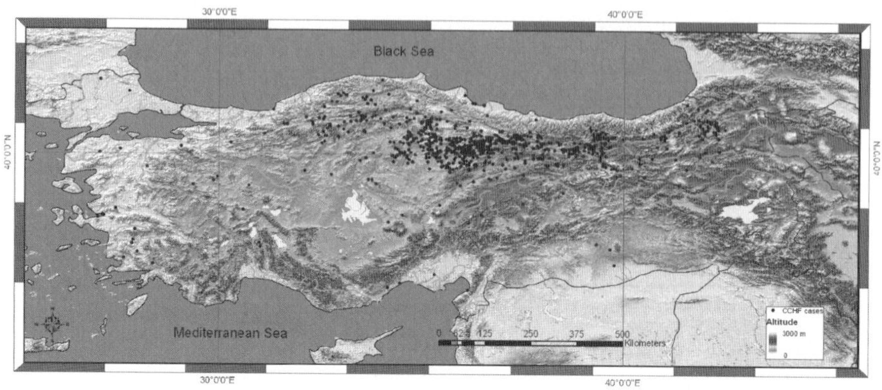

Fig. 6-3. CCHF distribution in Turkey. Dots represent the localities where confirmed cases of CCHF were recorded for the period 2003–2006.

Table 6-1. Demographic characteristics of the CCHF patients in Turkey

Characteristics	Result
Female gender	50%
Mean age	43
Mean days from symptoms to hospital	6 days
History of tick bite	40–60%
Farmer	90%
Number of infected health-care workers	4

was limited to only 19 patients. As mentioned previously, the majority of patients were farmers, who were working in agriculture and/or animal husbandry. The mean age of 43 reflects the working population. Only 40–60% of the confirmed cases had a history of the tick bite. However, it is likely that tick bites could have gone unnoticed, thus the lack of a tick bite history can not rule out infection (Table 6-1).

Nausea and vomiting, myalgia, headache, and fever are common symptoms. The beginning symptoms were typically severe flu-like. However, the body temperature rarely exceeds 40°C. Bleeding was seen in almost half of the patients. The most common bleeding type was epistaxis. Maculopapular rash was seen in one third of the patients (Table 6-2). All the patients in Turkey had thrombocytopenia, 80% had leukopenia, and only 30% had anemia.

In the diagnosis of CCHF, other endemic diseases should also be considered. The other endemic diseases that should be considered in differential diagnosis include leptospirosis, Lyme disease, rickettsiosis, brucellosis, and Q fever.

Table 6-2. Approximate rates of signs and symptoms of CCHF patients from Turkey. (From clinical studies performed in Turkey [6, 8, 11, 17, 21].)

Signs and symptoms	Percentage
Nausea-vomiting	80
Myalgia	70
Fever	75
Headache	75
Bleeding	48
Epistaxis	40
Hematemesis	26
Hematuria	16
Melena	14
Hemoptysis	8
Conjunctival injection	42
Maculopapular rash	35
Hepatomegaly	35
Diarrhea	34
Somnolence	20
Jaundice	10
Splenomegaly	10
Lymphadenopathy	15

Since 1999, about 500 CCHF cases were reported from Southern Federal District of Russia (Kalmykia, Stavropol, Astrakhan, Rostov, Volgograd, and Dagestan) [2]. Although there is no evidence yet, the relation between these two simultaneously apperaing outbreaks in Russia and Turkey should be investigated.

6.3. THE CCHFV STRAIN IN TURKEY

Several CCHF viral RNA sequences from both patients and ticks from Turkey have been determined. Sequence analysis of these samples showed that the CCHFV strain circulating in Turkey is closely related to those virus lineages from Kosovo and southwestern Russia [17]. In two tick survey studies, the genetic sequences of the viruses obtained from ticks collected in the epidemic region were found to be closely related to the Balkan strains [32, 34]. Preliminary results from another study also revealed that the circulating virus in Turkey is closest to the Balkan cluster [20] (Fig. 6-4).

6.4. TICKS OF TURKEY AND THEIR DISTRIBUTION IN THE CCHF ENDEMIC AREA

Among the 899 valid tick species recognized worldwide, 25 Ixodid and 7 Argasid species have been reported from Turkey. Especially, ticks in the genera *Rhipicephalus*, *Hyalomma*, and *Boophilus* are widespread (Table 6-3) and

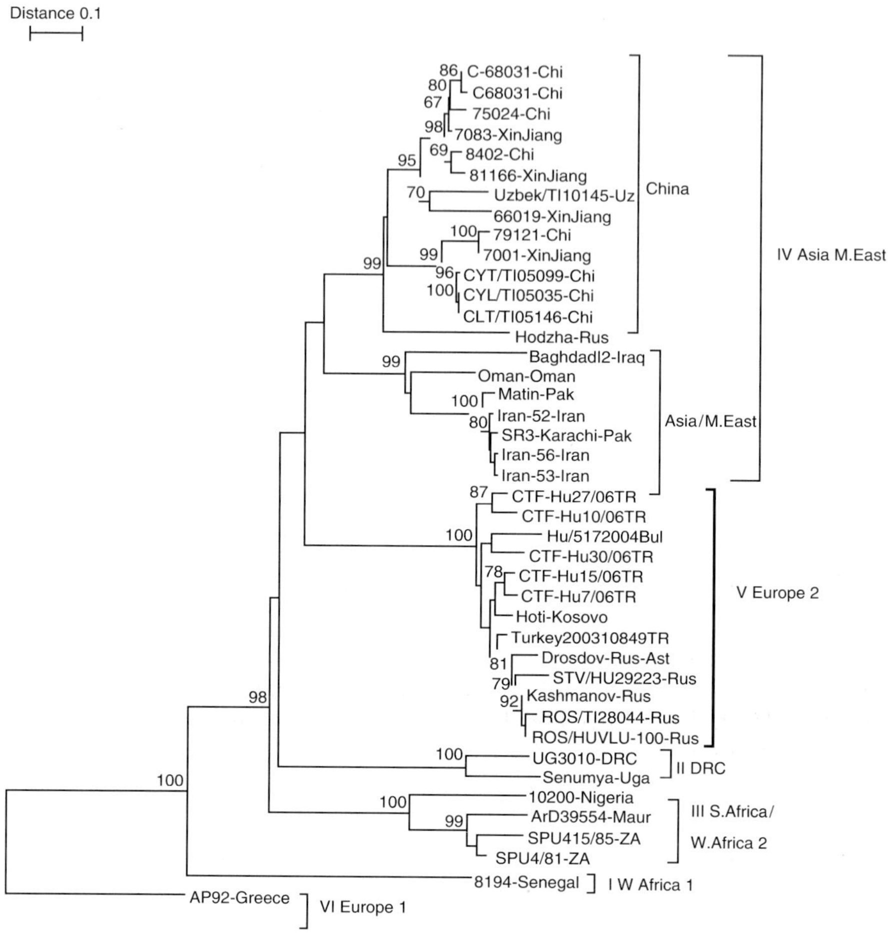

Fig. 6-4. Phylogenetic tree of the S segment RNA of CCHFV showing the placement of a strain originally isolated from a human patient from Turkey (Turkey 200310849TR) [20].

negatively influence the livestock industry in Turkey. Following two comprehensive studies [18, 19], numerous publications on the tick distribution and tick infestations of domestic animals in Turkey have resulted [4, 5, 28, 29, 35]. Since most of these studies were of the tick-borne diseases of the domestic animals in Turkey, little detailed information was available on the ecology of ticks in Turkey.

During the summer of 2005, field studies were conducted in 60 villages in the main epicenter of CCHF, around the province of Tokat. Preliminary results showed that 74% of cattle were tick-infested. Of the ticks collected, 83% were *Hyalomma* spp., with *Hyalomma marginatum marginatum* being predominant (95% of all *Hyalomma* spp.). Heavy infestation with *Hyalomma* spp. larvae was

Table 6-3. Common tick species affecting animal and human health in Turkey

Rhipicephalus bursa
Rhipicephalus turanicus
Rhipicephalus sanguineus
Hyalomma anatolicum anatolicum
Hyalomma anatolicum excavatum
Hyalomma detritum
Hyalomma matginatum marginatum
Boophilus annulatus
Haemaphysalis parva
Haemaphysalis punctata
Haemaphysalis sulcata
Dermacentor marginatus
Ornithodoros lahorensis
Argas persicus

found on a captured hare as well. CCHF antibodies were detected in 79% of sera collected from the same cattle (unpublished data). Molecular analysis showed that CCHFV RNA was present in three tick pools (two *H. margiantum margiantum* and one *Hyalomma detritum*) collected during this study [34]. In another study [32] carried out on sheep and cattle in the same area, 47% and 46% of the collected ticks were *Rhipicephalus bursa* and *H. marginatum marginatum*, respectively, and four tick pools (three *R. bursa* and one *H. marginatum marginatum*) were CCHFV RNA positive. In 2006, numerous ticks were collected from domestic animals throughout the country and 21 tick pools were CCHFV RNA positive by reverse transcription polymerase chain reaction (RT-PCR) [20]. Interestingly CCHFV positive ticks were found in some areas of the European region of Turkey where cases have not yet been reported. This may be an indication of possible virus dissemination towards the western part of the country and into the Balkans.

Until now *H. marginatum marginatum* never reached 5% of the collected ticks in any study in Turkey. The above results strongly indicate that *H. marginatum marginatum* ticks are involved in the outbreak and that there is high virus circulation in nature. This situation is quite similar to that seen in Crimea and the Southern Federal District of Russia where increased numbers of *H. marginatum marginatum* was recorded during CCHF outbreaks [16]. In recent years, increases in the *Hyalomma* spp. tick population have been noticed by local inhabitants and professionals in the CCHF-endemic areas of Turkey. On one occasion, three investigators were attacked by 140 unfed *H. marginatum marginatum* while walking a 30 m distance in a field near the CCHF epicenter (personal observations). Although local shepherds and agricultural workers are used to occasional tick bites during their daily activities, it has not been previously observed with such a magnitude.

The exact factors leading to the increase in the *Hyalomma* spp. tick population in Turkey are unknown. The number of days with a temperature of ≥5°C in April, and the daily mean temperature in April in the endemic region were reported to be increased in the years prior to the outbreak [9]. However, climate change is not necessarily the cause of the marked increased incidence of a variety of tick-borne diseases in many parts of Europe over the last two decades [26]. In general, CCHF outbreaks have developed on a background of favorable climatic factors and environmental changes beneficial for survival of large numbers of *Hyalomma* spp. ticks and of the hosts of both their immature and adult stages. In the former Soviet Union, environmental changes have resulted from neglect of agricultural lands, introduction of new settlers into infected areas, changing pasture patterns, converting floodplains to farmland, and flood control [16]. Interestingly a similar explanation was suggested for Turkey. According to this scenario [9, 10], prior to the initial outbreak in 2002, highlands in the region had been abandoned from pasturing between 1995 and 2001, and a strict hunting prohibition is still in effect for this region; this may have led to an increase in the population of wild animals and ticks. According to our personal observations and small-scale questionnaires, wildlife populations (especially hares and wild boar) were increased in the CCHF outbreak area, which supports the above scenario. There are no ordinary pastures for livestock in the region and cattle are grazed in the bush, sharing the same area with wildlife and their ticks.

6.5. SPATIAL DISTRIBUTION OF THE CCHF IN TURKEY

In order to determine the driving factors behind CCHF outbreaks, data on tick densities and tick infection rates are required to delineate areas of potential risk. However, distribution maps for vector ticks are not easily produced since most of the limited data on CCHFV-infected ticks from Turkey are from ticks feeding on domestic animals [32, 34] and not from questing ticks. In general, *H. marginatum marginatum* is considered the main vector of CCHFV in most parts of the Palaearctic region [16]. It was the main tick involved in outbreaks in Balkans, Crimea, and Southern Russia; the main factor leading to CCHF epidemics seems to be its increased abundance [16, 23]. Habitat pattern may also contribute to the risk of exposure to tick bites, as has been demonstrated for other tick-borne diseases. For instance, fragmented habitats have been incriminated as areas of higher risk exposure to infected ticks transmitting Lyme disease [3]. In any case, using geographical information systems and remote sensing, land cover, and vegetation indexes can now be mapped over large areas at high resolution.

Here, we define the spatial distribution of CCHF in Turkey since the onset of the CCHF epidemic in 2003. To this end, we mapped disease notifications and tick habitat suitability (HS) to determine whether there was significant variation in the risk for CCHF. Surface maps of standardized morbidity ratios were created and the spatial scan analysis method was used to test whether the number of cases and tick abundance correlated with the resident human population. By mapping the spatial relationships between CCHF risk, HS for ticks, and habitat fragmentation, we determined the spatial risk patterns associated with tick abundance and vegetation type.

6.5.1. Case notifications and spatial analysis of CCHF disease patterns

The MOH of Turkey compiled a list of CCHF cases at the village level, as confirmed by the Refik Saydam National Hygiene Center, according to their laboratory tests (both serology and polymerase chain reaction [PCR]) for the period 2003–2006. The mapping of CCHF disease was to the patient's place of residence, and we recognize that this may not necessarily be the same location at which the person was infected. However, the rural population in this region is commonly employed in agriculture, thus their movements are limited to areas close to their village or town of residence. Furthermore, adequate follow-up of the cases, including obtaining details of their normal habits through discussions with personal contacts, ensures that there is an adequate correlation between the patient's place of residence and the site of infection. The number of cases per 100,000 persons was obtained for each district. To facilitate further analysis, a grid of hexagonal polygons of 10 km radius was generated covering the entire country (3,803 hexagons total), and the number of cases and population values were allocated to each hexagon. CCHF disease patterns among the 3,803 hexagons were examined using spatial scan statistics [14]. This method was used to scan the entire country for significant disease clusters in a window size of 80 km. This method created a circular window at each hexagon centroid, and then tested the null hypothesis against the alternative hypothesis that there was an elevated risk of CCHF within, as compared to outside, the window. A likelihood ratio test statistic was then calculated for each window and its distribution under the null hypothesis was obtained by repeating the calculation on 999 random replications of the data set generated under the null hypothesis. The p-value of each cluster was obtained through using a Monte Carlo simulation. The null hypothesis of no cluster was rejected at a level of $p < 0.05$.

As of 2006, CCHF clinical cases in Turkey have been distributed along a narrow band covering most of the northern portion of the country. The human population in this region is highly heterogeneous, and this factor could explain, in part, the significant amount of variation in the observed disease rates. The 4-year CCHF-averaged incidence rate in Turkey was 3.37 cases per 100,000 person per year (py) and these values ranged from 0.12 (Konak) to 190.81 (Artova). There were large variations in the CCHF incidence rates among individual districts (Fig. 6-5) with seven districts above 100 cases per year and 23 below one case per year (of those with positive records). Figure 6-6 shows the spatial and temporal trends in CCHF incidence rates. While the origin of the current epidemic in Turkey is unknown, the first cases in 2003 were reported in

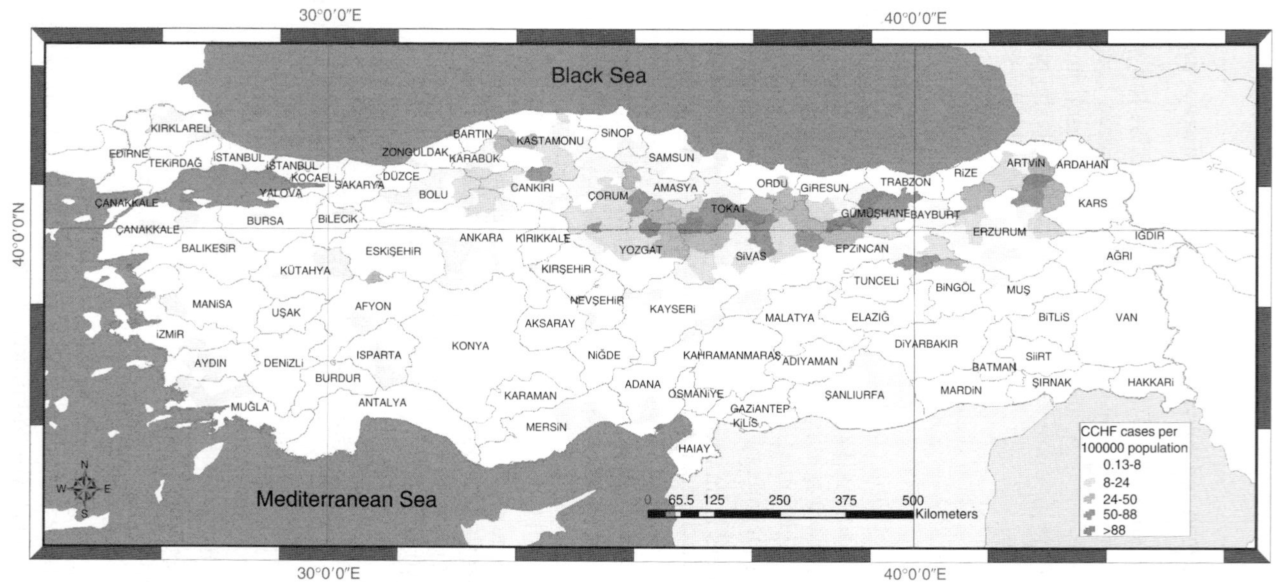

Fig. 6-5. Incidence rates of CCHF per 100,000 persons in Turkey, divided according to districts (administrative divisions) in the period 2003–2006. Contours represent the provinces in Turkey.

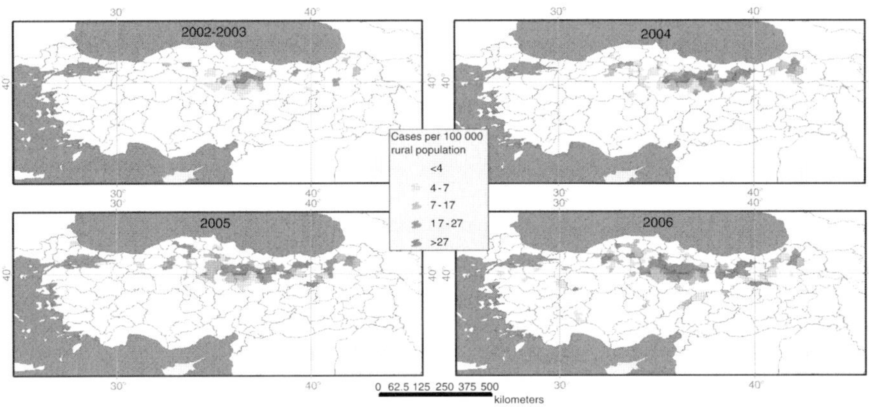

Fig. 6-6. Incidence rates of CCHF per 100,000 rural population in Turkey, divided according to districts (administrative divisions), as for the years 2003, 2004, 2005, and 2006. Contours represent the provinces in Turkey.

northern Turkey. Since then, the incidence of disease has increased with new foci in areas east of the original focus. While there has been an obvious trend towards increased incidence rates of CCHF and geographical spread of the disease in Turkey, whether these are due to actual dispersion of the disease or to an increased awareness is unknown.

In 359 hexagons, there were higher than expected CCHF case clusters with standard morbidity rates (SMRs) (number of cases/100,000 people) values ranging from 0 to 153. At this scanning window size (80 km), there were consistent patterns of significantly higher than expected numbers of CCHF cases in seven areas, involving a total of 40 districts as the central area of disease in 2003–2006 (Fig. 6-7).

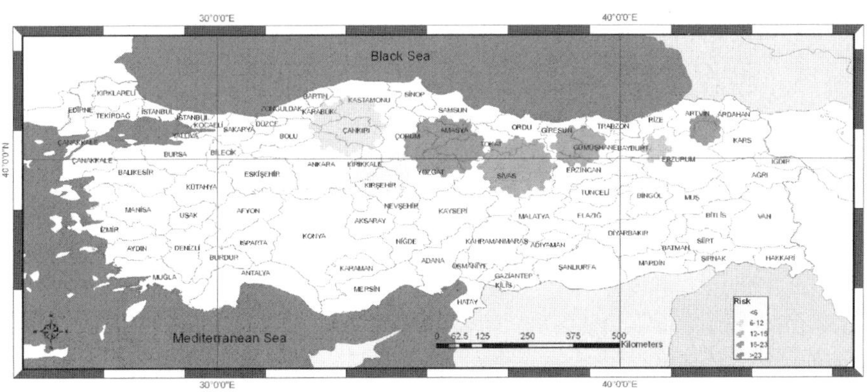

Fig. 6-7. Significant clusters of CCHF incidence, as detected by the use of the scan window algorithm at a window size of 80 km. Included are the contours of the clusters and the risk rates calculated for each one with significant risk rates ($p < 0.05$).

6.5.2. Climate data, vegetation features, and climate suitability for *H. marginatum marginatum*

Climate data (temperature and normalized difference vegetation index [NDVI]) were used to develop a predictive model of the HS for *H. marginatum marginatum* in Turkey. We used decadal (10-day) intervals of images, at a nominal 8 km resolution, taken over the region by the National Oceanic and Atmospheric Administration advanced very high resolution radiometer (NOAA-AVHRR) series of satellites for the period 1983–2000. Decadal images were converted into monthly averages and subjected to Fourier analysis. Fourier-derived values of both temperature and NDVI (i.e. yearly average, amplitude, and phase) were used to build the model as explained below. We used a data set with recent georegistered and accurate records of tick populations to allow comparison with contemporary climate. A total of 608 tick presences were recorded between 2000 and 2006 and selected as suitable for analysis. These records represent accurate determinations and were unambiguously referred to a pair of coordinates. We used a modeling algorithm based on presence-only data, called MaxEnt [24], which provided evaluation of the HS (ranging from 0 to 100) for ticks as defined by the set of environmental variables.

The models were developed with a random training set and checked against an evaluation set (50% of records each) and then evaluated against the remaining tick survey records. The evaluation of performance required the derivation of matrices of confusion that identified true positive, true negative, false positive, and false negative. From the confusion matrix we calculated the area under the curve (AUC) of a receiver operating characteristic (ROC) plot of sensitivity against (1-specificity) [31]. Sensitivity is defined as the proportion of true positives correctly predicted; whereas, specificity is the proportion of true negatives correctly predicted [12, 22].

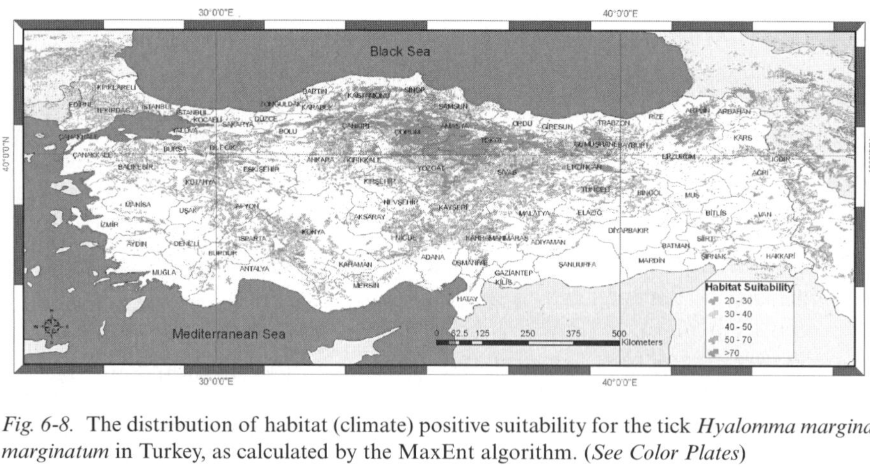

Fig. 6-8. The distribution of habitat (climate) positive suitability for the tick *Hyalomma marginatum marginatum* in Turkey, as calculated by the MaxEnt algorithm. (*See Color Plates*)

Figure 6-8 shows a map of the HS distribution for *H. marginatum marginatum* in Turkey. Variables involved in defining the environmental niche of the tick population were monthly minimum temperatures from April to September, monthly mean temperatures from May to October, monthly maximum temperatures from June to August, and monthly rainfall from May to September. All the variables defining the climate envelope of ticks were derived from climatic data collected from the late spring and summer. Approximately 62% of CCHF cases resided in areas of HS above 50, indicating a strong spatial correlation between favorable environmental features for tick populations and CCHF cases. Interestingly, positive HS was predicted to exist outside of the main foci of disease, underlining the existence of additional factors involved in the maintenance of disease foci.

6.5.3. Spatial relationships between CCHF disease risk, vegetation features, and HS for *H. marginatum marginatum*

Landsat imagery was used as a source for high-resolution vegetation data to calculate the patterns of habitat fragmentation for the entire country. Supervised classification was performed using the hybrid classification technique [7]. A total of 22 categories, including water, were extracted. Attention was paid to fragmentation in zones belonging to vegetation categories of cultured fields, bush or shrub, and forest, and to the distance of the case localities to these vegetation categories. In order to quantify these satellite-derived vegetation variables in the neighborhoods of each case location, and to compare with sites where no CCHF cases have been recorded, the images were overlaid with the hexagonal grid referred to previously. From these data, we determined whether there was a spatial association between CCHF risks (as obtained from spatial analysis) with the HS of the tick population and the patterns of the habitat (fragmentation of, and distance to, target categories). Risks for every statistically significant cluster, as obtained for each of the four scanning windows, were examined with a general lineal model (GLM) against the null hypothesis that the risks values were independent of either vegetation fragmentation, distance to fragmented habitats, or HS. Risk values were entered as dependent variables, with HS as continuous descriptor variables and vegetation features as discrete descriptor variables. The null hypothesis was rejected at values of $p < 0.05$.

The GLM regression model of the risk rates against the selected HS and vegetation-derived predictor variables showed a clear response: a highly significant relationship between HS and fragmentation and disease risks were observed in each hexagon ($p < 0.0001$). In contrast, the fragmentation of the agricultural-type categories, as a unique regression character, was not a good predictor of risks ($p = 0.12$). The distance of the case localities to these highly fragmented areas is not directly correlated with CCHF risk, alone or in combination with HS.

The abundance of ticks alone may not be the best indicator of disease risks because certain types of land cover may be of greater importance due to the activities of humans, resulting in higher levels of human–vector contact. Most of the villages where CCHF cases have been recorded are well within the area of high HS for ticks, and the highest rates of clinical cases were observed inside zones with high landscape fragmentation. Areas with high HS and fragmentation had variable, but positive risk rates, while sites where the tick population is expected to have low HS, or with low fragmentation, showed zero risk rates. These results are consistent with those for other tick-borne diseases. For example, in the USA, there was an effect of habitat fragmentation on the density of questing nymphal *Ixodes scapularis* ticks infected with the Lyme disease spirochete. The density of *I. scapularis* ticks was found to be higher in highly fragmented habitats, thus increasing the risk of contact between humans and infected ticks [3]. Clustering of tick-borne diseases may be the result of many interrelated factors. In our studies, we considered both tick abundance (as derived from the HS index) and landscape fragmentation as the main causes leading to an increased risk for CCHF. Another important parameter, however, the abundance of wildlife, was not included here because of the lack of adequate spatial data available related to wildlife in Turkey. It has been claimed that an increase in the abundance of wild hosts was responsible for most of the CCHF epidemics in the Eurasia [16]. Unpublished data from Turkey suggest that both wild boars and hare populations have increased considerably in the areas where most of the CCHF cases have occurred. Immature stages of *H. marginatum marginatum* ticks are known to feed on a wide range of ground-feeding birds and medium-sized mammals (e.g. hares), while adults prefer large animals (e.g. cattle, horses, wild boars) [15, 25, 27]. Therefore, the rise in the populations of these hosts could drive a disparate increase in tick numbers in areas where climate suitability is predicted to be maximum and landscape fragmentation is occurring, thus resulting in increased contact between humans and CCHFV-infected ticks. Predicting the future of a disease is not an exact science; however, based on the data presented here, CCHF cases will likely continue to occur in Turkey for the foreseeable future.

Acknowledgments

We thank the staff of the General Directorate of Primary Health Care of the Ministry of Health for their cooperation. In particular, we thank Dr. Ahmet Safran for his invaluable help in organizing the data. We are also grateful to the members of CCHF Scientific Committee (Turkish CCHF study group), jointly coordinated by Ministry of Health and Ministry of Agriculture and Rural Affairs in Turkey. We thank to Demir Serter, for providing the results of his serological studies and Olga Gimeno (University of Zaragoza, Spain), who participated in various aspects of the modeling analysis.

REFERENCES

1. The reports of Communicable Diseases Department (2006) The Ministry of Health of Turkey, Ankara
2. pro-med digests (2006): http://www.promedmail.org
3. Allan BF, Keesing F, Ostfeld BS (2003) Effect of forest fragmentation on Lyme disease risk. Conserv Biol 17:267–272
4. Arslan MO, Umur S, Aydin L (1999) Prevalence of Ixodidae on cattle in Kars province [In Turkish]. J Turk Parazitol 23:331–335
5. Aydin L (2000) Distribution and species of ticks on ruminants in the southern Marmara region [Turkish]. J Turk Parazitol 24:194–200
6. Bakir M, Ugurlu M, Dokuzoguz B, Bodur H, Tasyaran MA, Vahaboglu H (2005) Crimean-Congo haemorrhagic fever outbreak in Middle Anatolia: a multicentre study of clinical features and outcome measures. J Med Microbiol 54:385–389
7. Bourne JV, Graves M (2001) Classification of Land-cover types for the Fort Benning ecoregion using enhanced thematic mapper data. Strategic Environmental Research and development program, pp 1–9
8. Ergonul O, Celikbas A, Dokuzoguz B, Eren S, Baykam N, Esener H (2004) Characteristics of patients with Crimean-Congo hemorrhagic fever in a recent outbreak in Turkey and impact of oral ribavirin therapy. Clin Infect Dis 39:284–287
9. Ergonul O, Akgunduz S, Kocaman I, Vatansever Z, Korten V (2005) Changes in temperature and the Crimean-Congo hemorrhagic fever outbreak in Turkey. In: 15th European Congress of Clinical Microbiology and Infectious Diseases, Copenhagen. Clin Microbiol Infect 11(S2):360
10. Ergonul O (2006) Crimean-Congo haemorrhagic fever. Lancet Infect Dis 6:203–214
11. Ergonul O, Celikbas A, Baykam N, Eren S, Dokuzoguz B (2006) Analysis of risk-factors among patients with Crimean-Congo haemorrhagic fever virus infection: severity criteria revisited. Clin Microbiol Infect 12:551–554
12. Fielding AH, Bell JF (1997) A review of methods for the assessment of prediction errors in conservation presence/absence models. Environ Conserv 24:38–49
13. Gozalan A, Akin L, Rolain JM, Tapar FS, Oncul O, Yoshikura H, Zeller H, Raoult D, Esen B (2004) [Epidemiological evaluation of a possible outbreak in and nearby Tokat province]. Mikrobiol Bul 38:33–44
14. Hjalmars U, Kulldorf M, Gustafsson G, Nagarwalla N (1996) Childhood leukaemia in Sweden: using GIS and a spatial scan statistic for cluster detection. Stat Med 15:707–715
15. Hoogstraal H (1956) African Ixodoidea. I. Ticks of the Sudan (with special reference to Equatoria Province and preliminary reviews of the genera *Boophilus*, *Margaropus*, and *Hyalomma*), Department of the Navy, Bureau of Medicine and Surgery, Washington, DC, p 1101
16. Hoogstraal H (1979) The epidemiology of tick-borne Crimean-Congo hemorrhagic fever in Asia, Europe, and Africa. J Med Entomol 15:307–417
17. Karti SS, Odabasi Z, Korten V, Yilmaz M, Sonmez M, Caylan R, Akdogan E, Eren N, Koksal I, Ovali E, Erickson BR, Vincent MJ, Nichol ST, Comer JA, Rollin PE, Ksiazek TG (2004) Crimean-Congo hemorrhagic fever in Turkey. Emerg Infect Dis 10:1379–1384
18. Kurtpinar H (1954) Turkiye Keneleri: Morfoloji, biyoloji, yayilislari ve medical onemleri (tick fauna of Turkey: morphology, biology, distribution and medical importance), Ankara, p 112
19. Merdivenci A (1969) Turkiye Keneleri Uzerine Arastirmalar (investigations on the tick fauna of Turkey) [Turkish], Istanbul
20. Midilli K, Gargili A, Ergonul O, Sengoz G, Ozturk R, Bakar M, Jongejan F (2007) Crimean-Congo Haemorrhagic Fever in Istanbul (unpublished)
21. Ozkurt Z, Kiki I, Erol S, Erdem F, Yilmaz N, Parlak M, Gundogdu M, Tasyaran MA (2006) Crimean-Congo hemorrhagic fever in Eastern Turkey: clinical features, risk factors and efficacy of ribavirin therapy. J Infect 52:207–215

22. Pearce J, Ferrier S (2000) Evaluating the predictive performance of habitat models developed using logistic regression. Ecol Model 133:225–245
23. Petrova-Pointovskaya S (1947) Biological and ecological data on the tick *Hyalomma marginatum* Koch in the northwestern Crimean hemorrhagic fever focus (Russian, Translation 864, Medical Zoology Department, United States Naval Medical Research Unit No. 3, Cairo, Egypt). Nov Med 5:21–24
24. Phillips SE, Dudík M, Shapire RE (2004) A maximum entropy approach to species distribution modelling. In: 21st International Conference on Machine Learning, Banff, Canada, pp 13–19
25. Pomerantsev BI (1950) Fauna of USSR. Arachnida, vol IV, No. 2. Ixodid Ticks (Ixodidae). Zoological Institute of The Academy of Science USSR, New Series No. 41. Translated and published in English by The American Institute of Biological Sciences in 1959
26. Randolph SE (2004) Evidence that climate change has caused 'emergence' of tick-borne diseases in Europe? Int J Med Microbiol 293 (Suppl 37):5–15
27. Ruiz-Fons F, Fernandez de Mera IG, Acevedo P, et al. (2006) Ixodid ticks parasitizing Iberian red deer (*Cervus elaphus hispanicus*) and European wild boar (*Sus scrofa*) from Spain: geographical and temporal distribution. Vet Parasitol 140:133–142
28. Sayin F, Dumanli N (1982) Ticks (Ixodidae) of domestic animals in the province of Elazig, Turkey [Turkish]. Ankara Univ Vet Fak Derg 29:344–362
29. Sayin F, Dincer S, Karaer Z, Dumanli N, Cakmak A, Inci A, Yukari BA, Vatansever Z (1997) Status of the tick infestation of sheep and goats in Turkey. Parassitologia 39:145–152
30. Serter D (1980) Present status of arbovirus sero-epidemiology in the Aegean Region of Turkey. In: Vesenjak-Hirjan J, Porterfield JS, Arslanagic E (eds) Arboviruses in the Mediterranean Countries. Gustav Fisher Verlag, Stuttgart, NY, pp 155–161
31. Swets KA (1988) Measuring the accuracy of diagnostic systems. Science 240:1285–1293
32. Tonbak S, Aktas M, Altay K, Azkur AK, Kalkan A, Bolat Y, Dumanli N, Ozdarendeli A (2006) Crimean-Congo hemorrhagic fever virus: genetic analysis and tick survey in Turkey. J Clin Microbiol 44:4120–4124
33. Uzun R, Dokuzoguz B, Mentes H, Sen S, Gozalan A, Ergonul O (2003) The epidemiologic investigation report in Tokat Ministry of Health of Turkey, Tokat, Turkey
34. Whitehouse CA, Hottel H, Deniz A, Vatansever Z, Ergonul O, Paragas J, Garrison A, Kondig JP, Wasieloski LP (2006) Molecular detection of Crimean Congo hemorrhagic fever virus in ticks from Turkey. In: American Society of Tropical Medicine and Hygiene 55th Annual Meeting, Atlanta, GA, USA
35. Yukari BA, Umur S (2002) The prevalence of tick species (Ixodoidea) in cattle, sheep and goats in the Burdur region, Turkey [Turkish]. Turk J Vet Anim Sci 26:1260–1270

CHAPTER 7

EPIDEMIOLOGY OF CRIMEAN-CONGO HEMORRHAGIC FEVER IN THE BALKANS

TATJANA AVŠIČ-ŽUPANC, PH.D.

Institute of Microbiology and Immunology, Medical Faculty of Ljubljana, Zaloška 4, 1000 Ljubljana, Slovenia. Tel.: + 386 1 543 7450; Fax: + 386 1 543 7401; E-mail: tatjana.avsic@mf.uni-lj.si

7.1. INTRODUCTION

Crimean-Congo hemorrhagic fever (CCHF) is viral hemorrhagic fever with high significance in the Balkan Peninsula (Fig. 7-1), with a case fatality rate (CFR) of up to 40% and a high propensity for nosocomial spread. Epidemic outbreaks as well as sporadic cases have continuously been recorded in this area since 1952. There is strong evidence that *Hyalomma marginatum marginatum* ticks are implicated in the ecology of CCHF and serve as a principal vector of the virus in this region. Although the majority of documented cases have a history of tick bite, farmers having contact with the livestock have been found as a risk population for contracting the disease. Person-to-person transmissions, resulting in family outbreaks, as well as nosocomial transmissions in hospital settings, were frequently reported in the Balkans. Hospitalized patients, usually presented with a severe clinical course, were treated with supportive and replacement therapy. With regard to the availability of drugs and the country, antiviral drug ribavirin was effectively administered. Depending on the country, various preventive measures against the infection and spread of the disease are used. In general, after every outbreak or epidemic of CCHF, a better knowledge and awareness of the disease were obtained in general population, in high-risk groups and particularly among health-care workers.

7.2. KOSOVO

The first description of CCHF from Kosovo was a report of eight fatal cases in the village of Nishor of Suharekë [16]. However, the first recognized CCHF outbreak in former Yugoslavia was reported in 1970 among shepherds in the village of Čiflik on the border of Macedonia and Kosovo (Fig. 7-1). Based on

O. Ergonul and C. A. Whitehouse (eds.), Crimean-Congo Hemorrhagic Fever, 75–88.

Fig. 7-1. Map of the Balkan Peninsula.

epidemiological data, eight family members contracted the disease from the index case that was infected by tick bite. By retrospective analysis, using a complement fixation test and differential gel precipitation, a diagnosis of CCHF was confirmed in 10 out of 13 patients, two of whom died [31]. The results of serological examination of 97 sera from healthy residents of the village Čiflik revealed three positive individuals [23]. An outbreak of viral hemorrhagic fever was also noted in Kosovo during the period 1991–1992. Ninety-two sera from 76 hospitalized subjects suspected of CCHF were available for retrospective laboratory testing. Using enzyme immunoassays, we confirmed five CCHF cases by the detection of both IgM- and IgG-specific antibodies and further five cases (family members) by detection of specific IgG antibodies only. Ten years later, reverse transcription polymerase chain reaction (RT-PCR) was applied retrospectively to 19 stored serum samples from serologically confirmed CCHF cases. Viral RNA could be detected in all IgM-positive samples up to day 12 of illness. The sequence analysis of the partial S segment of the polymerase chain reaction (PCR) products revealed that the CCHF virus (CCHFV) responsible for the 1991–1992 outbreak in Kosovo is closely related to a CCHFV strain Drosdov [4].

The next CCHF outbreak in Kosovo occurred between June and November 1995. Based on the records of the Institute of Public Health in Pristine, Kosovo, the total number of CCHF patients in the 1995 epidemic was 65 with seven deaths. From 1996 to 2000, there were 33 sporadic cases with seven fatalities in Kosovo (Table 7-1) [8, 19].

Table 7-1. Confirmed cases of CCHF and HFRS in Kosovo from 1995 to 2005

Year	CCHF cases	CCHF fatal	Mortality rate (%)	HFRS cases
1995	65	7	10.8	24
1996	23	5	21.8	4
1997	0	0	0	0
1998	1	0	0	0
1999	7	2	28.6	1
2000	2	0	0	0
2001	30	7	23.3	2
2002	14	3	21.4	4
2003	8	0	0	3
2004	12	2	16.7	2
2005	5	1	20	2

During the spring and summer of 2001, the largest epidemic thus far occurred in this region. From 155 suspected CCHF cases, the diagnosis was confirmed in 30 patients. Among them, 28 patients were confirmed as acute CCHF by the presence of specific anti-CCHF IgM antibodies and/or positive PCR. Two additional cases were confirmed on clinical and epidemiological records – one was a contact with the index case and the second died in the emergency room with typical clinical signs of the disease. The CFR during the 2001 epidemic was 23.3% (Table 7-1). The mean age of the patients was 38 years (range from 8 to 76 years). Male patients were more often affected (70%). When 46 sera from healthy family members of the patients affected during the epidemic in 2001 were tested, seven were found IgG-positive and one IgM- and IgG-positive. Since the last epidemic on average ten sporadic cases are registered every year in the known endemic regions of CCHF in Kosovo (Table 7-1). CCHF cases in Kosovo were distributed among 18 municipalities with a high incidence in Klinë, Rahovec, Skënderaj, Malishevo, and Suharekë (Fig. 7-2) [19]. It is worth mentioning, that beside CCHF, sporadic cases of hemorrhagic fever with renal syndrome (HFRS) are diagnosed every year in Kosovo (Table 7-1.) [5]. But endemic regions of HFRS are different from those of CCHF [8].

7.2.1. Mode of transmission

Although, the main source of infection during the 2001 CCHF epidemic was tick bite (58%), there were five secondary cases reported [19]. From the second registered CCHF case during this epidemic, three hospital-acquired infections were confirmed and, in addition, two intrafamiliar cases occurred. One case in each group was fatal. No tertiary cases were detected. The viral RNA sequences obtained from the acute sera samples from these nosocomial cases were analyzed. Phylogenetic relationship determined that the causative agent was a CCHFV, which is closely related to the Drosdov strain [5]. Similar observation was reported by Papa and coworkers in a case from Kosovska Mitrovica in

Fig. 7-2. Map of Kosovo. Highlighted municipalities have the highest incidence of CCHF.

Kosovo which occurred during the same epidemic in 2001. The physician who had treated the index patient on admission to the hospital was infected [26]. In the CCHF outbreaks and epidemics described above, a large proportion of cases were found among health-care workers and relatives of patients. Therefore, prompt confirmation of the diagnosis in the first suspected case(s) is important for the treatment of patients, protection of medical staff, and prevention of larger outbreaks [6, 29]. Additionally, since the early clinical manifestations of CCHF and HFRS are virtually identical, rapid microbiological diagnosis of suspected cases is mandatory for distinguishing these two hemorrhagic fevers, mainly in the areas such as Kosovo, where both diseases coexist. In our experience, the combined use of real-time PCR for the detection of viral RNA and a serological assay for the detection of specific IgM antibodies are the approaches of choice for a rapid and specific diagnosis of acute CCHF [10]. A similar experience was reported by Drosten and coworkers who described a case from 2000 in Kosovo, with complete clinical, laboratory, and virological results. The viral RNA sequence obtained from the patient's serum sample drawn on day 3 of illness revealed the presence of a distinct strain of CCHFV circulating in the Kosovo Black Sea region [9].

7.2.2. Reservoirs and vectors

Three years after the 1970 outbreak, 269 ticks were collected from the cattle in the region where the disease had appeared and three strains of CCHFV were recovered. Two strains, designated "Čiflik 1" and "Čiflik 6", were isolated from the tick pools of *H. marginatum marginatum* and one strain "Čiflik 11" from the pool of *Ixodes ricinus* ticks [13]. During the 5-year period of 1973–1978, 691 sera were collected from the livestock in four different localities in Kosovo and Macedonia. The presence of CCHF antibodies was found on average in 14% of animals in a range from 2.3% to 32.6% in 1977 [14]. The highest prevalence of specific antibodies (32.6%) was found in sheep. While the prevalence in older cattle was 15.4%, calves were found positive in only 4.3% [24]. These data suggest that domestic animals, especially sheep and cattle, should be considered the principle host of adult ticks – the vectors of the virus. The ticks were collected from the domestic animals that were investigated for the presence of CCHF antibodies. Of 1,816 ticks examined, *H. marginatum marginatum* was found to be the most frequent in the region surveyed (58%), followed by the *Rhipicephalus bursa* (27.7%), *Boophilus calcaratus* (9.1%), *I. ricinus* (4.8%), and *Haemophysalis punctata* (0.4%) [24]. At the time of the 2001 epidemic the ticks were collected from pasturing cows owned by some patients. From the tick specimens collected, 267 were *H. marginatum marginatum*, four *Rhipicephalus sanguineus*, and one *Boophilus* sp. Among 272 ticks collected from 28 cows, 43 (15.8%) ticks of *H. marginatum marginatum* were determined to be infected with CCHFV. Amplicons obtained by using conventional RT-nested PCR were sequenced and the sequences from ticks were compared with the sequences from CCHF patients from different regions in Kosovo. The sequences from the ticks were identical, and the observed similarity between the sequences from ticks and patients was 98.6–100% [10]. These data again confirm that *H. marginatum marginatum* ticks are implicated in the ecology of CCHF and serve as a principal vector of the virus in this region. All investigated endemic regions of CCHF in Kosovo belong to the type of the Pond Caspian steppe, with dry, hot summers and cold winters, and abundant rainfall. This represents an excellent ecological niche for the ticks of the genus *Hyalomma*, whose seasonal dynamic coincides with the seasonal dynamic of CCHF in Kosovo [17, 19].

7.2.3. Treatment of patients

The signs and symptoms of CCHF in Kosovo resemble those described for the disease in other countries. However, there are some minor differences that were observed by the local physicians. For example, the incubation period after a tick bite is usually 7–10 days; whereas, the time to the onset of the disease in nosocomial cases is only 2–5 days. These data are just the opposite to what has been seen in the patients in South Africa [32]. Although inapparent infections and moderate forms are registered in each epidemic or yearly sporadic cases, a severe course of the disease is prevalent. During the 2001 epidemic, on average more than 50% of the patients showed hemorrhagic signs including hematomas (83.3%), petechiae (76.7%), epistaxis (70%), gingival hemorrhage (63.3%), conjunctival injection (63.3%), melena (43.3%), metrorrhagia (13.3%), and hematuria (3.3%). Other prominent clinical signs that were present in more than 80% of the same group of patients include fever, anorexia, vertigo, headache, hepatomegaly, elevated liver transaminases, and hypotension [19]. Similar clinical and/or laboratory observations were described for the sporadic cases from Kosovo [9, 26]. As mentioned above, Kosovo is an area where two viral hemorrhagic fevers coexist, CCHF and HFRS. Their seasonal occurrence is similar and the early clinical manifestations are virtually indistinguishable [1]. Hence, prompt and accurate diagnosis in suspected case(s) is needed, not only to prevent nosocomial spread of CCHF, but also to apply the adequate supportive and replacement therapies that are different for each of the disease. One of the most important reasons for a prompt and accurate diagnosis of CCHF is the specific treatment of patients. It has been shown that a broad-spectrum antiviral drug ribavirin was effective for treating the CCHFV infections [12, 18, 22, 37]. A drawback of ribavirin therapy is the need to administer the drug early in the course of disease, namely within 4 days after the onset of symptoms [22]. Given that specific IgM antibodies against the CCHFV are first detectable about 7 days after the onset of illness, a rapid and accurate diagnosis of CCHF can be made only by an adequate molecular method [10, 30]. Intravenous ribavirin (donation of ICN Pharmaceuticals) was used only in the last six severe CCHF patients at the time of the 2001 epidemic in Kosovo. Ribavirin was not applied during the epidemic peak due to the delayed approval by the Department of Health and Social Welfare. In all six patients, recovery was observed a short time after the initiation of treatment, without any side effects of the drug [1, 8]. Additionally, in a report of nosocomial transmission of CCHFV in a hospital in Belgrade in 2001, oral ribavirin was effective in treating the secondary case [26].

7.2.4. Preventive measures

The most often affected populations in CCHF endemic areas in Kosovo are the farmers who work outdoors and own cattle. After the 2001 epidemic, public health measures in the country included continuous information and education about the potential risks of the population who live in endemic areas by using widespread distribution of printed materials. This included personal protective measures such as regular examination of clothing and skin for ticks, their proper and safe removal, and the use of repellents. Furthermore, people were asked to take every precaution to avoid exposure to virus-contaminated animal blood or tissues. In addition, tick control was started in Kosovo by using acaricide treatment of livestock. But farmers are not consistent in using it adequately. During the 2001 epidemic in Kosovo, three health-care workers were affected as a result of a nosocomial transmission. Among them, two physicians in residence at ENT and a cleaning lady who died 10 days after the onset of the disease. None of them wore gloves. WHO Global Alert and Response Team who assisted in the response of the 2001 epidemic focused on the measures to improve infection control practice and to provide protective equipment and clothing with requisitions for appropriate barrier nursing at the Infectious Disease Clinic of the Pristine University Hospital. Since the 2001 epidemic, a total of 39 CCHF cases have been reported, with no nosocomial transmission.

7.3. BULGARIA

CCHF was first recognized in the country in early 1950 and became a notifiable disease in 1953. By retrospective analysis, at least 10 CCHF cases from three localities were hospitalized in the Burgass district from 1946 to 1952. In the period from 1953 to 1974, 1,105 sporadic cases were recorded with morbidity rate of 0.71% and mortality rate of 17.2% [17]. An immunization program was introduced in 1974 to protect health-care workers and military personnel in known endemic areas. Hence, from 1975 to 1996, the number of reported cases was reduced to 279, with a CFR of 11.4% [27]. Since 1997, a total of 170 sporadic cases have been reported to the Bulgarian Ministry of Health (MOH), 37 of them were fatal (Table 7-2) [7]. After the isolation of CCHFV from the blood of two patients in 1968, broad investigations of CCHF began in Bulgaria. The results from serologic survey showed the presence of asymptomatic infection in humans [34]. The degree of seropositive persons varied in different localities, but it was approximately 18% in humans tending cattle. In a 5-year study on CCHF

Table 7-2. Confirmed cases of CCHF in Bulgaria from 1997 to 2005

Year	No. of cases	No. of fatal cases	Mortality rate (%)
1997	20	4	20
1998	15	3	20
1999	5	2	40
2000	10	1	10
2001	18	5	27.8
2002	56	12	21.4
2003	14	2	14.3
2004	18	6	33.3
2005	14	2	14.3

in endemic regions, antibodies were detected in 9.1% of 580 people bitten by ticks, in 7.5% of 5,398 persons involved in raising cattle, and in 0.5% of other labor groups residing in these foci [17]. Most cases were reported from Plovdiv (central Bulgaria), Haskovo and Kardgali (southeastern Bulgaria), Shumen (northeastern Bulgaria), and Burgass (eastern Bulgaria) (Fig. 7-3) [34].

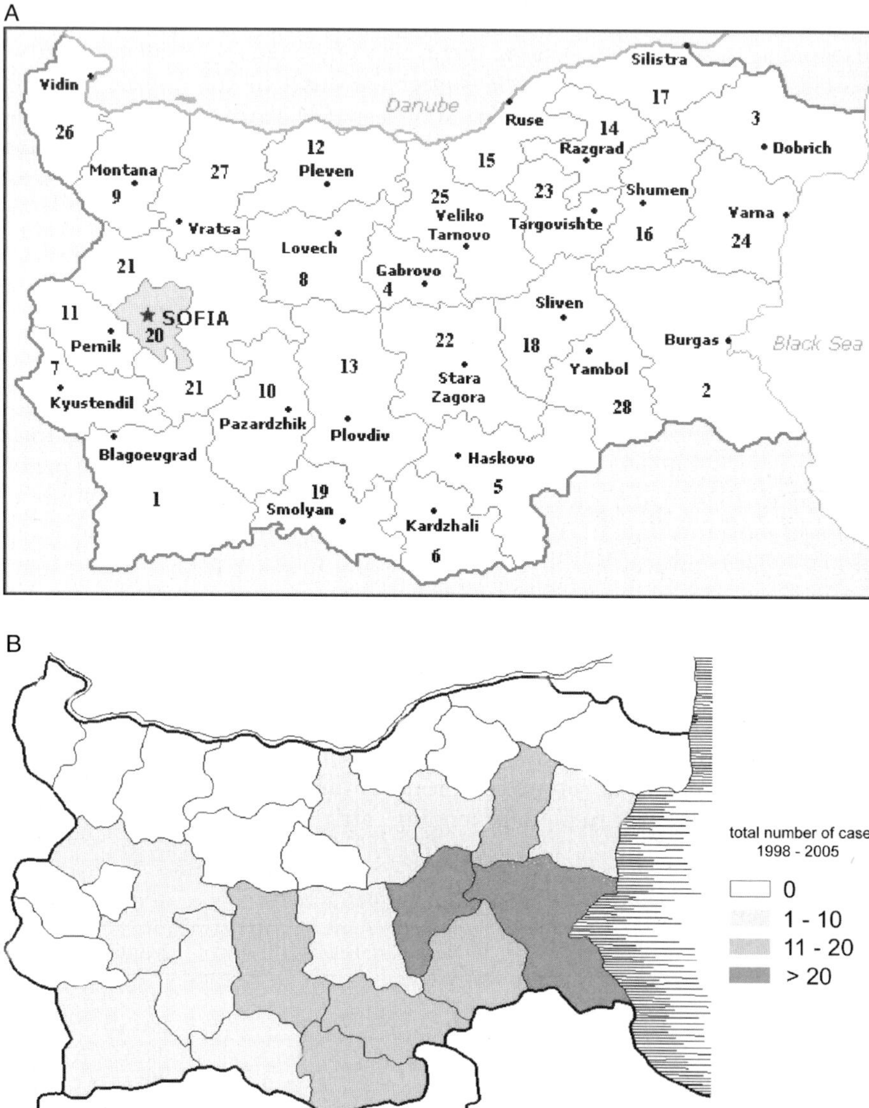

Fig. 7-3. (A) Map of Bulgaria showing the 28 provinces. (From www./Wikipedia.org.) (B) Geographical distribution of CCHF cases in a period from 1998 to 2005. (Courtesy of Dr. W. Monev, NCIPD, Sofia, Bulgaria.)

7.3.1. Mode of transmission

Most patients have been bitten by a tick; therefore, ticks are considered as the main source of infection. However, infections through a direct contact with CCHF patients also occurred. In the period from 1953 to 1965, 42 CCHF cases with 17 fatalities were caused by nosocomial infection in hospital [17]. There is no evidence of exposures of patients to the blood of infected animals or to the infection by the slaughter of livestock. Based on the available literature, there is no data of possible infection occurring in laboratory personnel.

7.3.2. Reservoir and vectors

Most of Bulgaria is an ecologically favorable environment for CCHFV circulation. Increased chances for human–tick contact and other ecological and climatic factors, during the 1950–1960 period of agricultural collectivization, provided the opportunity for a serious CCHF epidemic to develop. The seasonal dynamic of CCHF, appearing in April, reaching its peak in June, and disappearing in October is followed by a month of the seasonal dynamic of adult *H. marginatum marginatum* ticks [17]. Between 1968 and 1972, the CCHFV was isolated from human patients and from *H. marginatum marginatum*, *R. sanguineus*, and *Boophilus annulatus* ticks. In an early study in Bulgaria, it was shown that cattle were the chief host of adult *H. marginatum marginatum*. This tick species represented 90% of the 4,856 ticks taken from cattle, of the 1,278 ticks from horses, and of the 431 ticks from sheep. The results from this early investigation in Bulgaria showed that the prevalence of antibody to CCHFV in the sera of sheep is 28%, in cattle 47%, and in horses 82% [17]. When Levi performed entomological study in an endemic CCHF region in central Bulgaria (Pazardzhik), European hare, little owl, and blackbirds were found as the main hosts of the immature stages of the *H. marginatum marginatum* ticks. He further described that earlier detaching nymphs molt into adults in the hot Bulgarian fall and over winter before feeding; later detaching nymphs remain over winter in the fed state and molt the following spring [21]. Some other tick species, such as *Dermacentor marginatus*, *H. punctata*, and *I. ricinus*, were shown to maintain enzootic circulation of CCHFV in some areas. The virus has not been isolated from the common field mouse or other rodents and insectivores which are numerous in Bulgarian CCHF areas and are also known as tick hosts [17].

7.3.3. Treatment of patients

During the period from 1975 to 1996 mean age of patients registered was 52 years (range 11–79 years). Most patients were men (74%) because they were more frequently exposed to tick bites during the outdoor activities. The disease occurs in general from March to July which is in accordance with the ticks' activity. The main clinical symptoms include: fever, malaise, nausea, epistaxis, petechiae, and bleeding, mainly from gastrointestinal tract [27]. Pronounced laboratory findings were leukopenia, thrombocytopenia, and elevated liver enzymes. The Institute of Infectious and Parasitic Diseases in Sofia, Bulgaria, prepared a human immunoglobulin against CCHFV. The preparation was made of the plasma of donors boosted with one dose of vaccine against CCHFV. Forty units were defined as the potency of 1 mL of specific immunoglobulin preparation, with an immunodiffusion titer of 1:4. It was suggested to use a dose of 1,500 units for a passive immunization against CCHF and an administration of a single dose of 3,000 units for the therapy of CCHF [36]. There is very little, and often vague, information available on the efficacy of this immunoglobulin preparation. Namely, the authors claimed that the intravenous immunoglobulin preparation, designated as "CCHF-Venin", was successfully used in seven CCHF patients in the summer of 1989. The patients recovered quickly after the administration and their bleeding tendency ceased. No side effects were observed. However, the described study lacks a control group of patients [35]. There is no information available on the temporary use of the "CCHF-Venin" for the treatment of patients in Bulgaria.

7.3.4. Preventive measures

Various preventive measures are used in Bulgaria to protect against CCHF. The treatment of livestock with acaricide is a widespread practice in the country, although it may be impractical due to extensive farming conditions which prevail in the areas where *Hyalomma* ticks are the most prevalent. It seems that public health measures including information and education of the residents in high endemic areas about the potential risks of infection are effective [7]. Based on the registry of the MOH, it is obvious that the incidence of CCHF in Bulgaria decreased significantly after the introduction of immunization program for medical workers and military personnel in 1974 using the inactivated vaccine made by Vassilenko in 1970. The vaccine was subsequently given to 583 volunteers in 1970 and 1971; it was concluded that the vaccine was highly efficient [17]. This vaccine consists of mouse brain preparation inactivated by chloroform, heated at 58°C, and adsorbed on $Al(OH)_3$. The first two doses are given on days 0 and 30, the third dose is given 1 year later, and another, a booster dose, 5 years later [33]. As a result of specific immunization, the morbidity rate due to the CCHFV has visibly dropped. No infections have been reported from vaccinated military personnel since the immunization program began [20].

7.4. ALBANIA

In Albania, a country which is situated in the southwestern part of the Balkan Peninsula (Fig. 7-4), the first CCHF case was recognized in 1986. Since then, a number of cases have been reported each year. The majority of cases have

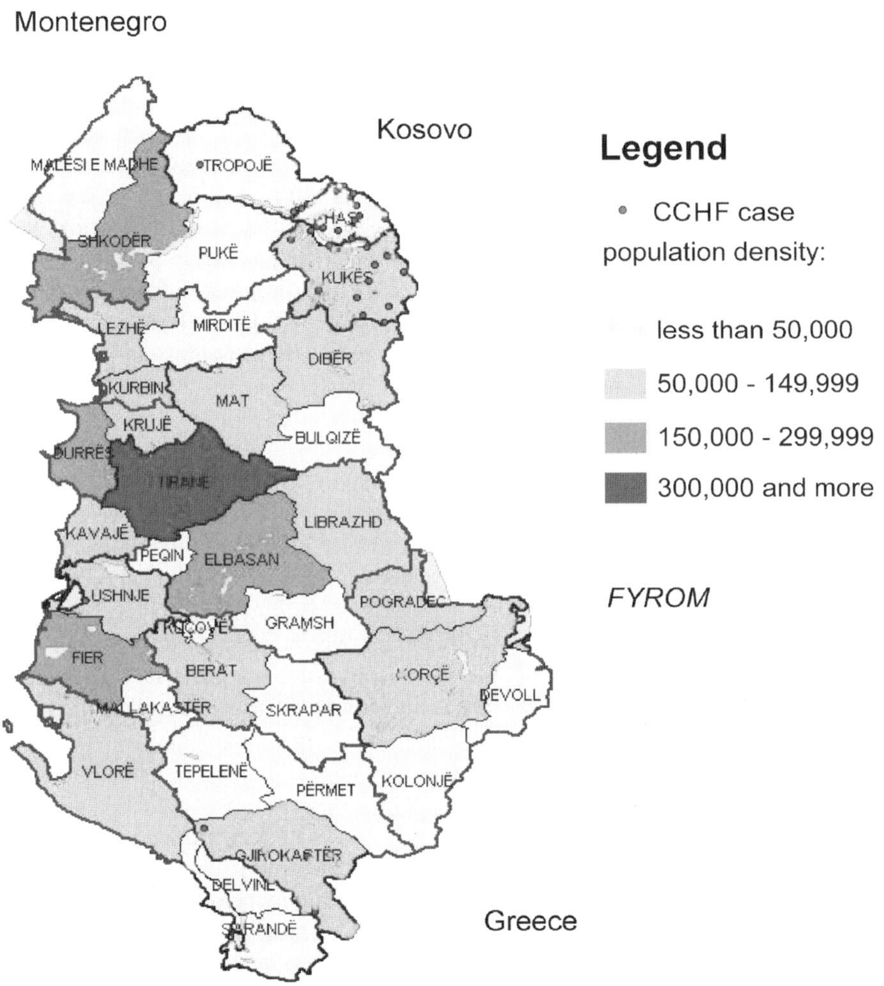

Fig. 7-4. Map of Albania showing the geographical distribution of CCHF cases in 2003 and 2004. (Courtesy of Dr. S. Bino, PHI, Tirana, Albania.)

been observed in Kukes area, in the northwestern part of the country. But sporadic CCHF cases have been reported throughout the country, including Tirana, Mirdita, Lezha, Gjirokastra, and Skapar [11]. From the end of May to September 2001, eight cases of CCHF were identified in Albania. Most of the cases (7 of 8) were recorded in the Kukes area, as was also the case in previous years. None of the patients died during this outbreak [25], although the mortality rate of the disease in the years 1985–1987 was as high as 16.6% [11].

7.4.1. Mode of transmission

For the 2001 outbreak, it is obvious that the main source of infection was tick bite or a lifestyle that included tending livestock. In the same outbreak, a cluster of three cases within a family was also reported [25]. However, person-to-person infections had been documented earlier in Albania. Furthermore, Eltari showed that, out of 233 sera from healthy population, 1.3% had antibodies to CCHFV [11]. During the 2001 community outbreak, one nosocomial infection was observed. The patient was a male nurse who was infected while performing an electrocardiogram on a patient, hospitalized with a diagnosis of acute hemorrhagic syndrome (that was later confirmed to be CCHF). The nurse's skin apparently intact was exposed to the patient's vomit (hematemesis) [15].

7.4.2. Reservoirs and vectors

Although there is no data available on the reservoirs or the vectors of CCHFV in Albania, we can presume that the ticks of the genus *Hyalomma* serve as the main vector. Similarly as elsewhere in southern Balkan Peninsula, owing livestock and performing outdoor pasture is a common lifestyle. During the outbreak of 2001, the tick population was extremely high. This was due to an optimal climatic condition, such as mild winter which allowed ticks to survive. In addition, the early arrival of spring accelerates tick activity. The same phenomenon was observed in 2001 in a neighboring country Kosovo [8].

7.4.3. Treatment of patients

The clinical presentation of the disease in Albania is in accordance with the one seen in the rest of the Balkan countries. In the above-mentioned outbreak, five out of eight patients had petechiae and severe hemorrhagic manifestations. Most of the cases displayed marked thrombocytopenia. The median age of the patients was 28 years (range 8–66 years). The disease lasted approximately 2 weeks [25]. When necessary, the patients received symptomatic treatment, consisting of replacement of blood components and use of corticosteroids [15]. Ribavirin was not used as postexposure prophylaxis or as a drug for treatment. No fatal case was reported in the 2001 outbreak.

7.5. GREECE

So far, CCHF has not been diagnosed in humans in Greece, although the virus was isolated from *R. bursa* ticks, collected in May 1975 from goats of a flock in Vergina village, 80 km west of Thessaloniki. During the 5-year period from 1971 to 1976, 118 tick pools were collected from goat, sheep, and cattle. The most abundant tick species was *R. bursa* (31 tick pools), followed by *R. sanguineus* (23 tick pools), *Ixodes gibosus* (16 tick pools), and *Hyalomma anatolicum* (15 tick pools), respectively. Using the agar gel precipitation assay and the above-mentioned CCHFV strain, designated as AP92, antibodies were detected in 139 of 422 (32.9%) goat sera and in 34 of 294 (11.6%) sheep sera collected in numerous locations in northern Greece [28]. Sixty-five sera from lifelong residents of the county, where the virus was recovered, were tested for the presence of antibodies against CCHF in 1980–1981. Four residents (6.1%) were found positive. The mean age of seropositive persons was 56 years (range 35–72 years) [3]. The results from a broad serosurvey study, which included 3,040 serum samples from apparently healthy residents from 26 of 54 counties in Greece, revealed an overall prevalence of 1.1% with a range from 0% to 6.3%. With an attempt to confirm the existence of the disease in Greece, 409 serum samples were taken from patients with clinical picture resembling CCHF. None of the patients were found positive [2]. The presence of CCHFV in ticks and seropositive individuals in Greece without any recognized disease is unique in the Balkan Peninsula.

REFERENCES

1. Ahmeti S (1991) Virusne hemoragijske groznice na Kosovu 1986–1989 – nova tumacenja patogeneze i klinickih manifestacija bolesti (Thesis in Croatian). In: Medical Faculty. University of Zagreb, Zagreb, p 130
2. Antoniadis A, Alexiou-Daniel S, LeDuc JW, Peters CJ (1988) Crimean-Congo Hemorrhagic Fever in Greece. In: 1st International Symposium on Hantaviruses and Crimean-Congo Hemorrhagic Fever Virus, Halkidiki, Greece, p 48
3. Antoniadis A, Casals J (1982) Serological evidence of human infection with Congo-Crimean hemorrhagic fever virus in Greece. Am J Trop Med Hyg 31:1066–1067
4. Avsic-Zupanc T, Ahmeti S, Petrovec M, Rossi CA (1999) Retrospective analysis of an outbreak of Crimean-Congo hemorrhagic fever in the Kosovo during 1991–1992. Am J Trop Med Hyg 61:318–319
5. Avsic-Zupanc T, Petrovec M, Duh D, Dedushaj I, Ahmeti S (2002) Description of nosocomial and intrafamiliar spread of CCHF in Kosovo during the 2001 epidemic. In: The World of Microbes. EDK, Paris, France, Paris, 27th July to 1st August 2002, Le Palais des Congres de Paris, p 87
6. Burt FJ, Leman PA, Smith JF, Swanepoel R (1998) The use of a reverse transcription-polymerase chain reaction for the detection of viral nucleic acid in the diagnosis of Crimean-Congo haemorrhagic fever. J Virol Methods 70:129–137
7. Christova I (2006) Epidemiology of CCHF in Bulgaria (personal communication)
8. Dedushaj I (2001) Epidemics of CCHF in Kosovo (personal communication)
9. Drosten C, Minnak D, Emmerich P, Schmitz H, Reinicke T (2002) Crimean-Congo hemorrhagic fever in Kosovo. J Clin Microbiol 40:1122–1123
10. Duh D, Saksida A, Petrovec M, Dedushaj I, Avsic-Zupanc T (2006) Novel one-step real-time RT-PCR assay for rapid and specific diagnosis of Crimean-Congo hemorrhagic fever encountered in the Balkans. J Virol Methods 133(2):175–179
11. Eltari E, Cani M, Cani K, Gina A (1988) Crimean-Congo hemorrhagic fever in Albania. In: 1st International Symposium on Hantaviruses and Crimean-Congo Hemorrhagic Fever Virus, Halkidiki, Greece, p 34
12. Fisher-Hoch SP, Khan JA, Rehman S, Mirza S, Khurshid M, McCormick JB (1995) Crimean Congo-haemorrhagic fever treated with oral ribavirin. Lancet 346:472–475
13. Gligic A, Stamatovic L, Stojanovic R, Obradovic M, Boskovic R (1977) The first isolation of the Crimean hemorrhagic fever virus in Yugoslavia. Vojnosanit Pregl 34:318–321

14. Gligic A, Stojanovic R, Obradovic M, Boskovic R (1980) Serological examination of Crimean-Congo hemorrhagic fever infections of domestic animals in natural foci. Zbl Bakt Suppl 9:263–266
15. Harxhi A, Pilaca A, Delia Z, Pano K, Rezza G (2005) Crimean-Congo hemorrhagic fever: a case of nosocomial transmission. Infection 33:295–296
16. Heneberg D et al. (1968) Crimean hemorrhagic fever in Yugoslavia. Vojnosanit Pregl 25:181–184
17. Hoogstraal H (1979) The epidemiology of tick-borne Crimean-Congo hemorrhagic fever in Asia, Europe, and Africa. J Med Entomol 15:307–417
18. Huggins JW (1989) Prospects for treatment of viral hemorrhagic fevers with ribavirin, a broad-spectrum antiviral drug. Rev Infect Dis 11 (Suppl 4):S750–S761
19. Humolli I (2003) Epidemiological and serological characteristics of CCHF epidemics in Kosova (Thesis in Albanian). In: Medical Faculty. University of Pristina, Pristina, p 110
20. Kovacheva T, Velcheva D, Katzarov G (1997) Studies on the morbidity of Congo-Crimean hemorrhagic fever before and after specific immunoprophylaxis (article in Bulgarian). Infectology 34:34–35
21. Levi V (1972) Seasonal activity of the ticks of the family Ixodidae in focus of Crimean hemorrhagic fever in Paradzjik region. Suvrem Med 23:44–50
22. Mardani M, Jahromi MK, Naieni KH, Zeinali M (2003) The efficacy of oral ribavirin in the treatment of Crimean-Congo hemorrhagic fever in Iran. Clin Infect Dis 36:1613–1618
23. Obradovic M, Gligic A (1981) Crimean-Congo hemorrhagic fever virus antibodies in people living in natural reservoirs. Vojnosanit Pregl 38:342–346
24. Obradovic M, Gligic A, Stojanovic R, Stamatovic L, Boskovic R (1978) Serological and arachno-entomological investigation of natural foci of Crimean hemorrhagic fever in various regions of Yugoslavia. Vojnosanit Pregl 35:253–256
25. Papa A et al. (2002) Crimean-Congo hemorrhagic fever in Albania, 2001. Eur J Clin Microbiol Infect Dis 21:603–606
26. Papa A, Bozovi B, Pavlidou V, Papadimitriou E, Pelemis M, Antoniadis A (2002) Genetic detection and isolation of crimean-congo hemorrhagic fever virus, Kosovo, Yugoslavia. Emerg Infect Dis 8:852–854
27. Papa A, Christova I, Papadimitriou E, Antoniadis A (2004) Crimean-Congo hemorrhagic fever in Bulgaria. Emerg Infect Dis 10:1465–1467
28. Papadopoulos O, Koptopoulo G (1980) Crimean-Congo hemorrhagic fever (CCHF) in Greece: isolation of the virus from *Rhipicephalus bursa* ticks and a preliminary serological survey. In: Vesenjak-Hirjan Jea (ed.) Arboviruses in the Mediterranean Countries. Gustav Fisher Verlag, Stuttgart, Yugoslavia, pp 117–121
29. Rodriguez LL et al. (1997) Molecular investigation of a multisource outbreak of Crimean-Congo hemorrhagic fever in the United Arab Emirates. Am J Trop Med Hyg 57:512–518
30. Shepherd AJ, Swanepoel R, Leman PA (1989) Antibody response in Crimean-Congo hemorrhagic fever. Rev Infect Dis 11 (Suppl 4):S801–S806
31. Stamatovic L, Panev D, Gerovski V, Miladi-Novic T, Grdanoski S, Radovic S (1971) Epidemija krimske hemoragične groznice. Vojnosanit Pregl 28:237–241
32. Swanepoel R et al. (1987) Epidemiologic and clinical features of Crimean-Congo hemorrhagic fever in southern Africa. Am J Trop Med Hyg 36:120–132
33. Todorov S, Kovacheva T, Velcheva D, Katzarov G (2001) Congo-Crimean hemorrhagic fever-prophylaxis and treatment (article in Bulgarian). Contemp Med 42:54–60
34. Vasilenko S et al. (1971) Investigations on Crimean-Congo hemorrhagic fever in Bulgaria II. Serological examinations of people and animals in endemic and nonendemic areas for CCHF (article in Bulgarian). Epidemiol Microbiol Infect Dis 8:150–156
35. Vassilenko SM, Vassilev TL, Bozadjiev LG, Bineva IL, Kazarov GZ (1990) Specific intravenous immunoglobulin for Crimean-Congo haemorrhagic fever. Lancet 335:791–792
36. Vassilev T, Valchev V, Kazarov G, Razsukanova L, Vitanov T (1991) A reference preparation for human immunoglobulin against Crimean/Congo hemorrhagic fever. Biologicals 19:57
37. Whitehouse CA (2004) Crimean-Congo hemorrhagic fever. Antiviral Res 64:145–160

CHAPTER 8

CRIMEAN-CONGO HEMORRHAGIC FEVER INFECTION IN IRAN

SADEGH CHINIKAR, PH.D.
Head of Laboratory of Arboviruses and Viral Haemorrhagic Fevers (National Center), Pasteur Institute of Iran, Tehran, Iran. E-mail: sadeghchinikar@yahoo.com

8.1. INTRODUCTION

8.1.1. History of crimean-congo hemorrhagic fever infection in Iran

Crimean-Congo hemorrhagic fever (CCHF) disease existed from many years in the northwest of Iran, in Ardebil, East and West Azerbaijan provinces with the local name "Kara-Mikh typhoid Fever". Kara-Mikh is a Turkish word which means black nail and it refers to the black spots appearing on the skin of the patients.

In 1974, Dr. Asefi studied clinically 60 patients with CCHF syndromes in Ardebil, Sarab, and Khalkhal counties of the Ardebil province [1]. In the years 1974–1975, Dr. Ardoin from the Pasteur Institute of Paris, with the collaboration of Dr. Younis Karimi from the Pasteur Institute of Iran, has studied clinically the disease in East Azerbaijan province [2]. In 1975, Dr. Saidi et al. have demonstrated the presence of antibodies against CCHF in the blood of human, domestic animals, and small mammals suspected for the disease in different regions of Iran mainly the region bordering the Caspian Sea and East Azerbaijan [23]. In 1978, professor Sureau from the Pasteur Institute of Paris with collaboration of the Pasteur Institute of Iran succeeded to isolate the virus of the disease from infected ticks in Khorassan province in northeast of Iran [25]. After that time, because there were no facilities, no research has been done on this disease in Iran and patients with hemorrhagic diseases have been misdiagnosed as they were previously when CCHF was considered to be "Kara-Mikh typhoid Fever" and unfortunately some of them died without diagnosis and efficient treatment. In summer 1999, suspected cases for this disease have been reported in Iran and with the sending of the sera of suspected patients from August 1999 to June 2000 to the National Center of Virology in South Africa, 7 Out of 34 persons have been diagnosed positive (Table 8-1).

O. Ergonul and C. A. Whitehouse (eds.), Crimean-Congo Hemorrhagic Fever, 89–98.

Table 8-1. The characteristics of CCHF-diagnosed patients in 1999

Cases	Sex	Age	Profession	Province	Positive test result
Case 1	Female	24	Housewife	Khuzestan	ELISA IgG+
Case 2	Male	32	Physician	Chaharmahal Bakhtiari	ELISA IgM+, IgG+, IF+
[a]Case 3	Female	26	Physician	Chaharmahal Bakhtiari	ELISA IgM+, IgG+, IF+, Virus culture+
[a]Case 4	Male	25	Livestock dealer	West Azerbaijan	ELISA IgG+
Case 5	Female	29	Housewife	Khuzestan	RT-PCR+
[a]Case 6	Male	55	Farmer	Khuzestan	Virus culture+
Case 7	Male	37	Worker	Khuzestan	ELISA IgM+, IgG+

[a] These patients have died

8.2. RESPONSE TO CCHF OUTBREAK

Following these finding, in 2000, CCHF was recognized as a major public health problem by sanitary authorities in Iran and thus the Laboratory of Arboviruses and Viral Haemorrhagic Fevers (National Center) has been founded in the Pasteur Institute of Iran which is a unique center for research and diagnosis of Arboviruses in the region and this laboratory is well equipped for the serological and molecular diagnosis of this disease and other arboviral and hemorrhagic fever diseases [3, 4, 8, 9]. The Laboratory of Arboviruses and Viral Haemorrhagic Fevers of the Pasteur Institute of Iran has provided the entire health system of the country with the collaboration of the Ministry of Health of Iran a protocol according to which human serum samples suspected of CCHF disease are sent from different provinces of Iran respecting the cold chain [10, 13, 24] (Fig. 8-1).

From each patient, three serum samples must be prepared.

1. Following the clinical diagnosis of the patient: for reverse transcription polymerase chain reaction (RT-PCR) (detection of a fragment in the S segment of the virus genome) and specific enzyme-linked immunosorbent assay (ELISA) for detection of anti-CCHF IgM and anti-CCHF IgG (if it exists)
2. 5–10 days after the onset of the first clinical signs for specific IgM and IgG ELISA
3. 10–15 days after the onset of the first clinical signs for specific IgM and IgG ELISA

In the survey carried out from 7 June 2000 until 6 September 2005 by our lab, 1,528 samples from 763 suspected patients from different provinces of Iran have been tested serologically (specific ELISA) and molecularly (RT-PCR) [5, 7]. The results have been regularly reported to the Promed organization [14–22] (Table 8-2).

Around 763 CCHF-suspected patients (which means they had a sudden fever, at least one sort of hemorrhage and thrombocytopenia: less than 100,000 platelets/mL), 295 (262 + 33) were confirmed cases which means they were either IgM-positive or RT-PCR-positive for CCHF.

○ Confirmed Case

□ Confirmed fatalities

Fig. 8-1. The CCHF-infected provinces of Iran.

Table 8-2. The CCHF cases by the year 2005

Issues	Number
Samples	1,528
Suspected patients	763
IgM+ patients	262
IgG+ patients	245
RT+ and IgM+ patients	47[a]
Only RT+ patients	33
Total confirmed cases	295 (262 + 33)
Total confirmed death	53

[a] These 47 patients are comprised in the 262 IgM+ patients

After analyzing the profession of the infected patients, it is obvious that the most exposed profession to the infection is profession in which the person is in contact with livestock (farmer, worker, and truck driver) or has to slaughter them (butcher and slaughterer). Many cases are housewives who lived in the countryside and had contact with livestock (Table 8-3).

Our results show that the most infected province of Iran is Sistan and Baluchistan in the southeast of the country. This province has long border with Afghanistan and Pakistan, infected countries from which illegal entrance of humans and livestock occur into Iran (Table 8-4).

Table 8-3. The age, sex, and profession of the confirmed patients

		Number of confirmed cases
Sex	Male	233
	Female	62
Age group (year)	0–20	60
	21–40	157
	41–60	58
	61–80	20
Profession	Farmer	61
	Worker	54
	Housewife	52
	Butcher	38
	student	27
	Slaughterer	16
	Truck driver	9
	Employee	6
	Jobless	6
	Teacher	4
	Soldier	3
	Animal seller	3
	Health worker	2
	Others	14

Table 8-4. The CCHF-infected provinces of Iran

Name of the province	Confirmed case	Confirmed fatal case
Sistan and Baluchistan	184	17
Isfahan	34	9
Fars	19	5
Tehran	9	3
Khuzistan	8	1
Golestan	7	0
Khorassan	7	2
Lorestan	4	2
Boushehr	4	1
Yazd	3	1
Hormozgan	3	2
Markazi	2	2
Qom	2	2
Semnan	2	2
Zanjan	1	1
East Azerbaijan	1	1
Hamedan	1	1
Gilan	1	1
Kordestan	1	0
West Azerbaijan	1	0
ChaharMahal and Bakhtiari	1	0

8.3. NOSOCOMIAL CASES

In Iran, there are nosocomial cases. As mentioned in previous paragraph, there was a lady physician who caught the disease from her husband, also a physician who contracted the disease after contact with a hospitalized patient, and died from CCHF in 1999 in Chahar Mahal Bakhtiari province.

In 2001, another physician also caught the disease from his patients (a father and son), both butchers hospitalized as CCHF-suspected in the city of Isfahan and later confirmed as CCHF-positive. The physician fortunately recovered from the disease.

8.4. COMMENTS ON THE CASE FATALITY RATE

Among the 295 confirmed cases, 53 died because of CCHF. The case fatality rate (CFR) is 17.9% which is much lower than the reported CFRs from different countries. This could be related to the rapid diagnosis by serological and molecular techniques performed in the Laboratory of Arboviruses and Viral Haemorrhagic Fevers of the Pasteur Institute of Iran (National Center) and also in part to the awareness of Iranian clinicians, who begin ribavirin as soon as after the suspicion of CCHF. There is a good interdepartment collaboration between our lab and the Center for Disease Control (CDC) of Iran for rapid case finding, diagnosis, and report of the results of the patients.

8.5. ECOLOGIC FEATURES OF THE ENDEMIC REGION IN IRAN

Most positive cases are from Sistan–Baluchestan province (southeast of Iran) which has long borders with Afghanistan and Pakistan, two countries in which CCHF is endemic. Sistan–Baluchestan is an arid region with a windy and dry climate. Due to the climate of this province, agriculture and livestock breeding is very difficult. The transport of livestock is seen in the borders of this province from neighboring countries such as Pakistan and Afghanistan. During the transport, livestock infected with this virus and also with ticks containing this virus easily enter into this province. Livestock market in Zahedan (capital of Sistan–Baluchistan province) is full of these livestock (cattle, sheep, goat, and camel). A significant number of humans also live in close proximity of the livestock so they are at a great risk to catch the disease.

8.6. THE PROPOSED SOURCE OF INFECTION

The proposed source of infection is the entrance of infected livestock from the eastern borders of the country and their dispatch throughout the country where numerous humans can catch the disease by handling them directly or their carcass and by manipulation of their blood and discharges. Direct infection of humans by tick bite is rare in Iran and has been observed only in few cases.

8.7. PHYLOGENETIC STUDY OF IRANIAN CCHFV

In a joint study by the Laboratory of Arboviruses and Viral Haemorrhagic Fevers (National Center), Pasteur Institute of Iran and the Center for Microbiological Preparedness, Swedish Institute for Infectious Disease Control, CCHFV genome fragments for nine Iranian patients infected during 2002 were examined genetically nucleotide sequencing of the S and M segments, encoding the nucleocapsid protein (NP) and the glycoproteins, respectively. The study revealed that the different isolates were closely related to each other with nucleotide identities exceeding 98% for both S and M segments.

Phylogenetic analysis using S-segment sequences demonstrated that the Iranian isolates formed a distinct lineage together with the Pakistanian strain Matin and the Madagascar strain ArMg851. The Iranian strains analyzed in this study are clearly separated from a previously published Iranian CCHFV strain ArTeh193-3, indicating that at least two genetic lineages of CCHFV could be cocirculating in Iran (Fig. 8-2). Using M-segment sequences, it was confirmed that the Iranian strains examined in this study formed a separate cluster (Fig. 8-3). Interestingly, on the M tree the Pakistanian strain Matin was located outside the cluster formed by the Iranian isolates. These observations might indicate that some kind of genetic exchange could occur between different strains of CCHFV. Genetic reassortment has previously been demonstrated both in vivo and in vitro within the Bunyaviridae family among different members of the genera Bunyavirus and Hantavirus. It is not known at present whether members within the genus *Nairovirus* have the same capacity. These findings could suggest a possibility for genetic reassortment within CCHFV, but further studies are needed to substantiate this hypothesis. The Iranian sequences reported in this study have been the deposited into GenBank nucleotide sequence database with the following accession numbers: for the CCHFV S-segment sequences; AY366373–AY366379, and for CCHFV M-segment sequences; AY366380–AY366387 [6].

8.8. TREATMENT

According to the protocol of the CDC of Iran, *suspected patient* is a patient who has sudden onset of fever, myalgia, bleeding, one of the epidemiological signs such as tick bite or hand crushing of ticks or contact with fresh blood or other tissues of infected domestic animals, direct contact with the blood, and excretions of a confirmed or suspected patient of CCHF. A *probable case* is a patient who is a suspected case with a thrombocytopenia (less than 150,000 platelets/mm^3) with leucopenia (less than 3,000 lymphocytes/mm^3) or leucocytosis (more than 9,000 lymphocytes/mm^3). A *definitive case* is a patient who is a probable case with positive serological test or RT-PCR-positive [26].

Immediately after the diagnosis of the probable case, the patient is treated 10 days with ribavirin (ribavirin is used in pill form in Iran and other forms of

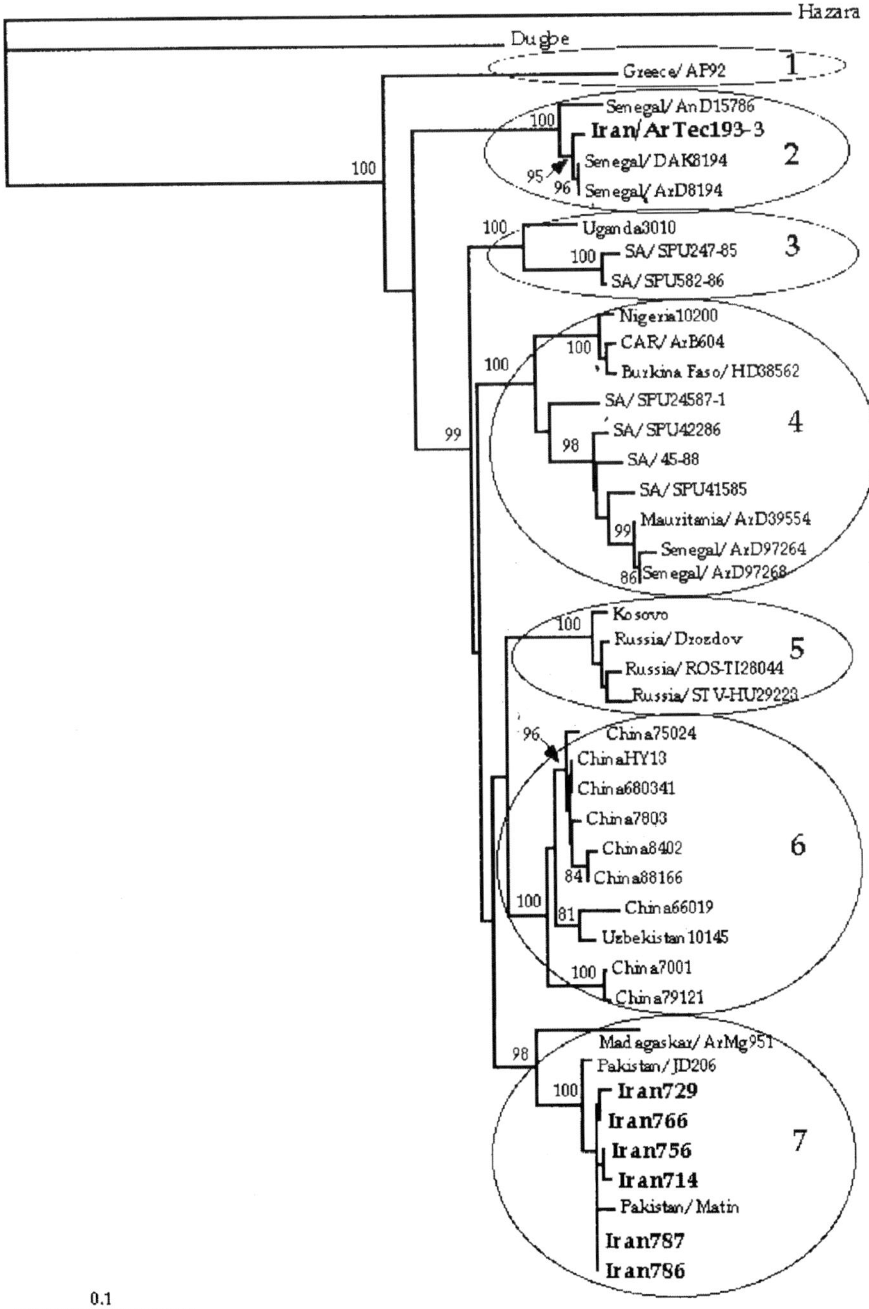

Fig. 8-2. Phylogenetic tree (neighbor-joining) calculated for Iranian CCHFV S-segment sequences.

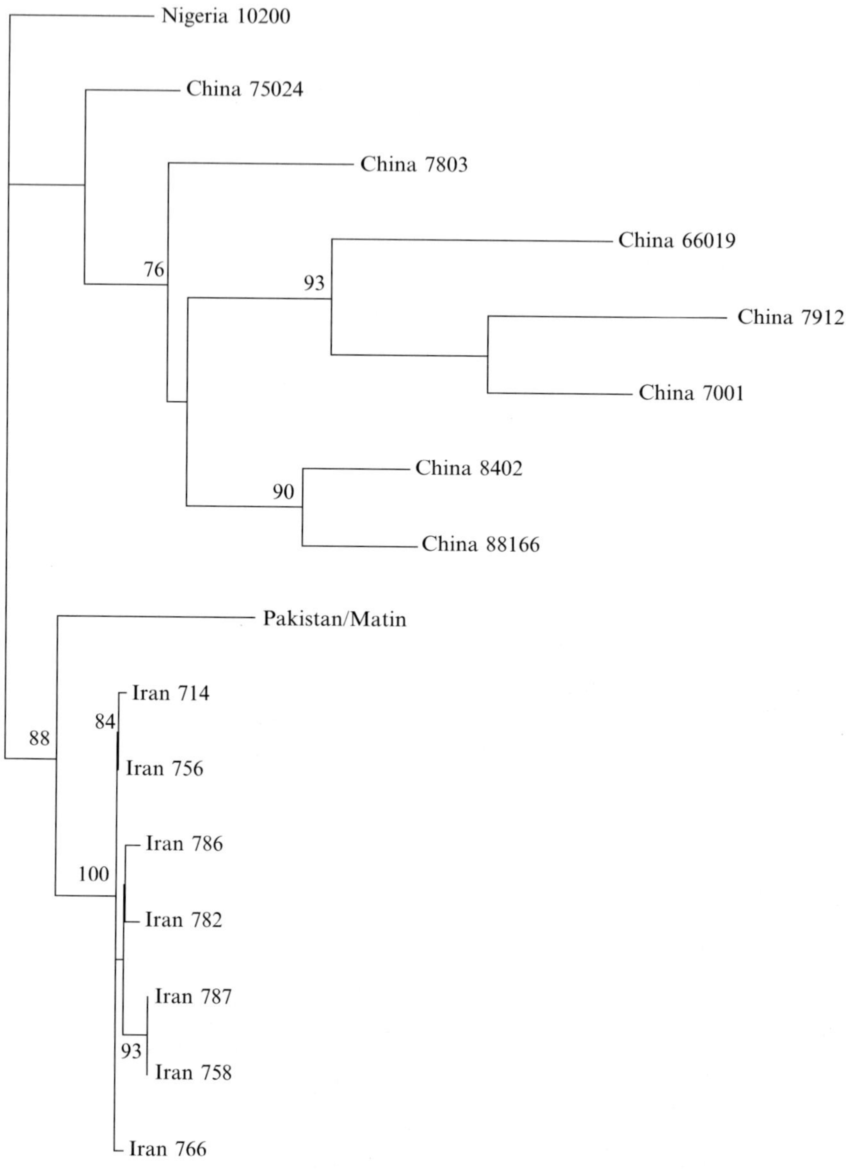

Fig. 8-3. Phylogenetic tree (neighbor-joining) calculated for Iranian CCHFV M-segment sequences.

ribavirin are inaccessible). According to the protocol of the CDC of Iran, the dose of ribavirin are recommended as follows:

- First 30 mg of ribavirin/kg of body weight
- Then 15 mg of ribavirin/kg of body weight each 6 h for 4 days
- At the end 7.5 mg of ribavirin/kg of body weight each 8 h for 6 days
- Also patients receive supportive therapy [11, 12, 26]

8.9. PREVENTION

The prevention measures performed in Iran are based on two axes: (1) public education and (2) ticks control. Concerning public education, seminars and speeches have been organized in collaboration with the Pasteur Institute of Iran and Ministry of Health, CDC of Iran, in the at-risk provinces of the country in the universities and health centers.

For the common people, mass media and several press articles have exposed the ways and risks of transmission CCHF and explained in detail the prevention measures and the ways to control it.

Concerning ticks control, all livestock are treated in industrial breeding center with insecticide showers, also in industrial slaughterhouses, persons are educated about the different ways of the transmission and the control of the disease, as slaughterers wear gloves, goggles, plastic gowns, and other preventive things, so the contamination risk is reduced.

In the traditional way of breeding livestock, there is less controls and preventive measures and livestock are infected with ticks carrying CCHF virus and there is more chance to contaminate oneself when slaughter these animals or handling their carcass and meat.

REFERENCES

1. Asefi V (1974) Clinical study of 60 human cases with haemorrhagic fever syndrome in East-Azarbaijan province (Iran). J Med Council 3:182–188 (in Persian)
2. Ardoin A, Karimi Y (1982) Foyer de purpura thrombocytopenique en Iran dans l'Azerbaidjan de l'est. Med Trop 42(3):319–26
3. Chinikar S (2003) Seroepidemiology of Crimean-Congo haemorrhagic fever in human and domestic animals in Iran by analyzing the quantities of specific IgM and IgG against the virus of the disease by ELISA method. J Vet Org 3:69–73
4. Chinikar S, Mehrabi-Tavana A, Mirahmadi R, Mazaheri V (2002) Crimean-Congo haemorrhagic fever. In: Ramezani A, Banifazl M, Mohraz M, Samar G (eds) Textbook of Infectious Diseases, vol 1. Porsina, Tehran, pp 705–710
5. Chinikar S, Fayaz A, Mirahmadi R, Mazaheri V, Mathiot C, Saron MF (2002) The specific serological investigation of suspected humans and domestic animals to have Crimean-Congo haemorrhagic fever in various parts of Iran using ELISA techniques. Hakim J (National Research Center of Medical Sciences) 4(4):294–300
6. Chinikar S, Persson SM, Johansson M, Bladh L, Goya M, Houshmand B, Mirazimi A, Plyusnin A, Lundkvist A, Nilsson M (2004) Genetic analysis of Crimean-Congo hemorrhagic fever virus in Iran. J Med Virol 73:404–411

7. Chinikar S, Mazaheri V, Mirahmadi R, Nabeth P, Saron MF, Salehi P, Hosseini N, Bouloy M, Mirazimi A, Lundkvist A, Nilsson M, Mehrabi Tavana A (2005) A serological survey in suspected human patients of Crimean-Congo hemorrhagic fever in Iran by determination of IgM specific ELISA method during 2000–2004. Arch Iranian Med 8(1):52–55
8. Garcia S, Chinikar S, Coudrier D, Billecocq A, Hooshmand B, Crance JM, Garin D, Bouloy M (2006) Evaluation of a Crimean-Congo hemorrhagic fever virus recombinant antigen expressed by Semliki forest suicide virus for IgM and IgG antibody detection in human and animal sera collected in Iran. J Clin Virol 35(2):154–159
9. Izadi SH, Holakouie Naieni K, Majdzadeh SR, Chinikar S, Rakhshani F, Nadim A, Hooshmand B (2003) Incidence of the infection of Crimean-Congo haemorrhagic fever in the Sistan–Baluchistan province: a serological study. Payesh J 2:85–93
10. Izadi SH, Holakouie Naieni K, Majdzadeh SR, Rakhshani F, Chinikar S, Nadim A, Hooshmand B (2003) Crimean-Congo haemorrhagic fever in Sistan and Baluchesatan province of Iran, a case–control study about epidemiological characteristics. J Vet Fac Univ Tehran 57(4):27–32
11. Mandell GL, Bennett JE, Dolin R (2000) Douglas and Bennette's Principles and Practice of Infectious Diseases, 5th edn. Churchill Livingstone, Philadelphia, 477–1861
12. McCormick JB, King IJ, Webb PA, et al. (1986) Lassa fever. Effective therapy with ribavirin. N Engl J Med 314:20–26
13. Mehrabi-Tavana A, Chinikar S, Mazaheri V (2002) The seroepidemiological aspect of Crimean-Congo haemorrhagic fever in three health workers: a report from Iran. Arch Iranian Med 15(4):255–258
14. Promed Digest, Crimean-Congo hemorrhagic fever in Iran, 20011023.2613, October 23 2001 Volume 2001: Number 263, http://www.promedmail.org, Accessed 15 September 2006
15. Promed Digest, Crimean-Congo hemorrhagic fever in Iran, 20020625.4596, June 25 2002 Volume 2002: Number 157. http://www.promedmail.org, Accessed 15 September 2006
16. Promed Digest, Crimean-Congo hemorrhagic fever in Iran, 20020827.5163, August 27 2002 Volume 2002: Number 233. http://www.promedmail.org, Accessed 15 September 2006
17. Promed Digest, Crimean-Congo hemorrhagic fever in Iran, 20030309.0580, March 9 2003 Volume 2003: Number 074. http://www.promedmail.org, Accessed 15 September 2006
18. Promed Digest, Crimean-Congo hemorrhagic fever in Iran, 20040114.0146, January 14 2004 Volume 2004: Number 020. http://www.promedmail.org, Accessed 15 September 2006
19. Promed Digest, Crimean-Congo hemorrhagic fever in Iran, 20040517.1324, May 17 2004 Volume 2004: Number 188. http://www.promedmail.org, Accessed 15 September 2006
20. Promed Digest, Crimean-Congo hemorrhagic fever in Iran, 20040906.2492, September 6 2004 Volume 2004: Number 355. http://www.promedmail.org, Accessed 15 September 2006
21. Promed Digest, Crimean-Congo Hemorrhagic Fever in Iran, 20050205.0397, February 5 2005 Volume 2005: Number 055. http://www.promedmail.org, Accessed 15 September 2006
22. Promed Digest, Crimean-Congo Hemorrhagic Fever in Iran, 20050907.2647, September 7 2005 Volume 2005: Number 397. http://www.promedmail.org, Accessed 15 September 2006
23. Saidi S, Casals J, Faghih A (1975) Crimean hemorrhagic fever-Congo (CHF-C) virus antibodies in man, and in domestic and small mammals in Iran. Am J Trop Med Hyg 24(2):353–57
24. Shirani M, Asmar M, Chinikar S, Mirahmadi R, Piazak N, Mazaheri V (2004) Detection of CCHF virus in soft ticks (Argasidae) by RT-PCR method. J Infec Dis Trop Med 9(24):11–15
25. Sureau P, Klein JM, Casals J, Digoutte J.P, Salaun JJ, Piazak N, Calvo MA (1980) Isolement Des Virus Thogoto, Wad Medani, Waowarie Et De La Fievre Hemorragique De Crimée-Congo En Iran À Partir De Tiques D'animaux Domestiques. Ann Virol (Inst Pasteur) 131E:185–200
26. Suspected, probable and definitive CCHF cases defined by Center of Disease Control of Iran, Ref: 4.17481/24.10.1379 (in Persian)

CHAPTER 9

CRIMEAN-CONGO HEMORRHAGIC FEVER IN RUSSIA AND OTHER COUNTRIES OF THE FORMER SOVIET UNION

A.M. BUTENKO, PH.D.[1] AND G.G. KARGANOVA, PH.D.[2]

[1]*D.I. Ivanovsky Institute of Virology, Russian Academy of Medical Sciences, Moscow, Russia*

[2]*M.P. Chumakov Institute of Poliomyelitis and Viral Encephalitides, Russian Academy of Medical Sciences, Moscow, Russia*

9.1. HISTORICAL ASPECTS

In the Compendium of the Sheikh of Khorezm by Dzhurzhoni in the 12th century, written in the Tajik language, the physician Zayn ad-Din abu Ibrahim Ismacil ibn Muhamad al-Husayini al-Jurjani described a hemorrhagic disease, now considered to have been Crimean-Congo hemorrhagic fever (CCHF), from the area that is presently Tajikistan [29]. The signs were presence of blood in the urine, rectum, gums, vomit, sputum, and abdominal cavity. The arthropod causing the disease was said to be tough, small, related to a louse or tick, and normally parasitizing a blackbird. Treatment that was sometimes ineffectual was application of *bodzkhar* and essence of red sandalwood at the site of the bite and feeding the patient fresh goat milk together with butter, *khot'ma* flowers (Malvaceae) and leaves or essence of *khovre* and essence of flax seeds, chicory, and gourd.

CCHF was also recognized for centuries under at least three names by indigenous peoples of southern Uzbekistan [18, 30]. In the opinion of Drobinsky [24], cases of CCHF had probably been seen in Romanian hospitals in occupied Crimea in 1942. Sipovsky [51] described 18 cases of peculiar gastrointestinal bleeding that occurred in the former Stalinabad (Dushanbe, Tajikistan Republic). This disease was called acute infectious capillary toxicosis, and *post factum* analysis clearly shows that this illness and CCHF resemble each other.

Crimean hemorrhagic fever, as originally named, was first fully described by Chumakov and coauthors [14, 15] who analyzed the data collected from the first outbreak of human disease in 1944, when about 200 members of the Soviet military were infected while assisting peasants in war-devastated Crimea (Ukraine SSR).

O. Ergonul and C. A. Whitehouse (eds.), Crimean-Congo Hemorrhagic Fever, 99–114.

9.2. HISTORY OF STUDIES ON THE ETIOLOGY OF CCHF

Chumakov and colleagues established the viral etiology of CCHF by inducing disease in psychiatric patients for whom pyrogenic therapy was prescribed. An infectious agent was found in both the blood of CCHF patients and in suspensions of *Hyalomma plumbeum* (*Hyalomma marginatum marginatum*) ticks. These authors showed its ability, after being filtered through bacterial filters, to cause disease with typical clinical features of CCHF in humans. Multiple species of laboratory animals, including adult white mice, guinea pigs, monkeys, and cats were found to be refractory to infection. Later, in 1955, Chumakov and coauthors similarly showed the viral etiology of CCHF in the Astrakhan region of Russia. A decade later several reports of the isolation of CCHF virus (CCHFV) using different types of cell cultures appeared in the literature [16]. These isolated viruses turned out to be nonpathogenic for newborn white mice and golden hamsters, and the attribution of these isolates to CCHFV was questionable.

Intracerebral inoculation of newborn white mice was first used for virus isolation in the USSR in 1967. Obviously at that time, this model was very useful for isolation of numerous arboviruses from around the world. Two groups of virologists from the Laboratory of Hemorrhagic Fevers, Institute of Poliomyelitis and Viral Encephalitides (USSR Academy of Medical Sciences), headed by Chumakov, were involved in this work. One group (Butenko and colleagues) worked in Astrakhan and another (Shalunova and colleagues), in Moscow. As a result, nine strains of CCHFV were isolated and identified. One strain (Drozdov) was isolated from a CCHF patient from the Astrakhan region; seven strains were recovered from patients in the neighboring Rostov region and one from Samarkand in Central Asia [10, 20].

Evidence of the role of these agents in the etiology of CCHF included regular detection of seroconversion of specific antibodies in patients with typical CCHF and virus isolation from patient's blood during the acute period of disease. Using serological tests, CCHFV was differentiated from nine spontaneous murine viruses, Omsk hemorrhagic fever virus, the virus of hemorrhagic fever with renal syndrome, and ten other arboviruses. The prominent American scientist Harry Hoogstraal, in his review of CCHF [30] wrote "... it was only in 1967, when Soviet workers first used the generally accepted newborn white mouse (NWM) inoculation technique for CCHF virus isolation and study, that the etiologic agent could be characterized antigenically, physiochemically, and morphologically." In 1968, the CCHFV (Drozdov strain) was transferred to the Yale Arbovirus Research Unit (YARU). In October 1968, Jordi Casals from YARU wrote in a letter to Chumakov, "as you can see, it appears that by complement-fixation your strain Drozdov is indistinguishable from three strains of Congo virus – Ug K 2/61, Congo 3010 and Pak JD 206. The fact that the serum pool from persons who have had Crimean hemorrhagic fever, which serum you sent me in your first shipment of materials, reacts with Congo virus (strain 3010) antigen with the same titer, 1:8, as it does with the Drozdov antigen, is of crucial importance and significance. We all believe here at YARU that these results are very exciting; whether or not Drozdov strain and Congo virus will turn out to be identical by neutralization test remains to be seen. I am planning to run the neutralization tests in 4 or 5 days and will let you know the result. If your strains of Crimean hemorrhagic fever virus turn out to be indistinguishable or very close to Congo virus, then the latter becomes a pathogen of exceptional interest, not just to you in the Soviet Union where you already knew it in the form of a very serious human illness, but also to the other areas in the world where it has been recognized heretofore."

Soon, both Russian and American investigators obtained additional data on identity of the Crimean and Congo viruses [12, 21]. Hoogstraal, in his 1979 review [30] declared, "The Drozdov strain of CCHF virus, isolated by this method from a patient (Drozdov) in Astrakhan, became the now-famous prototype CCHF strain for much experimental work in the USSR and abroad. Following the publication of Congo virus in 1967 [54], some have argued that the common name of the virus should be Congo virus, but Soviet authorities have insisted that the long recognized name Crimean hemorrhagic fever virus should be retained. As a compromise between 'unofficial' historical antecedents and 'official' Registry criteria, J. Casals et al. [13] suggested CHF-Congo virus as an accepted common name." For more on the controversy over the naming of CCHFV, see Chapter 3.

Major results of intensive research during the years since 1967 have included the development of cell culture techniques for virus isolation, which has allowed the identification of the virus in endemic regions of Europe, Asia, and Africa. This has also allowed for the identification of biological and morphological properties of the virus making it possible to classify it as a member of the family *Bunyaviridae* [23]. Plaque formation, hemagglutination, interference phenomena, and some peculiarities of cultivation of the virus in cell cultures were discovered. The spectrum of pathogenicity of CCHFV for laboratory and domestic animals was studied, as well as possibilities of laboratory techniques for serological diagnosis and experimental investigations. Serological methods, of course, have been highly beneficial in the areas of CCHFV laboratory diagnostics and research, for example, by placing CCHFV in the genus *Nairovirus*, and together with Hazara virus isolated from *Ixodes redikorzevi* ticks from Pakistan, forming the CCHFV antigenic group [2, 3].

9.3. EPIDEMIOLOGY

Although many of the fundamental epidemiological parameters of CCHF in the southern territories of the USSR and Bulgaria have been identified previously, after 1967 new data radically expanded our knowledge of the geographical distribution of natural foci, host reservoirs, and tick species associated with CCHFV. The virus is distributed in the following regions in Europe: Russian Federation (Astrakhan, Rostov, Volgograd regions, Kalmykia,

Krasnodar and Stavropol territories, Dagestan, Ingushetia), Ukraine (Crimea, Lugansk region) Azerbaijan, Armenia, Georgia (?), Bulgaria, Greece, Hungary (?), Republics of former Yugoslavia, Albania, France (?), and Portugal. In Asia, the virus occurs in all the former Soviet Union republics of Central Asia, Kazakhstan, China (western provinces), Afghanistan, Pakistan, India, Iran, Iraq, United Arab Emirates, Kuwait, and Turkey.

9.4. INCIDENCE OF DISEASE

The total number of registered CCHF cases in the world is about 5,000, with more than 200 cases in the Crimea, 1,154 in the Russian Federation (Fig. 9-1), some 700 in Central Asia, more than 550 cases in Uzbekistan, about 150 in Tajikistan, a single case in Kyrgyzstan, about 200 in Kazakhstan, 1,500 cases in Bulgaria, nearly 50 in territories of the former Yugoslavia (mainly in Kosovo), 40 cases in the Middle East countries and Pakistan, over 1,000 cases in Turkey, and approximately 200 cases in Africa – half of them reported from the Republic of

Fig. 9-1. Map of the Russian Federation showing the total number of CCHF cases for 1948–2005.

South Africa. The situation in the latter country is interesting, because, unlike other African countries, clinical and epidemiological features in South Africa strongly resemble those in classical foci in Central Asia; mortality rates are up to 30%, modes of transmission include ticks bites, direct contact with crushed infected ticks, contacts with infected human, bovine and sheep blood.

9.4.1. Incidence of CCHF in the countries of the former Soviet Union

Three hundred thirty-nine cases of CCHF were registered in the Astrakhan region of Russia in the period from 1953 to 2005; 377 cases in the Rostov region from 1963 to 2005; 263 cases in the Stavropol region from 1953 to 2005. In 1948, 18 persons became ill in the Krasnodar region. From 2000 to 2005, 102, 41, and 10 cases occurred in Kalmykia, the Volgograd region, and Dagestan, respectively [8, 9]; four cases were identified in 2004 from Ingushetia [49]. The most cases (230 patients) occurred in the Astrakhan region from 1953 to 1967. Only nine cases were registered from 1970 to 1983 and only a single case in 1984–1999. In the following 4-year period ending in 2004, 50 persons became ill. An additional 37 cases were recognized in 2005.

After the first description of CCHF in the Rostov region in 1963, the number of human cases in the region during the following 8 years (to 1970) was 338. The highest incidence of disease was recorded in 1968 (138 cases; 23.4 per 100,000 population). In the other years of the same 8-year period (1963–1970), incidence varied between 8.9 and 17.0 cases per 100,000 population. Twenty-eight cases were reported in 2001–2004, with an additional ten cases in 2005.

The number of cases recorded from the Stavropol region from 1953 to 1968 was 25. There were one and two cases in 1970 and 1972, respectively. However, from 1999 to 2005, the total number of cases increased to 237 [9]. Eighteen cases were reported in the Krasnodar region in 1948 [33]; however, no other information of additional cases from this region is available.

During the initial outbreak of CCHF on the Crimean peninsula in 1944, approximately 200 cases of disease were recorded. Only small outbreaks and sporadic cases have been seen there in the years that followed [8, 9]. Additionally, three laboratory-confirmed cases were reported from the Lugansk region in Ukraine in 1969 [32].

The only case of CCHF in Armenia was reported in 1974 [50]. Local physicians reported 72 cases of CCHF in the Chimkent region of Kazakhstan from 1948 to 1975, although 49 patients were reported in the entire Republic from 1965 to 1982. An outbreak involving 90 patients occurred in 1989 in the Kyzyl-Orda, Chimkent, and Djambul regions of Kazakhstan; morbidity rates ranged from 0.01 to 0.09 per 100,000.

From 1948 to 1963, in Uzbekistan, a total of 525 CCHF cases were reported. However, from 1973 to 1983, only 28 cases were reported. Ninety-three CCHF cases were reported in Tajikistan from 1943 to 1970. By 1974, the number of cases reached 121, and during 1975–1983, an additional 23 cases were seen. Morbidity rates varied from 0 to 0.27 per 100,000 (in 1975) [8, 9].

The hospital-acquired outbreak in Turkmenistan in 1946 involved seven persons [41]. In the years that followed, only sporadic cases occurred. Two cases were recorded in 1953, and one in 1971, in the Osh region of Kyrgyzstan [31].

As shown by seroepidemiological studies, the level of naturally acquired specific immunity to CCHFV in all endemic territories is extremely insignificant, and so, the overwhelming majority of both local and newly arrived population is susceptible to CCHF. The morbidity in all of the endemic regions over the years has been very sporadic. Even during the significant outbreaks, only single cases were noticed in the same territories, and not every year. For example, during the epidemic period from 1963 to 1971, in the Rostov region, 169 settlements in 18 administrative regions were affected by CCHF with a mean number of reported patients from 1.0 to 1.42 [1].

9.5. SEASONAL ACTIVITY

Not surprisingly, the seasonality of CCHF cases corresponds to the months of the year when the tick vectors are most active. The first cases of CCHF during the initial outbreak in 1944–1945 in the Crimean peninsula were noticed in April and the last in September, with the highest number occurring in July (53% of all cases) [30, 34]. The earliest CCHF case in the Astrakhan region was reported in March, with the last one reported in August. In fact, the majority of cases occurred between the first week of May and the second week of June [8, 9, 35, 52]. The first CCHF cases in the Rostov region appeared at the end of May and the morbidity reached its maximum (as in Astrakhan region) at the end of May/beginning of June. The last cases were usually seen by the end of August. Only two patients acquired the disease in September in the Rostov region. The data of 1963–1969 shows the seasonal dynamics of CCHF morbidity in the Rostov region as follows: 0.9% of all cases were in April, 34.2% in May, 41% in June, 17.9 in July, 5.3% in August, and 0.6% in September. Two hundred fifty-one patients were recorded in May and June during the period from 1963 to 1970, comprising 72.4% of all cases (338) [1, 44].

Based on data from 1950–1969, CCHF cases in the Samarkand region of Uzbekistan were noticed year-round, nevertheless, the majority of infections occurred during the summer months (June, July, and August). During 1948–1975 in the Chimkent region of Kazakhstan, the first cases of the disease were seen in January, which was highly unusual. The distribution of CCHF cases by season was as follows: 61.4% summer, 24.2% autumn, 11.6% spring, and 2.8% in the winter (January) [8, 9, 30].

9.6. EPIDEMIOLOGY

The age range of the majority of the CCHF cases in the Rostov region was 20–60 years. The risk groups in the Rostov region, as well as in other endemic regions, comprising up to 70% of all CCHF cases, included milkmaids, cattle farm workers, agricultural workers, and housewives. Regional epidemiologists often stated that there is no strict association of CCHF with a particular trade in the rural area. Nevertheless, it was clear that the incidence of disease was highest among those who work in the open steppe environment [1, 44].

Likewise, in the Astrakhan region, approximately 80% of the patients were 20–60 years old; again, most were agricultural worker, and males (54.2%) were more likely to become infected than females (45.8%) [35, 52].

Adult rural residents made up 68% of all those infected with CCHFV in Kazakhstan, being cattle breeders and farm hands. Two persons who lived in an urban environment also became infected; one worked on the slaughtering floor of a meatpacking factory, and the other was a miner, who occasionally visited the endemic area. It is noteworthy that the numbers of shepherds and health-care workers infected with CCHFV were relatively high (38.8% and 16.3%, respectively) in Kazakhstan [53].

In Tajikistan, the breakdown of CCHF patients by profession was as follows: 28% agricultural workers, 19% shepherds, 14% housewives, 13% health-care workers, 11% farm machine operators, 8% teachers or students, and 6% dairy farm workers [42].

The CCHF patients in Uzbekistan ranged from 2 to 74 years old, with the majority (83%) being between 15–50 years old. Sixty percent of the patients were farm workers, and 9% were school children brought to the fields to help gather the harvest [40].

9.7. MODES OF TRANSMISSION

As stated elsewhere, tick bites are the most common route of transmission of CCHFV to humans. In the original Crimean outbreak in 1944, 87.8% of patients reported that they had been bitten by a tick several days before onset of the disease; human-to-human transmission was not reported [14, 15, 24].

In the Astrakhan region, approximately 30% of CCHF patients have reported being bitten by a tick or have found one crawling on their body or clothes. A case from May 1962 of the simultaneous infection of both husband and wife who sheared sheep has been reported. That spring the weather in the Astrakhan region was dry and hot, and the fields were very dusty. Neither husband nor wife reported a tick bite, and, thus, it was suggested that the disease was acquired by inhalation of infected dust particles; although undetected tick bite or exposure to infectious blood or body fluids during the process of shearing is also a possibility. Such a possibility is highlighted by the case of a nurse at a regional hospital, who became infected following manipulation of the blood of CCHF patients. She had severe eczema of the hands, and presumably became infected through the multiple lesions on her hands [52].

CCHF has a long history of nosocomial spread (see Chapter 22). Three cases of nosocomial infection are known from the Rostov region, all of which were medical personnel who become infected from the blood of CCHF patients [30, 34]. In 1946, a CCHF patient in Turkmenistan was a source of infection for six other persons, medical attendants, and other patients hospitalized in the same ward (six of the patients died).

The transmission of CCHFV through contact with the blood of infected cows and sheep was repeatedly reported from the countries of Central Asia (Tajikistan and Uzbekistan); slaughtering-infected cattle posed the highest risk [8, 9].

Two cases of laboratory infections are known from Russia. Manipulation of infected material during viral isolation, preparation of viral antigen, and serologic analyses resulted in infection by direct contact through the small skin injuries, accidental skin puncture by syringe needles, or inhalation of aerosols. Both of the laboratory-acquired cases were very severe and one patient ultimately died [26].

The analysis of the data from Russia and other former Soviet Union countries suggest that the major risk factors for CCHFV transmission to humans are residence in endemic regions, exposure to tick vectors (such as agricultural and cattle-breeding workers, sheep herders, and dairy farmers), direct contact with CCHF patients, and working with the virus or virus-infected materials in the laboratory (for more on CCHF risk factors, see Chapter 21).

9.8. MORTALITY RATES

Various authors have reported the mortality rates of CCHF during the initial outbreak in Crimea in 1944 as 8–11%; whereas, in succeeding years, when only small outbreaks or sporadic cases were reported, up to 30% of the cases were fatal. A possible explanation for this disparity is the underreporting of milder cases [17, 30, 34]. Mortality rates in the Astrakhan region, in the epidemic period from 1953 to 1967, varied from 12% to 16%. Eleven of 25 (44%) CCHF patients from 1953–1968 in Stavropol region died [34, 35]. The average mortality rate in Kazakhstan was 32.6% (1965–1982); however, interestingly, very high mortality has also been seen in the Chimkent region of that country when patients were infected by the blood of CCHF patients (62.5%). Likewise, in Tajikistan, transmission of CCHFV by infected blood resulted in a 50% case fatality rate.

More recently, mortality rates in endemic regions of Russia (southern territories of European Russia) have significantly decreased when compared with previous years. For instance, in 2004, there were six fatalities of 68 patients (8.8% mortality), and in 2005, six of 130 cases (4.6%) had fatal outcomes. Such a decrease may be the result of three main causes: (1) wide distribution and availability of specific diagnostic methods that lead to laboratory confirmation of multiple milder cases without, or with only slight, hemorrhagic syndrome, (2) utilization of ribavirin for the treatment of patients, and (3) predominant circulation of less pathogenic viral strains in endemic regions.

9.9. ARTHROPOD VECTORS OF CCHFV

Approximately 30 species and subspecies of Ixodid ticks are ecologically associated with CCHFV virus (see Chapter 12). *H. marginatum marginatum* is the most important tick vector in the European part of Russia. In Central Asia, many other ticks in the genus *Hyalomma* are known to harbor the virus (see Table 12-1 in Chapter 12 for a complete list) [18, 30].

9.10. WARM-BLOODED HOSTS OF CCHFV

Natural hosts and reservoirs of CCHFV include hedgehogs, hares, ground squirrels, jerboas, and some species of rodents and ungulate animals (see Chapter 13). These animals may develop constant viremia, high enough to transmit the virus to feeding ticks. Adult ungulates, immunized by their first encounter with CCHFV, loose the ability to produce a high-titer viremia and thus became a dead-end host.

As many as 70% of the European hares in the Astrakhan region, which were examined for the presence of specific antibodies to CCHFV, were found to be positive. Similar data were obtained from the Rostov region and from Bulgaria. Experimental work showed the ability of hares to be the source of virus infection for *H. marginatum marginatum* larvae, which feed on them during the period of viremia [4, 19, 56]. Subcutaneously inoculated animals did not develop clinically evident infection, but viremia was prolonged (from day 1 to 10 postinoculation).

H. plumbeum (*H. marginatum marginatum*) tick larvae were found to become infected while feeding on experimentally infected big-eared (Asian) hedgehogs (*Hemiechinus auritis*); moreover, they were also able to transmit the virus transstadially. Neither the big-eared nor the European (*Erinaceus europaens*) hedgehogs presented any clinical signs of disease. Virus could not be detected in the blood of European hedgehogs 5–13 days postinoculation; whereas the big-eared hedgehog produced high-titered (10^4) viremia 4–6 days postinfection [5].

Small ground squirrels (*Citellus pygmaeus*) infected subcutaneously at the age of 4–6 weeks retained CCHFV without marked clinical signs. The virus was regularly recovered from blood and parenchymatous organs 2–7 days after inoculation. In some individuals, virus could be detected in the kidneys and brain. The amount of virus needed to produce viremia in ground squirrels was determined to be 10 mouse LD_{50} (50% lethal dose), and studies have shown that artificially infected ground squirrels are capable of transmitting CCHFV to feeding larval and nymphal ticks of multiple species [6].

9.11. CLIMATIC INFLUENCE ON THE ACTIVITY AND DISTRIBUTION OF CCHFV IN THE NORTHERN PART OF ITS NATURAL FOCUS

Worldwide, the occurrence of CCHF coincides with the natural foci of *Hyalomma* species ticks. In Russia, this is primarily *H. marginatum marginatum*, which has a northern geographic limit of 48° north latitude. Epidemic potential and disease incidence are strongly dependent on the abundance of the main tick vector (adult *H. marginatum marginatum*) in the spring and summer. And this, in turn, is dependent on the climatic conditions during the winter months in the northern limits of its habitat (for a more general discussion of the effects of climate on tick-borne diseases, see Chapter 14).

Since 1891, significant warming has been observed in the Astrakhan and Rostov regions. In the period from 1963 to 1968, the number of reported CCHF patients in the Rostov region increased (11 in 1963, 24 in 1964, 27 in 1965, 38 in 1966, 61 in 1967, and 131 in 1968) proportionally to the average seasonal rates of adult *H. marginatum marginatum* tick abundance. Likewise, beginning in 1995, the six winters that followed were mild, with increased average seasonal rates of *H. marginatum marginatum* tick abundance. This led to an increase in CCHF cases in 2001–2003. The number of administrative regions considered endemic for CCHF also increased from 2 to 14, as well as the overall area of the endemic regions (from 600 to 16,000 km^2). This expansion of endemic regions has been suggested to be due to the dissemination of virus-infected nymphs and larvae of *H. marginatum marginatum* by rooks (*Corvus frugilegus*) and other birds. In addition, perceptible warming in the last few years has led to the increase of epizootic and endemic activities of CCHF in the Rostov region, synchronously with that in the Astrakhan region.

The emergence of CCHF in the Volgograd region in 2000 is of special interest. This is evidence of the spreading of the natural focus of CCHFV, possibly as a result of global warming, and formation of abundant local populations of *H. marginatum marginatum* ticks, much further north than the known endemic regions [11].

9.12. MOLECULAR EPIDEMIOLOGY

In recent years, a large body of genetic data has become available for many CCHFV strains isolated from around the world, with many strains isolated from Russia and other countries of the former Soviet Union represented [22, 27, 28, 38, 39, 45, 55].

Attempting to analyze the molecular epidemiology of CCHFV strains isolated from the European part of Russia and the Central Asian republics of the former Soviet Union, we compiled all available sequences in GenBank. Based on the phylogeny of all three segments, strains from the European part of Russia grouped together with strains isolated from southern Europe. In addition, our data supported those of others [7, 25, 28, 43, 55] regarding the phylogeny of strains isolated from the Astrakhan, Rostov, Volgograd, and Stavropol regions. S-segment sequences of these strains reliably grouped with viruses isolated from Bulgaria, Kosovo, Albania, and Turkey. Similarly, 12 fragments of the L segment of strains isolated in the southern regions of Russia form a phylogenetic group with a Bulgarian isolate [38, 39].

Phylogeny based on complete S segments show that viruses isolated from China, Tajikistan, Kazakhstan, and Uzbekistan group together (Asia 2); whereas, isolates from the Stavropol, Rostov, Volgograd, and Astrakhan regions, along with Bulgarian isolates form a distinct group (Europe 1) (Fig. 9-2). Thus,

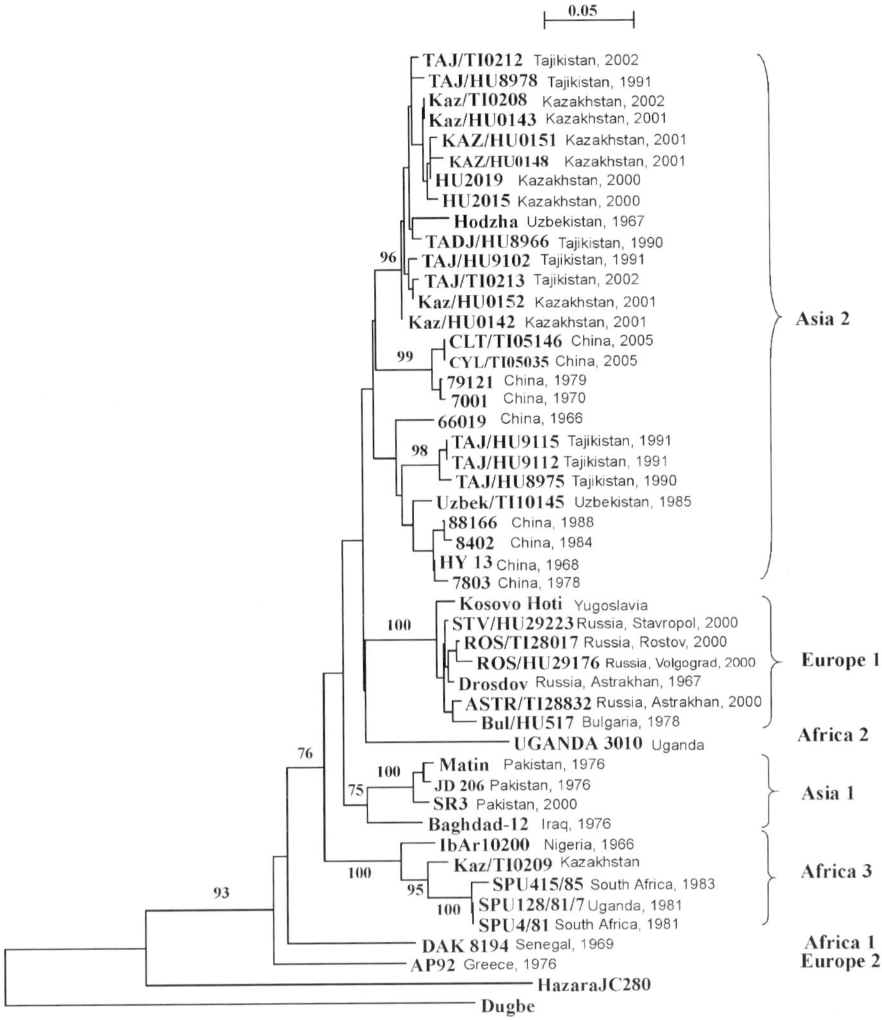

Fig. 9-2. Phylogenetic tree of a 294-nucleotide region of the S segment of CCHFV. Tree was constructed with CLUSTAL W software (version 1.83) using the neighbor-joining algorithm. Dugbe and Hazara viruses were included as out-groups. Bootstrap values from 1,000 replicates are shown at each branch point (only those of 70% and greater are shown).

one may find at least two genotypes of CCHFV circulating in Russia and the Central Asian Republics, one of which is genetically close to viruses circulating in China. For a more general discussion of the molecular epidemiology and phylogeny of CCHFV, see Chapter 5.

9.13. TREATMENTS USED IN RUSSIA AND OTHER COUNTRIES OF THE FORMER SOVIET UNION

Along with symptomatic therapy, specific antiviral treatment with 200 mg of oral ribavirin has been used for CCHF patients in the endemic regions of Russia. The loading dose of 30 mg per kg (2,000 mg for a 70 kg person) was established by experimental studies performed in the Stavropol region. A maintenance dose of 600 mg twice a day for those under 75 kg, and 500 mg twice a day for the persons over 75 kg, was also established, with a treatment length of 5–10 days. Maleyev et al. [37] presented data on their treatment of 20 CCHF patients. In 18 patients (90%), ribavirin was administered during the first 4 days after the onset of the disease. Clinical signs of hemorrhage were evident in only five patients; they presented with localized postinjection hematomata and petechiae. All patients demonstrated marked hypocoagulation. High fever resolved during the 2nd day after ribavirin administration. Blood tests performed 48 h after the initiation of therapy revealed white blood counts rising to normal levels in all patients. Two patients who developed nasal bleedings, gingival hemorrhages, and postinjection hematomata received hemostatic therapy, together with ribavirin, from the 2nd day of the disease. No fatalities were observed in this group of patients.

Ribavirin-Meduna (15 mg per kg) was also applied for the treatment of CCHF patients in the Astrakhan region (the clinical hospital of the Astrakhan Medical Academy). Normally, a 2-dose per day regimen of 1,000 mg was used (400 mg in the morning and 600 mg in the evening) and continued for up to 10 days. As a supplemental treatment, 0.5 g endogenous interferon (Cycloferon) was given on days 1, 2, 4, 6, 8, 10, and 12 after onset of disease [36].

Thus, the data from the Stavropol and Astrakhan regions suggest that the use of ribavirin is an effective treatment of CCHF, especially when administered before hemorrhagic manifestations (usually before the 5th day of the disease). For more on the use of ribavirin and other potential treatments for CCHF, see Chapter 20.

9.14. EPIDEMIOLOGICAL SURVEILLANCE AND INVESTIGATION IN ENDEMIC REGIONS

Epidemiological surveillance plays an important role in the overall control strategy of CCHF in Russia and other FSU countries. Various research institutions, Centers of Sanitary and Epidemiological Surveillance, Plague Control Institutions and Stations, and veterinary institutions collaborate on the surveillance of CCHF in Russia. It is compulsory for health authorities to inform their respective ministries of health on each suspected CCHF case. Isolation arrangements and a strict antiepidemic regimen for CCHF patients and their contacts are mandatory. Epidemiological investigation of each case or outbreak is carried out for the identification of the source of infection, transmission routes, and the identification of persons at risk of contracting the disease.

Editor's comments

Since 2005, there has been a marked expansion of CCHF cases in several regions of Russia, as evidenced by the following ProMED-Mail (www.promedmail.org) reports from 2006:

Date: 24 June 2006

As of 8 June 2006 there has been a marked expansion in the distribution of CCHF cases: new cases have been detected in the Zymovnikovskiy, Tsymlyanskiy, and Tselinniy districts of the Rostov region where no cases have been observed in recent years [46].

Date: 10 August 2006

[Thus far in 2006] a severe deterioration in the epidemiological situation for CCHF has been observed in the Southern Federal District of Russia. As of 8 August 2006, the Federal Service for Surveillance of Consumer Rights and Human Well-being reported that 192 cases of CCHF had been recorded in the southern federal district. This figure exceeds that for the corresponding period of (2005) by 43%. The greatest deterioration has occurred in the Republic of Kalmykia (65 cases of CCHF recorded) and in the Rostov region (53 cases of CCHF). CCHF has been diagnosed in 15 patients in the Astrakhan area, in 16 patients in the Volgograd area, and in 3 patients in the Republic of Dagestan. So far in 2006, five patients have died as a consequence of CCHFV infection, compared with four fatalities in the corresponding period of 2005. Two fatal cases were recorded in the Republic of Kalmykia, and one each in the Stavropol, Rostov, and Astrakhan regions [47].

Date: 31 August 2006

Forty-one cases of Crimean-Congo hemorrhagic fever (CCHF), including one fatality, have been recorded in 12 districts of the Stavropol region, according to the Territorial Management of Rospotrebnadzor (Federal Service on Surveillance of Consumer Rights and Human Well-being) in Stavropol. These figures are worse than those of 2005, when 38 cases were recorded in 15 areas of the region. The 334 people admitted to hospital with suspected CCHF exceeded by 46% of the number admitted to hospital in 2005 on suspicion of CCHF infection [48].

REFERENCES

1. Badalov ME, Koimchidi EK, Semenov MYa, Karinskaya GA (1971) Crimean hemorrhagic fever in Rostov Region [In Russian]. Sb Tr Inst Polio Virusn Encefalitov Akad Med Nauk SSSR 19:167–173
2. Begum F, Wisseman CL, Casals J (1970) Tick-borne viruses of West Pakistan. II. Hazara virus, a new agent isolated from *Ixodes redikorzevi* ticks from the Kaghan Valley, W. Pakistan. Am J Epidemiol 92:192–194
3. Begum F, Wisseman CL, Casals J (1970) Tick-borne viruses of West Pakistan. IV. Viruses similar to, or identical with, Crimean hemorrhagic fever (Congo-Semunya), Wad Medani, and Pak *Argas* 461 isolated from ticks of the Changa Manga Forest, Lahore district, and Hunza, Gilgit Agency, W. Pakistan. Am J Epidemiol 92:197–202

4. Berezin VV, Chumakov MP, Stolbov DN, Butenko AM (1971) On the problem of natural hosts of Crimean hemorrhagic fever virus in Astrakhan Region [In Russian]. Sb Tr Inst Polio Virusn Encefalitov Akad Med Nauk SSSR 19:210–216
5. Blagoveshchenskaya NM, Donets MA, Zarubina LV, Kondratenko VF, Kuchin VV (1975) Study of susceptibility to Crimean hemorrhagic fever virus in European and long-eared hedgehogs [In Russian]. Tezisy Konf Vop Med Virus, Moscow, pp 269–270
6. Blagoveshchenskaya NM, Vyshnivetskaya LK, Gusarev AF (1972) Investigation of susceptibility in little susliks (*Citellus pygmeus*) to CHF virus [In Russian]. Tezisy 17 Nauchn Sess Inst Posvyasch Aktual Probl Virus Profilakt Virus Zabolev, Moscow, p 356
7. Burt FJ, Swanepoel R (2005) Molecular epidemiology of African and Asian Crimean-Congo haemorrhagic fever isolates. Epidemiol Infect 133:659–666
8. Butenko AM (2005) Crimean hemorrhagic fever. Part I [In Russian]. *RAT-info* 2:52–56
9. Butenko AM (2005) Crimean hemorrhagic fever. Part II [In Russian]. *RAT-info* 3:45–48
10. Butenko AM, Chumakov MP, Bashkirtsev VN (1968) Isolation and investigation of Astrakhan strain ("Drozdov") of Crimean hemorrhagic fever virus and data on serodiagnosis of this infection [In Russian]. Mater 15 Nauchn Sess Inst Polio Virus Encephalitov (October 1968), Moscow, pp 88–90
11. Butenko AM, Larichev VF (2004) The influence of the climate on the activity and distribution of Crimean hemorrhagic fever (CHF) in the northern part of its distribution. In: Izmerov NF, Revich BA, Korenberg EI (eds) Climate Change and Public Health in Russia in the XXI century. Adamant, Moscow
12. Casals J (1969) Antigenic similarity between the virus causing Crimean hemorrhagic fever and Congo virus. Proc Soc Exp Biol Med 131:233–236
13. Casals J, Henderson BE, Hoogstraal H, Johnson KM, Shelokov A (1970) A review of Soviet viral hemorrhagic fevers, 1969. J Infect Dis 122:437–453
14. Chumakov MP (1945) A new tick-borne virus disease - Crimean hemorrhagic fever [In Russian]. In: Sokolov AA, Chumakov MP, Kolachev AA (eds) Crimean Hemorrhagic Fever (Acute Infectious Capillary Toxicosis). Izd Otd Primorskoi Armii, Simferopol, pp 13–43
15. Chumakov MP (1946) Crimean hemorrhagic fever (acute infectious capillary toxicosis). Short reports [In Russian]. Krymskiy Oblastnoy Otdel Zdravookhraneniya "Krymizdat" Simferopol, pp 1–27
16. Chumakov MP (1965) A short story of the investigation of the virus of Crimean hemorrhagic fever [In Russian]. Sb Tr Inst Polio Virusn Encefalitov Akad Med Nauk SSSR 7:193–196
17. Chumakov MP (1974) On 30 years of investigation of Crimean hemorrhagic fever [In Russian]. Sb Tr Inst Polio Virusn Encefalitov Akad Med Nauk SSSR 22:5–18
18. Chumakov MP (1979) Crimean hemorrhagic fever [In Russian]. In: Chumakov MP (ed.) Viral hemorrhagic fevers. Meditsina i zdravookhranenie, Moscow, pp 10–33
19. Chumakov MP, Butenko AM, Rubin SG (1969) Questions on the ecology of Crimean hemorrhagic fever virus. Mater 5 Simp Izuch Roli Pereletn Ptits Rasprostr Arbovirus, Novosibirsk, pp 222–229
20. Chumakov MP, Butenko AM, Shalunova NV, Mart'ianova LI, Smirnova SE, Bashkirtsev IuN, Zavodova TI, Rubin SG, Takachenko EA, Karmysheva Via, Reingol'd VN, Popov GV, Savinov AP (1968) New data on the virus causing Crimean hemorrhagic fever [In Russian]. Vopr Virusol 13:377
21. Chumakov MP, Smirnova SE, Tkachenko EA (1970) Relation between strains of Crimean hemorrhagic fever and Congo viruses. Acta Virol 14:82–85
22. Deyde VM, Khristova ML, Rollin PE, Ksiazek TG, Nichol ST (2006) Crimean Congo hemorrhagic fever virus genomics and global diversity. J Virol 80:8834–8842
23. Donets MA, Chumakov MP, Korolev MB, Rubin SG (1977) Physicochemical characteristics, morphology and morphogenesis of virions of the causative agents of Crimean hemorrhagic fever. Intervirology 8:294–308
24. Drobinsky IR (1945) Epidemiology and diagnosis of Crimean hemorrhagic fever [In Russian]. In: Sokolov AA, Chumakov MP, Kolachev AA (eds) Crimean Hemorrhagic Fever (Acute Infectious Capillary Toxicosis). Izd Otd Primorskoi Armii, Simferopol, pp 49–68

25. Drosten C, Minnak D, Emmerich P, Schmitz H, Reinicke T (2002) Crimean-Congo hemorrhagic fever in Kosovo. J Clin Microbiol 40:1122–1123
26. Gaidamovich SYa, Butenko AM, Leschinskaya EV (2000) Human laboratory acquired arbo-, arena-, and hantavirus infections. J Am Biol Safety Assoc 5:5–11
27. Hewson R, Chamberlain J, Mioulet V, Lloyd G, Jamil B, Hasan R, Gmyl A, Gmyl L, Smirnova SE, Lukashev A, Karganova G, Clegg C (2004) Crimean-Congo haemorrhagic fever virus: sequence analysis of the small RNA segments from a collection of viruses world wide. Virus Res 102:185–189
28. Hewson R, Gmyl A, Gmyl L, Smirnova SE, Karganova G, Jamil B, Hasan R, Chamberlain J, Clegg C (2004) Evidence of segment reassortment in Crimean-Congo haemorrhagic fever virus. J Gen Virol 85:3059–3070
29. Hoogstraal H (1966) Ticks in relation to human diseases caused by viruses. Ann Rev Entomol 11:261–308
30. Hoogstraal H (1979) The epidemiology of tick-borne Crimean-Congo hemorrhagic fever in Asia, Europe and Africa. J Med Entomol 15:307–417
31. Karas' FR, Risaliev DD, Vargina SG (1976) Crimean hemorrhagic fever foci in southwestern climatic region of Kyrgyzia [In Russian]. Tezisy Dokl Vses Konf Prirodn Ochag Bolez Chelov Zhivot, Omsk, p 128
32. Karinskaya GA, Badalov ME, Primakov SV (1970) Detection of new Crimean hemorrhagic fever foci (CHF) in Rostov and Luga Oblast [In Russian]. Mater 3 Oblast Nauchh Prakt Konf, pp 108–110
33. Koval'sky GN, Rybkina LG (1957) Infections of the hemorrhagic fever group in the steppe region of Krasnodar Oblast [In Russian]. Sb Nauchn Tr Kuban Med Inst, p 15
34. Lebedev AD, Pak TP, Birulya NB, Meliev AM, Berezin VV, Badalov ME (1977) Ecological geography of Crimean Congo hemorrhagic fever virus. Part I. [In Russian]. Itogi Nauki i Tekhniki 8:122–187
35. Lebedev AD, Pak TP, Birulya NB, Meliev AM, Berezin VV, Badalov ME (1977) Ecological geography of Crimean-Congo hemorrhagic fever virus. Part II. [In Russian]. Itogi Nauki i Tekhniki 9:185–234
36. Maleyev VV, Galimzyanov KhM, Butenko AM, Cherenov IV (2004) Crimean hemorrhagic fever [In Russian]. Izdatelstvo Astrakhanskoy Meditsinskoy Akademii, Moscow-Astrakhan, pp 1–119
37. Maleyev VV, Sannikova IV, Yuryev YuP (2004) Ribavirin "Meduna" in the treatment of patients of Crimean-Congo hemorrhagic fever [In Russian]. Medicofarma, Moscow, pp 3–7
38. Meissner JD, Seregin SS, Seregin SV, Vyshemirskii OI, Samokhvalov EI, Lvov DK, Netesov SV, Petrov VS (2006) A variable region in the Crimean-Congo hemorrhagic fever virus L segment distinguishes between strains isolated from different geographic regions. J Med Virol 78:223–228
39. Meissner JD, Seregin SS, Seregin SV, Yakimenko NV, Vyshemirskii OI, Netesov SV, Petrov VS (2005) Complete L segment coding-region sequences of Crimean-Congo hemorrhagic fever virus strains from the Russian Federation and Tajikistan. Arch Virol 151:465–475
40. Meliev AM (1967) A contribution to epidemiology of hemorrhagic fever in Uzbekistan [In Russian]. Zh Mikrobiol Epidemiol Immunobiol 44: 93–97
41. Mikhailov GI (1946) On the epidemiology of an acute infectious hemorrhagic disease. Klin Med (Moscow) 24:67–69
42. Pak TP, Mikhailova LI (1973) Crimean hemorrhagic fever in Tajikistan [In Russian]. Dushanbe, Irton, pp 1–154
43. Papa A, Bino S, Llagami A, Brahimaj B, Papadimitriou E, Pavlidou V, Velo E, Cahani G, Hajdini M, Pilaca A, Harxhi A, Antoniadis A (2002) Crimean-Congo hemorrhagic fever in Albania, 2001. Eur J Clin Microbiol Infect Dis 21:603–606
44. Perelatov VD, Vostokova KK (1971) Epidemiology of Crimean hemorrhagic fever in Rostov Region [In Russian]. Sb Tr Inst Polio Virusn Encefalitov Akad Med Nauk SSSR 19:174–179
45. Platonov AE, Karan LS, Yazyshina SB (2005) Molecular identification of Crimean-Congo hemorrhagic fever virus in human clinical cases in Southern Russia. Abstract Book of International Conference on Emerging Infectious Diseases, 2005. Atlanta, CDC 24-3-2002

46. ProMED-mail. Crimean-Congo hem. fever – Russia (Southern Federal District). ProMED-mail 2006; 24 June: 20060624.1760. http://www.promedmail.org. Accessed 13 January 2007
47. ProMED-mail. Crimean-Congo hem. fever – Russia (Southern Federal District). ProMED-mail 2006; 10 Aug: 20060810.2242. http://www.promedmail.org. Accessed 13 January 2007
48. ProMED-mail. Crimean-Congo hem. fever – Russia (Stavropol). ProMED-mail 2006; 31 Aug: 20060831.2479. http://www.promedmail.org. Accessed 13 January 2007
49. Schelkanov MIu, Kolobukhina LV, Moskvina TM, Aushev ID, Kartoev AA, Kelli EI, Merkulova, LN, Grenkova EP, Samokhvalov EI, Petriaev VG, Serobian AG, Klimova EA, Galkina IV, Malyshev NA, Aristova VA, Slavskii AA, Luk'ianova NA, Deriabin PG, Gromashevskii VL, Efremenko VI, Onishchenko GG, L'vov DK (2005) Detection of the circulation of Crimean-Congo hemorrhagic fever virus in the piedmont steppes of the North Caucasus [In Russian]. Vopr Virusol 50:9–15
50. Semashko IV, Chumakov MP, Karapetyan RM (1974) First isolation of the CHF virus in Armenia from the blood of the patient with Crimean hemorrhagic fever [In Russian]. Sb Tr Inst Polio Virusn Encefalitov Akad Med Nauk SSSR 7:271–278
51. Sipovsky PV (1944) Atypical cases of gastro-intestinal hemorrhage [In Russian]. Klin Med (Moscow) 22:64–67
52. Stolbov DN, Butenko AM, Egorova PS, Leschinskaya EV, Chumakov MP (1965) Crimean hemorrhagic fever (CHF) in Astrakhan Oblast [In Russian]. Sb Tr Inst Polio Virusn Encefalitov Akad Med Nauk SSSR 7:271–278
53. Temirbekov ZhT, Dobritsa PG, Kontaruk VM (1971) Investigation of Crimean hemorrhagic fever in Chimkent region of the Kazakh SSR [In Russian]. Sb Tr Inst Polio Virusn Encefalitov Akad Med Nauk SSSR 19:160–166
54. Woodall JP, Williams MC, Simpson DI (1967) Congo virus: a hitherto undescribed virus occurring in Africa. II. Identification studies. East Afr Med J 44:93–98
55. Yashina L, Petrova I, Seregin S, Vyshemirskii O, Lvov D, Aristova V, Kuhn J, Morzunov S, Gutorov V, Kuzina I, Tyunnikov G, Netesov S, Petrov V (2003) Genetic variability of Crimean-Congo haemorrhagic fever virus in Russia and Central Asia. J Gen Virol 84:1199–1206
56. Zgurskaya GN, Berezin VV, Smirnova SE (1975) Threshold levels of blood infectiousness for *Hyalomma plumbeum* tick during viremia in hares and rabbits caused by CHF virus [In Russian]. Tezisy Konf Vop Med Vir, Moscow, pp 146–147

CHAPTER 10

CRIMEAN-CONGO HEMORRHAGIC FEVER IN THE XINJIANG UYGUR AUTONOMOUS REGION OF WESTERN CHINA

MASAYUKI SAIJO, M.D., PH.D.

Department of Virology 1, National Institute of Infectious Diseases, Tokyo, Japan, 4-7-1 Gakuen, Musashimurayama, Tokyo 208-0011, Japan. Tel.: +81-42-561-0771 (ext. 320); Fax: +81-42-561-2039; E-mail: msaijo@nih.go.jp

10.1. INTRODUCTION

Certain regions of China are well known as Crimean-Congo hemorrhagic fever (CCHF)-endemic areas, particularly the western part of the Xinjiang Uygur Autonomous Region (Xinjiang) in western China. Xinjiang is unique in terms of CCHF virus (CCHFV) infections because this region forms the eastern boundary of CCHF endemic region and borders Pakistan, Afghanistan, Tajikistan, and Kyrgyzstan. These neighboring countries also have CCHF-endemic areas. The study of CCHFV infections in this region can, therefore, provide important insights into CCHFV infections.

10.2. HISTORICAL ASPECTS OF CCHF IN XINJIANG, CHINA

10.2.1. CCHF known as "Xinjiang Hemorrhagic Fever"

An outbreak of a disease similar to that in Crimea was first documented in Bachu County in the westernmost region of Xinjiang in 1965 (Fig. 10-1). The disease was named Xinjiang hemorrhagic fever. The etiological agent for Xinjiang hemorrhagic fever was isolated from the blood of patients and from pasture ticks by the Hygiene and Epidemic Prevention Station of Xinjiang [32] and was named Xinjiang hemorrhagic fever virus. Since the first documented outbreak of Xinjiang hemorrhagic fever in 1965, Bachu County has been struck by Xinjiang hemorrhagic fever almost every year. In 1985, it was first reported that the Xinjiang hemorrhagic fever virus was indistinguishable from CCHFV by serological analyses, indicating that Xinjiang hemorrhagic fever is similar to

O. Ergonul and C. A. Whitehouse (eds.), Crimean-Congo Hemorrhagic Fever, 115–130.

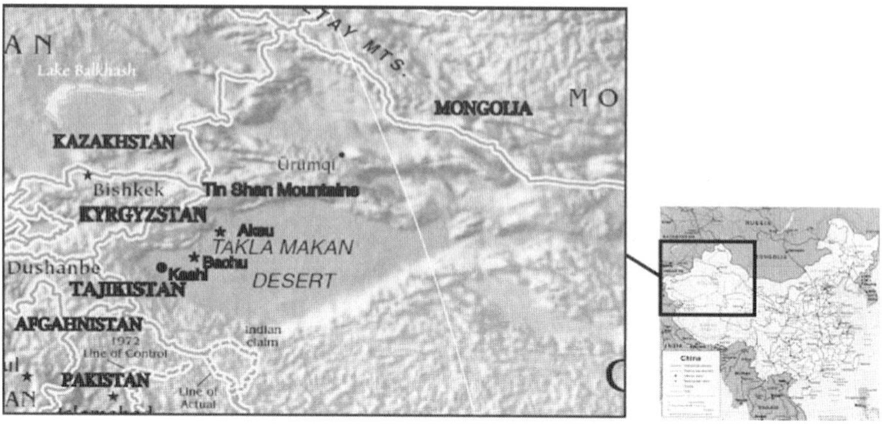

Fig. 10-1. Geography of Xinjiang, China. Patients with CCHF in Xinjiang have been reported in Bachu and Aksu Counties. The neighboring countries such as Kazakhstan, Kyrgyzstan, Afghanistan, and Pakistan also have endemic regions of CCHF.

CCHF [32]. It has now been confirmed by serological and molecular analyses that Xinjiang hemorrhagic fever virus is a CCHFV. Although the disease, Xinjiang hemorrhagic fever, was first recognized as a clinical entity in 1965, there is no doubt that the disease had been present in the region since antiquity.

10.2.2. History of CCHFV isolation in China

Since the first isolation of CCHFV in Xinjiang in 1966, CCHFV isolates have been recovered by inoculating materials collected from patients, ticks, sheep, and long-eared jerboa into the brains of suckling mice (Table 10-1). The latest isolation of CCHFV from a patient occurred in 1988. Although the number of CCHFV isolates is relatively small, studies on CCHFV infections in the region have progressed thanks to the isolation of the virus from patients, ticks, and vertebrates in the region.

10.2.3. Recent progress of studies on CCHF in Xinjiang, China

The first research articles written in Chinese that described Xinjiang hemorrhagic fever were published in 1983 [11, 28, 29]. Since 2000, collaborative research on CCHFV infections has been initiated by China and Japan, and recombinant protein-based diagnostic systems for CCHF have been developed [16, 17, 20, 24, 25]. Field studies on CCHF in Bachu County in Xinjiang were then initiated using these newly developed diagnostic systems. Furthermore, molecular epidemiological studies of CCHFV infections in this region were also carried out [14, 26]. Studies on CCHFV infections in Xinjiang have progressed dramatically in recent years.

Table 10-1. Description of nucleotide sequences of the partial S segment of CCHFV isolates in the Xinjiag Uygur Autonomous Region used in the present study

CCHFV	Origin	Place of isolation (county)	Year of isolation	Accession no. for S-segment sequence	Accession no. for M-segment sequence
66019	Human	Bachu	1966	AJ010648	AB069669
68031	Sheep	Bachu	1968	M86625	Not reported
HY-13	Tick (*Hyalomma asiaticum asiaticum*)	Bachu	1968	U88413	AY900145
7001	Human	Bachu	1970	AF415236	AB069670
75024	Human	Aksu	1975	AF362080	AB069671
7803	Human	Bachu	1978	AF354296	AB069672
79121	Long-eared jerboa	Bachu	1979	AF358784	AB069673
8402	Tick (*H. asiaticum asiaticum*)	Bachu	1984	AJ010649	AB069674
88166	Human	Bachu	1988	AY029157	AB069675

10.3. LIFE CYCLE OF CCHFV IN XINJIANG, CHINA

10.3.1. Vector and host

CCHFV is maintained in nature in Xinjiang through cycles of asymptomatic infection between tick (*Hyalomma asiaticum asiaticum*) and mammals including livestock such as sheep and goats (Fig. 10-2).

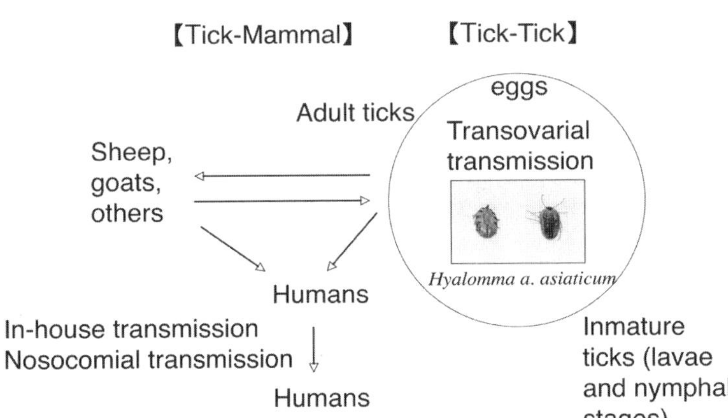

Fig. 10-2. Life cycle of CCHFV and infection routes of CCHFV to humans in Xinjiang. The main tick vector in the region is *Hyalomma asiaticum asiaticum*. Inhabitants in Xinjiang depend on sheep for nutrition, economic well-being, and religious activities. Inhabitants are usually infected with the CCHFV through close contact with the viremic tissues of sheep or through tick bites.

10.3.1.1. Ticks

CCHFV strains 8402, 68013, and HY-13 were isolated from *H. asiaticum asiaticum* ticks. The partial genome of the S segment of CCHFV was amplified from *H. asiaticum asiaticum* ticks, collected from the endemic area during the outbreak seasons in 2001 and 2002 [26]. The tick species, *H. asiaticum asiaticum*, certainly plays a key role in CCHF outbreaks in Xinjiang; however, little information on tick species that are associated with CCHFV infections in Xinjiang have been collected to date. Further studies are needed on the maintenance of CCHFV in nature in Xinjiang and on the relationship between the habits of these ticks with CCHF outbreaks in the region.

10.3.1.2. Vertebrates

Since the first documented outbreak of CCHF in Xinjiang in 1965, CCHFV isolates have been recovered not only from patients but also from sheep and long-eared jerboa [32] (Table 10-1). It was reported that 37 out of 125 sheep (30%) in the Bachu County in Xinjiang showed a positive reaction in a complement fixation assay for the detection of antibodies to CCHFV [32]. Recently, we conducted a similar study and found that approximately 60% of adult sheep showed a positive reaction in indirect immunofluorescent and immunosorbent assays for the detection of immunoglobulin G (IgG) antibody to CCHFV [24]. These results strongly suggest that sheep are commonly infected with CCHFV. *Hyalomma* tick infestations are also common in sheep in the region (Fig. 10-3), and these two factors indicate that sheep are important to the life cycle of CCHFV in Xinjiang. Unfortunately, no seroepidemiological studies on CCHFV infections in other mammals in the region have yet been conducted. The fact that CCHFV was isolated from a small mammal such as long-eared jerboa suggests that small mammals may also play an important role in the maintenance of CCHFV in nature in the region. It is suggested that CCHF outbreaks among people in the region are associated with CCHFV infections in sheep based on the fact that sheep in the region are commonly infected with CCHFV and that the residents in the CCHF-endemic region have close contact with sheep in their daily life. Epidemiological studies on CCHFV infections among vertebrates other than sheep are needed to clarify the ecology of the CCHFV in nature in the region and the risk to residents to CCHFV infection in Xinjiang.

10.4. CLINICAL ASPECTS OF CCHF IN XINJIANG, CHINA

10.4.1. Patients

Patients with CCHF have been reported only in Bachu and Aksu counties in Xinjiang (Fig. 10-1). In 2001 and 2002, field studies on CCHFV infections in Bachu County in Xinjiang were conducted using virological examinations. Forty-six and 12 patients diagnosed as having CCHF based on clinical symptoms such as fever, backache, joint pain, bleeding from body orifices, and/or

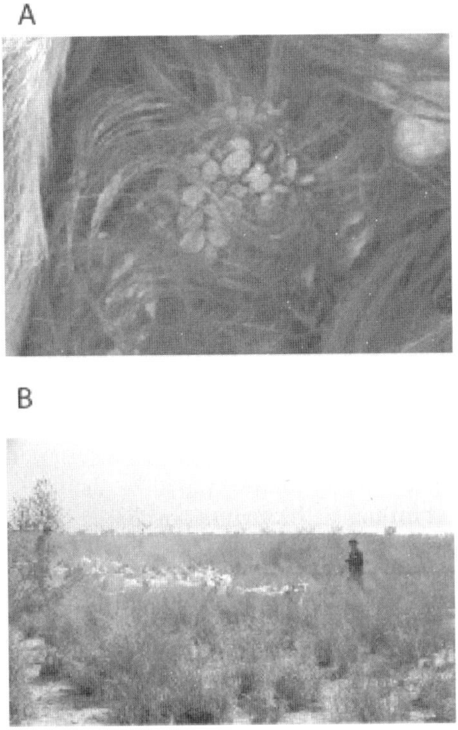

Fig. 10-3. Picture shows ticks (*Hyalomma asiaticum asiaticum*) infesting sheep (*arrow* indicates ticks (A) and with shepherds (B) in the rural and desert areas in Bachu County, Xinjiang.

purpura were registered with the local health authority in 2001 and 2002, respectively. However, one third of the 46 patients and one half of the 12 patients registered in 2001 and 2002, respectively, were confirmed to have CCHFV infections by virological assays such as serology, CCHFV antigen detection and reverse transcription polymerase chain reaction (RT-PCR) for amplification of virus genome. In the 2001 outbreak, three males died of hemorrhagic symptoms and multiple organ failure possibly due to CCHFV infection. From the 16 of the patients with CCHF diagnosed through virological examination, information on age, sex, clinical manifestations, and occupation were available for analyses. Male to female ration was 10:6. Age of the patients varied from 4 to 70 years with an average age of 31. Seven and four out of the 16 were farmers and shepherds, respectively. Of the six female patients, three had had close contact with sheep through raising sheep. The youngest patient was a 4-year-old girl and her mother also suffered from CCHF. The girl developed fever on the 5th day after the onset of infection in her mother. The partial genome of the CCHFV amplified from the girl was identical to that from her mother. It is speculated that the girl may have been infected with CCHFV through contact with her mother [18].

The area of CCHF outbreak in Bachu County is restricted to some small villages. Although no outbreaks of CCHF have been reported in urban areas where commercial activities associated with sheep and lambs, such as slaughter, marketing and preparation for human consumption in restaurants, are conducted, there is a great potential risk of CCHF outbreaks in such areas.

10.4.2. Clinical manifestations

The symptoms of patients with CCHF range in severity from fever only or fever with flu-like symptoms to hemorrhage with multiple organ failure resulting in death. All patients developed fever and joint pains. Orbital pain, backache, and headache are common symptoms in patients. One patient, a 28-year-old shepherd, with severe symptoms of hemorrhage from gingiva, nostrils, and rectum was reported [25] (Fig. 10-4). He eventually recovered without any sequelae. In severe cases such as this, elevation of liver enzymes is often seen. Furthermore, oliguria was a common symptom and it was presumably associated with renal failure caused by the direct influence of CCHFV infection or by indirect influence through hypovolemic shock.

10.4.3. Treatment

Basically, maintenance therapies such as hydration, blood transfusion, and other specific therapies, i.e. administration of diuretics and/or antibiotics (if necessary), should be initiated as soon as possible. However, it is usually difficult for patients with CCHF to receive such treatment due to populations in the endemic region being economically disadvantaged and having limited access to a regional hospital. Once patients with suspected CCHF are hospitalized, they should be treated with of ribavirin as well as supportive therapies.

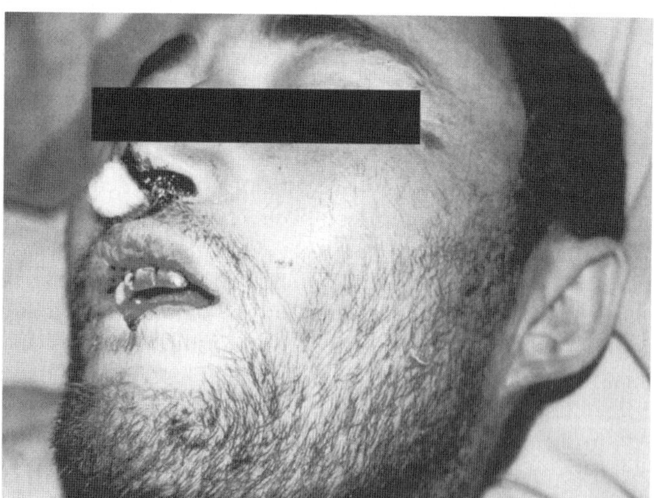

Fig. 10-4. Hemorrhagic symptoms in a patient with CCHF. The clinical course and virological data for this patient have been reported [25]. This [Au1] patient was treated with an intravenous administration of ribavirin and recovered without any sequelae.

10.4.4. Infection route for CCHFV in humans

Most patients with CCHF in Xinjiang were shepherds or farmers. The residents living in the endemic area usually have close contact with sheep in their daily lives and some inhabitants even share their houses with the sheep they are raising. Sheep in the endemic area were closely observed for tick infestations during the endemic season, and it was quite easy to find sheep infested with ticks (Fig. 10-3). The people in the region in Bachu County appear to be infected with CCHFV through close contact with tissues, blood or bodily fluid of viremic sheep, including lambs, or the bite of CCHFV-infected ticks (mainly *H. asiaticum asiaticum*). Cultivating cotton is also a major economic activity in the area. When farmers are working in the cotton fields or on farms on which vegetables are grown, the chance of infection with CCHFV through tick bite is increased. Although there have not yet been any reported cases of nosocomial and in-house CCHF outbreaks in Xinjiang except for one case in which a child might have been infected from her mother, the human-to-human transmission of CCHFV might be one of major routes of CCHFV infection in humans [18]. The mode of infection of CCHFV in people within their region should be clarified in order to develop an efficacious strategy for reduction of patients with CCHF.

10.5. PREVENTION

10.5.1. Education

The reduction of the number of patients with CCHF in the CCHF-endemic regions in Xinjiang is of great importance. Most patients with CCHF in this region are aged between 15 and 40 years, indicating that they are of a working age. For this reason, preventive measures are important. Surveillance of CCHFV infections, development of diagnostics for CCHF, and assessment of risk factors in CCHFV, and so on, must be undertaken.

The risk factors for CCHFV infections must be clarified for each region in which outbreaks occur in order to establish an efficacious strategy to limit CCHF outbreaks. The life of the residents is closely associated with the raising of sheep. When people work on farms, they have a high risk of tick bite, thus increasing the risk of CCHFV infections. The economic activities associated with raising sheep and work in the field may pose a great risk of CCHFV infections. It is speculated that the number of patients can be reduced if proper education for residents on preventive measures of CCHFV infections are implemented. As discussed below, CCHF outbreaks occur in spring from March to July in the region. Therefore, such education should be implemented especially during the endemic season. However, prevention of CCHF in the region is unfortunately very difficult as the lives of many of the inhabitants depend on very activities that pose a risk of CCHFV infection.

10.5.2. Development of diagnostic systems for CCHF

CCHFV is classified as a biosafety level-4 pathogen, indicating that infectious virus must be manipulated in a high-containment (BSL-4) laboratory. This fact suggests that the development of antibody-detection systems using authentic CCHFV antigens is difficult in an institute without a BSL-4 laboratory. To overcome this difficulty, recombinant CCHFV-nucleoprotein (NP)-based antibody detection systems have been developed [13, 16, 17, 19, 20, 24, 25]. The recombinant CCHFV-NP was expressed in a recombinant baculovirus system from the cDNA of S segment of the CCHFV Chinese strain 8402 and then purified. A recombinant CCHFV-NP-based enzyme-linked immunosorbent assay (ELISA) for the detection of IgG and IgM antibodies was developed and shown to have high sensitivity and specificity [16, 20, 24, 25]. A recombinant NP-based ELISA for the detection of IgG antibodies to CCHFV in sheep sera [24] and, a recombinant CCHFV-NP-based indirect immunosorbent assay have also been developed [17]. Furthermore, a CCHFV antigen detection sandwich ELISA was developed using a novel monoclonal antibody [19]. The advantage of these recombinant protein-based diagnostic systems is that these diagnostics can be employed in regional institutes without a BSL-4 laboratory.

As the CCHF outbreak area in Xinjiang is a remote and economically disadvantaged region, it is difficult to equip local laboratories with these diagnostic systems for CCHF. However, to combat outbreaks of CCHF and reduce the mortality and morbidity of CCHF in such regions, precise diagnosis of CCHF is necessary. The establishment of diagnostic systems for CCHFV infections in a regional laboratory near the site of CCHF outbreaks is, therefore, an issue to be settled in the future.

10.5.3. Assessment of risk factors for CCHFV infection

The assessment of risk factors for CCHFV infection in the endemic area in Xinjiang is still to be performed. Most people in the endemic region are Muslim and their life depends heavily on raising sheep. They also cultivate cotton and vegetables in fields where ticks including *H. asiaticum asiaticum* are abundant. There is no doubt that raising sheep is one of the occupational activities with a high risk of infection. The slaughter of lambs and sheep is also a job with a high risk of CCHFV infections. CCHFV were isolated from ticks, *H. asiaticum asiaticum*, and CCHFV genomes were amplified by nested RT-PCR in 4 of 16 tick pools (a total of 80 ticks) and in 3 of the 65 tick pools (a total of 280 ticks) collected in the 2001 and 2002 outbreak seasons, respectively. Manipulation of ticks infesting sheep (including lambs) may increase the risk of CCHFV infection. Therefore, in order to prevent CCHF outbreaks in the CCHF-endemic region, clarification of risk factors for CCHFV infections is needed (See Chapter 21).

10.6. EPIDEMIOLOGY OF CCHF IN XINJIANG, CHINA

10.6.1. Geographical prevalence of CCHFV

Patients with CCHF have been reported from Central Asian countries that have borders with Xinjiang, China, such as Kazakhstan, Kyrgyzstan, Tajikistan, Afghanistan, and Pakistan. No patients with CCHF have been reported within China excluding the Xinjiang region. Xinjiang is divided into northern and southern Xinjiang by the Tian Shan Mountains (Fig. 10-1). All patients in Xinjiang were from the southern region and most patients were reported in a small village near Bachu County in the Kashi district (Fig. 10-1).

10.6.2. Seroepidemiology of CCHFV infections in the endemic region

Seroprevalence to CCHFV was investigated in residents living in the CCHF-endemic village near Bachu County in 2001. Serum samples were collected from 70 residents of various ages under informed consent. The antibody to CCHFV was detected by the recombinant NP-based IgG-ELISA [16] in approximately one fourth of participants. Furthermore, 60% of the participants over 50 years of age showed a positive reaction. There was no significant difference in the antibody positive rate to CCHFV between males and females. Together with that fact that 60% of the sheep in the region showed a positive reaction in the IgG-ELISA [24], these results indicate that residents living in the village within Bachu County are at high risk of CCHFV infection.

10.6.3. Environment of the region in which CCHF is endemic in Xinjiang

10.6.3.1. Environment

The CCHF endemic regions, Bachu and Aksu counties, in Xinjiang are located on the Silk Road in the Taklamakan desert, which is arid and receives little rain throughout the year (Figs. 10-1 and 10-3).

10.6.3.2. Lifestyle of people in the endemic region

More than 80% of people living in Xinjiang are Muslim, as are most residents in the CCHF endemic region within Xinjiang, and their lives are closely associated with the cultivation of domestic livestock of sheep. They depend on primary industry such as raising sheep and farming cotton and vegetables. Some of the adult males work as shepherds (Fig. 10-3). Their socioeconomical status is usually not high. Therefore, the residents usually live in the same accommodation in which the sheep are raised. *H. asiaticum asiaticum* ticks are abundant in the region, and it is not difficult to find these ticks moving on the ground.

10.6.3.3. Economic activities associated with CCHF outbreaks in the endemic region

The economic activities associated with sheep such as cultivating sheep, shearing, slaughter, and the cooking of sheep meat, pose a risk of CCHFV infection. Furthermore, as sheep are usually infested with ticks and are routinely sold at supermarkets in urban areas as well as being slaughtered for meat, there is a possibility that CCHF outbreaks will occur not only in the remote villages, but also in urban areas in the region. Nosocomial infections of CCHF among healthcare workers have been reported in several countries [2–4, 9, 27]. Although no nosocomial outbreaks of CCHF have been reported in Xinjiang, the management of febrile patients poses a risk of CCHFV infections among the healthcare workers in the region. They must always be careful in the treatment of febrile patients to prevent infection.

10.6.4. Seasonality of CCHF outbreaks in Xinjiang

Most patients contract CCHF in spring, between March and July, each year. Very few cases are observed at other times of the year. There may be risk factors for residents in the regions for infection with CCHFV in this season. It is expected that the virus load in the region may increase in spring. As indicated above, sheep are commonly infected with CCHFV and ticks harbor the CCHFV. The number of newborn lambs increases in spring in the region. Newborn sheep that are negative for CCHFV antibodies are naïve to CCHFV infections, suggesting that the viremia of the CCHFV occurs in the newborn sheep once they are infected with the CCHFV through tick bites. Therefore, the CCHFV load is expected to increase in the endemic region in spring each year, according to the increase in the number of newborn sheep. Furthermore, ticks become active in spring in the region. Based on these facts, the increase in the number of patients with CCHF may be due to both factors, the increase in CCHFV load in sheep, especially newborn sheep, and the increase in tick activity in the region.

10.7. MOLECULAR EPIDEMIOLOGY OF CCHF IN XINJIANG, CHINA

10.7.1. Molecular epidemiology of CCHF determined with genetic information of the viral isolates

The molecular epidemiological studies on CCHFV infections have been conducted with genetic information from the partial S segment of CCHFV isolates [5, 7–10, 15, 22, 23, 30]. The phylogenetic analyses based on the nucleotide sequence of the partial S segment of CCHFV isolates around the world indicates that the Chinese CCHFV isolates as well as the Kazakhstan and Uzbekistan strains form a group independent from those consisting of the other CCHFV isolates from other part of the world (Fig. 10-5a). Even the Pakistani

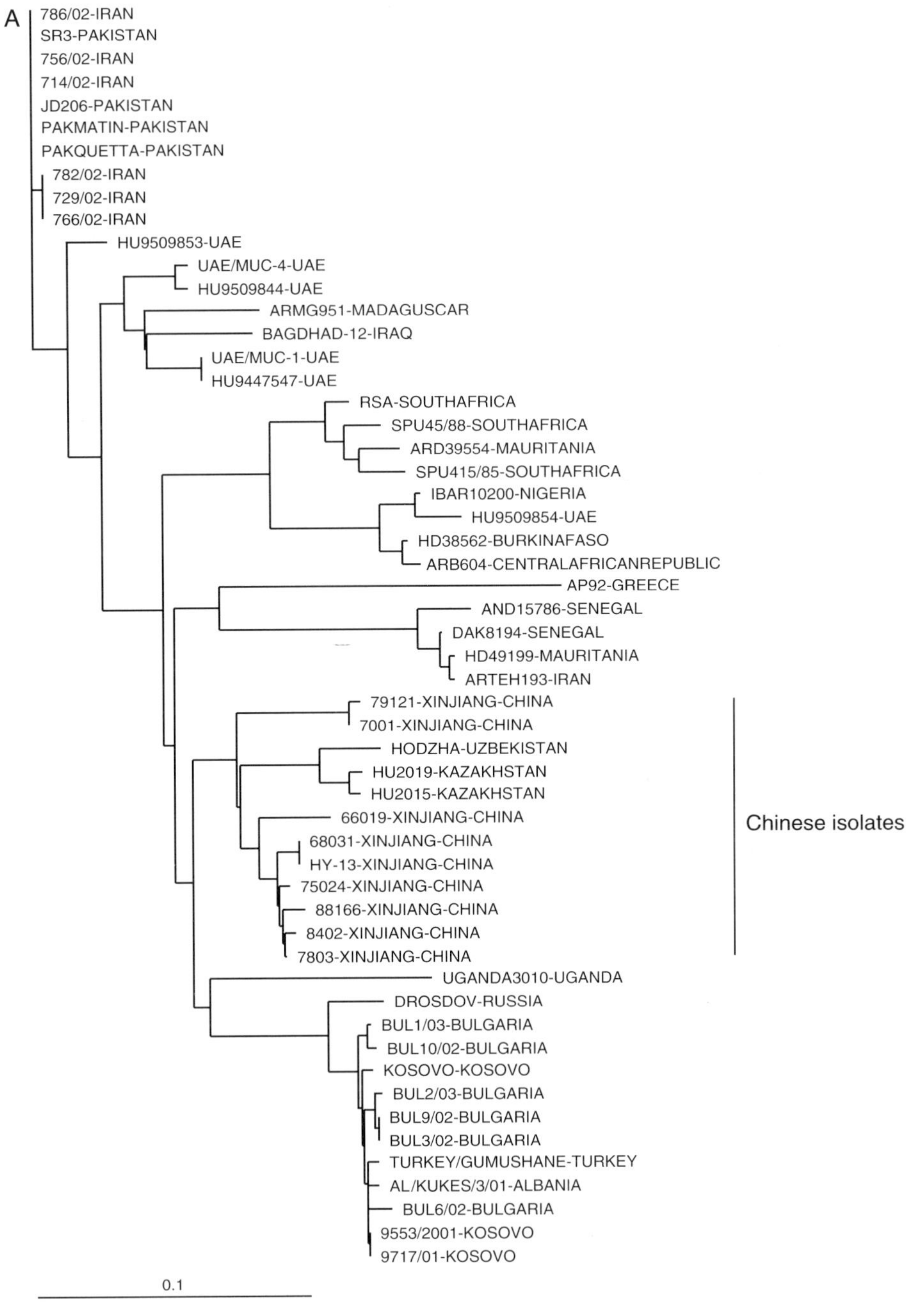

Fig. 10-5.

(Continued)

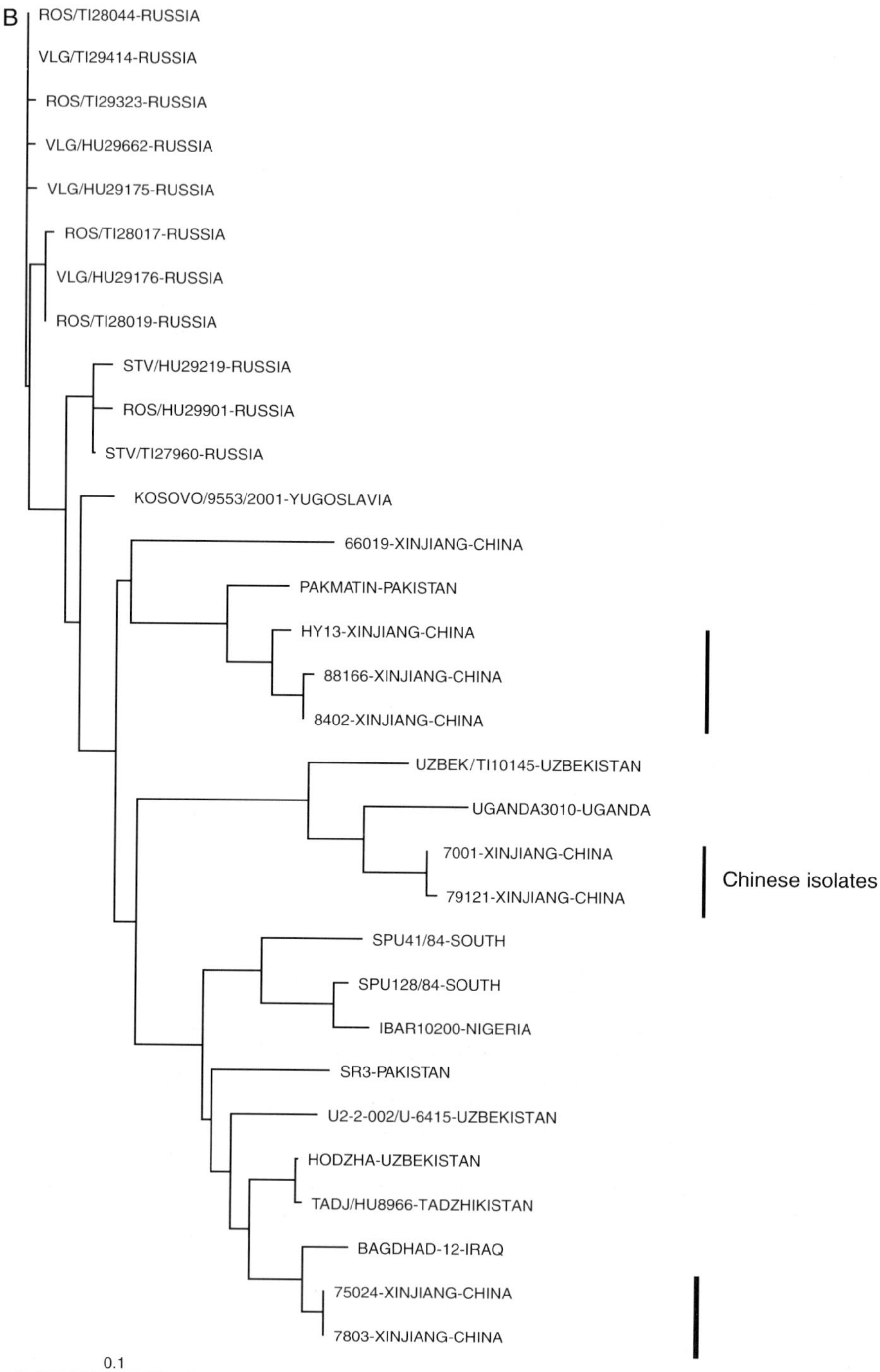

Fig. 10-5.

(*Continued*)

CCHFV isolates form a different clade from that of the Chinese isolates. The CCHFV in Xinjiang have evolved in an independent manner possibly associated with the evolution of CCHFV-related tick species. The Chinese virus isolates 7001 and 79121 have a close relationship with the Kazakhstan CCHFV in the evolutional event.

Recently, genetic information on the M segment of CCHFV isolates including those of the Chinese virus isolates has become available. It is indicated that there is a hypervariable region and a relatively conserved region in terms of nucleotide sequence within the M segment [1, 14, 21]. Phylogenetic analyses have also been carried out based on nucleotide sequence of the conserved region of the partial M segment [5, 14, 31]. In contrast to the phylogenetic relationship analyzed using the partial S segment (Fig. 10-5a), the Chinese isolates of 66019, HY-13, 88166, and 8402, the Chinese isolates of 7803 and 75024, and the Chinese isolates of 7001 and 79121 form independent clades with the Pakistani isolate, Matin, the Pakistani isolate, SR3, the Tajikistani isolate, TADJ/HU8966, the Uzbekistani isolates, Hodzha and U2-2-002/U-6415, the

Fig. 10-5. cont'd. Phylogenic relationship of CCHFV Chinese isolates determined by the neighbor-joining method using the nucleotide sequence of the partial S segment (A) and the partial M segment (B). The accession number of the isolates used in the phylogenetic analyses for the partial S segment excluding those of Chinese isolates are as follows (Country, accession number): Pakquetta (Pakistan, U75677), Pakmatin (Pakistan, U75678), JD206 (Pakistan, U88414), 714/02 (Iran, AY366376), 756/02 (Iran, AY366378), 786/02 (Iran, AY366374), HU9509853 (UAE, U75672), 729/02 (Iran, AY366375), 766/02 (Iran, AY366373), 782/02 (Iran, AY366377), UAE/MUC-1 (UAE, S82580), HU9447547 (UAE, U75670), ArMg951 (Madagascar, U15024), UEA/MUC-4 (UAE, S82581), HU9509844 (UAE, U75668), IbAr10200 (Nigeria, U75674), HU9509854 (UAE, U75671), HD38562 (Burkina Faso, U15093), ArB604 (South Africa, U15092), SPU 45/88 (South Africa, U84637), ArD39554 (Mauritania, U15089), RSA (South Africa, U75675), HD49199 (Mauritania, U15023), ArTeh193 (Iran, U15022), DAK8194 (Senegal, U88411), AnD15786 (Senegal, U15020), AP92 (Greece, U04958), BUL6/02 (Bulgaria, AY550256), AL/Kukes/3/01 (Albania, AF449482), Turkey (Turkey, AY508485), BUL10/02 (Bulgaria, AY550258), 9553/2001 (Kosovo, AF428144), 9717/01 (Kosovo, AF428145), BUL9/02 (Bulgaria, AY550257), BUL3/02 (Bulgaria, AY550255), BUL2/03 (Bulgaria, AY550254), Kosovo (Kosovo, AF404507), BUL1/03 (Bulgaria, AY550253), Drosdov (Russia/Astrahan, U88412), UGANDA3010 (Uganda, U88416), HU2019 (Kazakhstan, AF362746), HU2015 (Kazakhstan, AF362744). The accession number of the isolates used in the phylogenetic analyses for the partial M segment excluding those of Chinese isolates are as follows: ROS/TI29323 (Russia/Rostov, AF401650), VLG/TI29414 (Russia/Volgograd, AY179961), VLG/HU29662 (Russia/Volgograd, AY093622), ROS/TI28044 (Russia/Rostov, AF401651), ROS/TI28019 (Russia/Rostov, AF401649), VLG/HU29176 (Russia/Volgograd, AY093621), ROS/TI29017 (Russia/Rostov, AF401648), VLG/HU29175 (Russia/Volgograd, AY093620), ROS/HU29901 (Russia/Rostov, AY093625), STV/TI27960 (Russia/Stavropol, AF401647), STV/HU29219 (Russia/Stavropol, AY093624), Uzbek/TI10145, Uzbekistan, AY093627), Matin (Pakistan, AF467769), IbAr10200 (Nigeria, U39455), CRI538199 (Pakistan, AJ538199), Tadi/HU8966 (Tajikistan, AY179962), CRI538197 (Iraq, AJ538197).

Iraqi isolate, Baghdad-12, and with the Uzbekistani isolate, Uzbek/TI110145, respectively (Fig. 10-5b). However, the Russian isolates of the CCHFV form an independent clade from those of other CCHFV isolates including the Chinese isolates. These results suggest that the CCHFV in Xinjiang has a close relationship with CCHFV in the neighboring countries such as Pakistan, Tajikistan, and Uzbekistan.

Xinjiang in China is a unique and important region for the study of CCHFV infections as it forms the eastern border of CCHFV infections and has long historical ties with the neighboring countries in terms of transportation, communications, and economic activities, which might have an influence on the evolutional events of CCHFV in the region. The discrepancy between the phylogenetic relationships based on the partial S segment and that based on the partial M segment may be due to a difference in the mutational rate between the S and M segments and/or more dynamic evolutional events such as recombination and reassortment [6, 12]. Further study on the molecular epidemiology of the CCHFV will provide a deeper insight into the evolutional events of the CCHFV.

10.8. SUMMARY

Xinjiang hemorrhagic fever, a form of CCHF, was first identified to be caused by the CCHFV in the early 1980s when the CCHFV was isolated from patients, small mammals and ticks in Xinjiang. These scientific achievements led us to understand the features of CCHFV infections in the region more precisely. However, there are still many questions to be resolved: the ecology of CCHFV, risk factors of CCHFV infection, surveillance of CCHF outbreaks, molecular epidemiology of CCHFV infections, effective preventive measures for CCHF outbreaks, efficacious treatment for patients with CCHF, development of diagnostics for CCHF that can be easily carried out at the site of the outbreak, development of efficacious vaccines, and so on. We strongly hope that these subjects will be resolved in the near future.

Acknowledgments

The author acknowledges Dr. Tang Qing, Chinese Center for Disease Control and Prevention for Viral Infections, Beijing, China, and Dr. Buyondong Shimayi, Director of Bachu Center for Disease Control and Prevention, for their great contribution to the field study on CCHFV infections in the Xinjiang Uygur Autonomous Region. The author also acknowledges all the staff and collaborators for their studies on CCHFV infections. Furthermore, the author acknowledges Dr. Shigeru Morikawa, Dr. Masahiro Niikura, Dr. Akihiko Maeda, Dr. Momoko Ogata, and Dr. Ichiro Kurane, Department of Virology 1, National Institute of Infectious Diseases, Tokyo, Japan, for their contributions to the study of CCHFV infections.

REFERENCES

1. Ahmed AA, McFalls JM, Hoffmann C, Filone CM, Stewart SM, Paragas J, Khodjaev S, Shermukhamedova D, Schmaljohn CS, Doms RW, Bertolotti-Ciarlet A (2005) Presence of broadly reactive and group-specific neutralizing epitopes on newly described isolates of Crimean-Congo hemorrhagic fever virus. J Gen Virol 86:3327–3336
2. Altaf A, Luby S, Ahmed AJ, Zaidi N, Khan AJ, Mirza S, McCormick J, Fisher-Hoch S (1998) Outbreak of Crimean-Congo haemorrhagic fever in Quetta, Pakistan: contact tracing and risk assessment. Trop Med Int Health 3:878–882
3. Athar MN, Baqai HZ, Ahmad M, Khalid MA, Bashir N, Ahmad AM, Balouch AH, Bashir K (2003) Short report: Crimean-Congo hemorrhagic fever outbreak in Rawalpindi, Pakistan, February 2002. Am J Trop Med Hyg 69:284–287
4. Burney MI, Ghafoor A, Saleen M, Webb PA, Casals J (1980) Nosocomial outbreak of viral hemorrhagic fever caused by Crimean Hemorrhagic fever-Congo virus in Pakistan, January 1976. Am J Trop Med Hyg 29:941–947
5. Chinikar S, Persson SM, Johansson M, Bladh L, Goya M, Houshmand B, Mirazimi A, Plyusnin A, Lundkvist A, Nilsson M (2004) Genetic analysis of Crimean-Congo hemorrhagic fever virus in Iran. J Med Virol 73:404–411
6. Hewson R, Gmyl A, Gmyl L, Smirnova SE, Karganova G, Jamil B, Hasan R, Chamberlain J, Clegg C (2004) Evidence of segment reassortment in Crimean-Congo haemorrhagic fever virus. J Gen Virol 85:3059–3070
7. Iashina LN, Petrov VS, Petrova ID, Gutorov VV, Kazakov SV, Ospanov KS, Karimov SK, Tiunnikov GI, Seregin SV, Kuzina II, Babkin IV, Netesov SV (2002) Genetic identification of the Crimean-Congo hemorrhagic fever virus during epidemic outbreak in Kazakhstan in 2000. Mol Gen Mikrobiol Virusol 4:31–35
8. Iashina LN, Petrov VS, Vyshemirskii OI, Aristova VA, Moskvina TM, L'Vov DK, Petrova ID, Gutorov VV, Tiunnikov GI, Kuzina II, Samokhvlov EI, Seregin SV, Netesov SV (2002) Characteristics of Crimean-Congo hemorrhagic fever virus circulating in Russia and Central Asia. Vopr Virusol 47:11–15
9. Karti SS, Odabasi Z, Korten V, Yilmaz M, Sonmez M, Caylan R, Akdogan E, Eren N, Koksal I, Ovali E, Erickson BR, Vincent MJ, Nichol ST, Comer JA, Rollin PE, Ksiazek TG (2004) Crimean-Congo hemorrhagic fever in Turkey. Emerg Infect Dis 10:1379–1384
10. Khan AS, Maupin GO, Rollin PE, Noor AM, Shurie HH, Shalabi AG, Wasef S, Haddad YM, Sadek R, Ijaz K, Peters CJ, Ksiazek TG (1997) An outbreak of Crimean-Congo hemorrhagic fever in the United Arab Emirates, 1994–1995. Am J Trop Med Hyg 57:519–525
11. Li F (1983) Characteristics of Xinjiang hemorrhagic fever virus. III. Electron microscopic observation of Xinjiang hemorrhagic fever virus. Zhonghua Liu Xing Bing Xue Za Zhi 4:135–136
12. Lukashev AN (2005) Evidence for recombination in Crimean-Congo hemorrhagic fever virus. J Gen Virol 86:2333–2338
13. Ma B, Hang C, Zhao Y, Wang S, Xie Y (2002) Use of a novel baculovirus vector to express nucleoprotein gene of Crimean-Congo hemorrhagic fever virus in both insect and mammalian cells. Zhonghua Shi Yan He Lin Chuang Bing Du Xue Za Zhi 16:249–252
14. Morikawa S, Qing T, Xinqin Z, Saijo M, Kurane I (2002) Genetic diversity of the M RNA segment among Crimean-Congo hemorrhagic fever virus isolates in China. Virology 296:159–164
15. Rodriguez LL, Maupin GO, Ksiazek TG, Rollin PE, Khan AS, Schwarz TF, Lofts RS, Smith JF, Noor AM, Peters CJ, Nichol ST (1997) Molecular investigation of a multisource outbreak of Crimean-Congo hemorrhagic fever in the United Arab Emirates. Am J Trop Med Hyg 57:512–518
16. Saijo M, Qing T, Niikura M, Maeda A, Ikegami T, Prehaud C, Kurane I, Morikawa S (2002) Recombinant nucleoprotein-based enzyme-linked immunosorbent assay for detection of immunoglobulin G antibodies to Crimean-Congo hemorrhagic fever virus. J Clin Microbiol 40:1587–1591

17. Saijo M, Qing T, Niikura M, Maeda A, Ikegami T, Sakai K, Prehaud C, Kurane I, Morikawa S (2002) Immunofluorescence technique using HeLa cells expressing recombinant nucleoprotein for detection of immunoglobulin G antibodies to Crimean-Congo hemorrhagic fever virus. J Clin Microbiol 40:372–375
18. Saijo M, Tang Q, Shimayi B, Han L, Zhang Y, Asiguma M, Tianshu D, Maeda A, Kurane I, Morikawa S (2004) Possible horizontal transmission of Crimean-Congo hemorrhagic Fever virus from a mother to her child. Jpn J Infect Dis 57:55–57
19. Saijo M, Tang Q, Shimayi B, Han L, Zhang Y, Asiguma M, Tianshu D, Maeda A, Kurane I, Morikawa S (2005) Antigen-capture enzyme-linked immunosorbent assay for the diagnosis of crimean-congo hemorrhagic fever using a novel monoclonal antibody. J Med Virol 77:83–88
20. Saijo M, Tang Q, Shimayi B, Han L, Zhang Y, Asiguma M, Tianshu D, Maeda A, Kurane I, Morikawa S (2005) Recombinant nucleoprotein-based serological diagnosis of Crimean-Congo hemorrhagic fever virus infections. J Med Virol 75:295–299
21. Sanchez AJ, Vincent MJ, Nichol ST (2002) Characterization of the glycoproteins of Crimean-Congo hemorrhagic fever virus. J Virol 76:7263–7275
22. Schwarz TF, Nsanze H, Longson M, Nitschko H, Gilch S, Shurie H, Ameen A, Zahir AR, Acharya UG, Jager G (1996) Polymerase chain reaction for diagnosis and identification of distinct variants of Crimean-Congo hemorrhagic fever virus in the United Arab Emirates. Am J Trop Med Hyg 55:190–196
23. Tang Q, Gao D, Zhao X, Han L, Hang C (2002) Study on the molecular biology of hemorrhagic fever virus in Xinjiang. Zhonghua Liu Xing Bing Xue Za Zhi 23:449–452
24. Tang Q, Saijo M, Lei H, Niikura M, Maeda A, Ikegami T, Xinjung W, Kurane I, Morikawa S (2003) Detection of immunoglobulin G to Crimean-Congo hemorrhagic fever virus in sheep sera by recombinant nucleoprotein-based enzyme-linked immunosorbent and immunofluorescence assays. J Virol Methods 108:111–116
25. Tang Q, Saijo M, Zhang Y, Asiguma M, Tianshu D, Han L, Shimayi B, Maeda A, Kurane I, Morikawa S (2003) A patient with Crimean-Congo hemorrhagic fever serologically diagnosed by recombinant nucleoprotein-based antibody detection systems. Clin Diagn Lab Immunol 10:489–491
26. Tang Q, Zhao XQ, Wang HY, Simayi B, Zhang YZ, Saijo M, Morikawa S, Liang GD, Kurane I (2005) Molecular epidemiology of Xinjiang hemorrhagic fever viruses. Zhonghua Shi Yan He Lin Chuang Bing Du Xue Za Zhi 19:312–318
27. van Eeden PJ, Joubert JR, van de Wal BW, King JB, de Kock A, Groenewald JH (1985) A nosocomial outbreak of Crimean-Congo haemorrhagic fever at Tygerberg Hospital. Part I. Clinical features. S Afr Med J 68:711–717
28. Yan YC (1983) Characteristics of Xinjiang hemorrhagic fever virus. II. Physicochemical properties of Xinjiang hemorrhagic fever virus. Zhonghua Liu Xing Bing Xue Za Zhi 4:132–134
29. Yan YC (1983) Studies on the characteristics of Xinjiang hemorrhagic fever virus. I. Serological relationship between Crimian-Congo hemorrhagic fever virus and Xinjiang hemorrhagic fever virus. Zhonghua Liu Xing Bing Xue Za Zhi 4:129–131
30. Yashina L, Petrova I, Seregin S, Vyshemirskii O, Lvov D, Aristova V, Kuhn J, Morzunov S, Gutorov V, Kuzina I, Tyunnikov G, Netesov S, Petrov V (2003) Genetic variability of Crimean-Congo haemorrhagic fever virus in Russia and Central Asia. J Gen Virol 84:1199–1206
31. Yashina L, Vyshemirskii O, Seregin S, Petrova I, Samokhvalov E, Lvov D, Gutorov V, Kuzina I, Tyunnikov G, Tang YW, Netesov S, Petrov V (2003) Genetic analysis of Crimean-Congo hemorrhagic fever virus in Russia. J Clin Microbiol 41:860–862
32. Yen YC, Kong LX, Lee L, Zhang YQ, Li F, Cai BJ, Gao SY (1985) Characteristics of Crimean-Congo hemorrhagic fever virus (Xinjiang strain) in China. Am J Trop Med Hyg 34:1179–1182

CHAPTER 11

CRIMEAN-CONGO HEMORRHAGIC FEVER IN SOUTH AFRICA

FELICITY J. BURT, PH.D.[1], JANUSZ T. PAWESKA, BVSc, DVSc[2], AND ROBERT SWANEPOEL, BVSc, DTVM, PH.D.[2]

[1]*Department of Virology, Faculty of Health Sciences, University of the Free State, Box 339, Bloemfontein 9300, South Africa*

[2]*Special Pathogens Unit, National Institute for Communicable Diseases, Private Bag X4, Sandringham 2131, South Africa*

11.1. INTRODUCTION

Crimean-Congo hemorrhagic fever (CCHF) is a tick-borne viral zoonosis which occurs widely in Africa, eastern Europe, and Asia within the distribution range of ticks of the genus *Hyalomma.* A disease named Crimean hemorrhagic fever was first observed in the Crimean Peninsula in 1944, and the causative agent which was isolated in 1967, was found to be identical to Congo virus isolated in 1956 from a febrile child in the Belgian Congo (now Democratic Republic of Congo – DRC); hence the names Crimean and Congo are used in combination [8, 10, 11, 49]. By 1979, the virus was known to occur in many countries in eastern Europe and Asia, but the furthest south that evidence of infection had been found in Africa was the detection of antibodies in cattle sera in Tanzania [21]. Case fatality rates recorded in Eastern Europe and Asia varied from 15% to 40%, but in Africa only 1/15 known human infections had been fatal. Nevertheless, suggestions that African strains of the virus were less pathogenic for humans than Eurasian strains were rejected on the grounds that observations had been too limited [21].

In February 1981, the first case of CCHF was recognized in South Africa, and there was speculation that the virus had been introduced recently by translocation of infected tick vectors on migrating birds [20]. However, antibody to CCHF virus was found to be widely distributed in the sera of livestock and wild vertebrates in South Africa, Zimbabwe, and Namibia, including sera which had been in storage for many years [2, 44, 52, 54, 56]. Moreover, by the end of March 2006 a total of 180 cases of the disease had been confirmed, including 161 infections which had occurred within South Africa, plus two imported from DRC and Tanzania, and

O. Ergonul and C. A. Whitehouse (eds.), Crimean-Congo Hemorrhagic Fever, 131–141.

17 infections which had occurred in Namibia. There had been 49/180 (27%) fatalities [20, 42, 51, 54–57]. Thus, it appears that CCHF virus had been in southern Africa long before its presence was realized, with greater awareness and the availability of a specific diagnostic service leading to regular recognition of about 5–20 cases of the disease each year since 1981. It is also clear that the disease in Africa is no less severe than that which occurs in Eastern Europe and Asia.

11.2. EPIDEMIOLOGY

CCHF virus has been isolated from at least 30 species of ticks, but in most instances there is no proof that the ticks are capable of serving as vectors since virus isolated from engorged ticks may merely have been present in the blood meal obtained from a viraemic host [21, 52, 53, 55, 59, 64]. Nevertheless, the ability to transmit infection has been demonstrated for ixodid ticks of several genera, and transovarial transmission of the virus from adult females to the succeeding generation of larval ticks has been shown to occur in a few members of the *Hyalomma*, *Dermacentor*, and *Rhipicephalus* genera. However, the coincidence in distribution of CCHF virus and *Hyalomma* ticks implies that members of this genus are the most important vectors of the virus [21, 46, 48, 52–54, 59, 64]. Ixodid ticks have three stages or instars in their life cycle: larvae, nymphs, and adults, each of which attaches to a host and feeds to repletion before molting to the next stage and feeding again, either on the same host or on separate hosts. Many *Hyalommas* behave as two-host ticks: the larvae remain attached and molt in situ before feeding as nymphs on the same host, and then detach to molt into adults which feed on a second host.

Three species of *Hyalomma* occur in South Africa, *Hyalomma glabrum* (formerly thought to be identical to the Asian *Hyalomma marginatum turanicum*), *Hyalomma marginatum rufipes*, and *Hyalomma truncatum* [1, 25–27, 32]. These xerophilic ticks are most prevalent in the semiarid central and western inland regions of the country, and sparse or absent in the moister eastern and southern coastal areas [32, 60]. The preferred hosts of the immature stages of the *Hyalommas* are small mammals up to the size of hares, and ground-feeding birds, while the preferred hosts of the adult ticks are large vertebrates such as cattle, eland antelope, African buffaloes, giraffes, zebras, rhinoceroses, and ostriches [19, 28, 29, 36, 61, 63]. Sheep and goats are also frequently infested, but seldom harbor large numbers of ticks [17, 18, 30, 39].

CCHF virus causes inapparent infection or mild fever and viremia which lasts for up to a week, with maximum recorded titers of infectivity ranging from $10^{2.7}$ to $10^{4.2}$ mouse intracerebral 50% lethal doses/mL ($MICLD_{50}$/mL), in cattle, sheep, goats, and small mammals such as rodents and hares [21, 45, 48, 52, 53, 59, 64]. Ticks become infected while feeding on the hosts during viremia and are able to transmit the virus when they feed on a second host during the succeeding instar of their life cycle. It has been calculated that transovarial transmission of virus from infected females to larval ticks does not occur with sufficient frequency to ensure perpetuation of the virus in the absence of amplification of infection in vertebrate hosts [21, 64]. Consequently, it can be deduced that infection of immature ticks on small vertebrates constitutes a more important amplifying mechanism for circulation of the virus than infection of adult ticks on large hosts.

Passerine birds and domestic chickens have been found to be refractory to experimental infection with CCHF virus, and guinea fowl developed transient viremia of low intensity, so it is considered unlikely that birds can infect ticks directly [21, 43]. However, it has been shown that infection can be transferred between infected and noninfected ticks feeding together on hosts through the so-called phenomenon of nonviremic transmission which is thought to involve factors present in tick saliva. Moreover, since birds carry immature ticks they could serve to disseminate virus which has been transmitted transovarially in the ticks, both locally and on long-range migratory routes [23, 24, 65, 66]. Antibody to CCHF virus was found to be present in ostrich sera in southern Africa, and following the occurrence of two outbreaks of the disease among workers in ostrich abattoirs it was shown that these birds develop viremic infection similar to that in domestic ruminants [58].

Antibody to CCHF virus is highly prevalent in domestic and wild animals in South Africa, with lower prevalences occurring along the southern coast and the extreme northeast where *H. truncatum* is the sole representative of the three species of *Hyalomma* known to occur in the country [2, 25–27, 32, 43, 44, 50, 63]. Antibody prevalences are highest in the sera of large mammals such as cattle, eland antelope, African buffaloes, giraffes, zebras, rhinoceroses, and ostriches which are the preferred hosts of adult ticks, while among small mammals antibody is most prevalent in hares, but also occurs in ground-frequenting birds, such as guinea fowl and rodents [2, 43].

Mechanisms for the dissemination of ticks and hence virus, which must have operated in Eurasia and Africa for millennia, include the migration of large numbers of birds on a north–south axis annually, carrying immature ticks [21–24]. In addition, ticks have historically been dispersed between continents by transportation of livestock, and there is recent evidence of CCHF outbreaks in the Near East being associated with importation of animals from Africa [34].

Analysis of the genetic diversity of CCHF isolates based nucleotide sequences determined for a region of the S segment of the genome produced evidence for the existence of three groups of the virus: A, B and C, with group A containing two distinct lineages [7, 35, 40]. One lineage of group A circulates throughout Africa; the second circulates in Asia and Madagascar, while group B circulates in southern and West Africa, and subtype C is represented by a unique isolate from Greece. Although there are nucleotide differences of up to 18% between southern African isolates within groups A and B, the majority of the nucleotide changes are nonsynonymous, and the high degree of amino acid homology could explain the antigenic similarity of the nucleocapsid protein of CCHF isolates throughout the distribution range of the virus [9, 12]. Thus, despite the potential which exists for dispersal of the virus between Africa and Eurasia, it appears that circulation of the virus is largely compartmentalized within the two land masses, and the evolution of CCHF virus strains within specific geographic regions is probably a consequence of association with particular tick vector species. [7, 14, 34, 35, 37, 40, 41].

Humans commonly acquire CCHF infection from tick bite, or from contact of broken skin or mucus membranes with fresh blood or other tissues of infected livestock or human patients, and although some patients are unable to recall specific incidents constituting exposure to infection, it is invariably found that they have lived in or visited an environment where exposure was possible. Most cases of the disease in South Africa have occurred in North West, Free State, and Northern and Western Cape Provinces, where the relatively arid climate suits *Hyalomma* ticks [25–27, 31, 32, 56, 57]. Human disease has been diagnosed in most months of the year, but there has been a slight preponderance of cases in February–March and October–November, when adult *Hyalommas* manifest peak questing activity [38]. The majority of patients tend to be adult males engaged in the livestock industry, such as farmers, labourers, slaughtermen, and veterinarians, but infection sometimes occurs in town dwellers who have been exposed to infection in rural settings [21, 64, 56, 57].

Among the 180 cases of CCHF recorded in southern Africa from February 1981 up to the end of March 2006, the largest group of 80 (44%) cases arose from known tick bite or the squashing of ticks; a similar number of 72 (40%) cases arose from known or potential contact with fresh blood or other tissues of livestock and/or ticks; 7 (4%) nosocomial infections arose from contact with blood or fomites of known CCHF patients, while in 21 (12%) instances there was no direct evidence of contact with livestock or ticks, but the patients lived in or visited a rural environment where such contact was possible [20, 33, 42, 51, 54–57, 62]. Males constituted 150/180 (83%) of all cases of the disease diagnosed.

It was observed in South Africa and the former USSR that humans do not have to be bitten by ticks, and can become infected from merely squashing ticks between the fingers, with the virus presumably gaining entry through skin wounds such as torn cuticles or through contact with mucus membranes [21, 56]. People are not always aware of having been bitten, and in one instance a semiengorged tick, from which virus was isolated, was found attached between the toes of a comatose and moribund patient [51]. However, many people involved with livestock are familiar with *Hyalomma* ticks, which have distinctive brown and white bands on their legs, and patients in southern Africa are often able to verify that they have been bitten by ticks fitting this description [51, 56].

Apart from nosocomial and laboratory infections, and those which occur in abattoirs, the disease has never been recorded in urban residents without an explicit history of exposure to ticks or fresh tissues of farm or wild animals, often encountered on hiking or hunting trips. In two exceptions where people became infected within an urban environment, one person is thought to have been exposed to ticks brought in with garden manure obtained from cattle pens on a farm, and the other slaughtered a sheep in town for a wedding feast [51].

In view of the serological evidence that infection of livestock occurs on a wide scale, it is surprising that so few human infections are diagnosed and this raises the possibility that many human infections are asymptomatic or mild and pass unnoticed. However, the low prevalence of antibody detected in surveys conducted on rural residents and the sparse evidence of infection encountered among cohorts of cases of the disease indicate that a high proportion of CCHF infections are symptomatic and come to medical attention [15, 55, 56]. Reasons for the low incidence of human infection probably include the fact that humans are not the preferred hosts of *Hyalomma* ticks, particularly immature ticks, and are infrequently bitten in comparison with livestock. Moreover, viremia in livestock is short-lived and of low intensity compared to that which occurs in other zoonotic diseases such as Rift Valley fever, which is more readily acquired from contact with infected tissues. Although people become infected from slaughtering animals in rural environments or in urban abattoirs, there has been no indication that CCHF virus constitutes a public health hazard in meat processed and matured according to normal health regulations [21, 55, 56, 58]. This may be related to the fact that the sharp drop in pH which occurs during the maturation of meat in abattoirs would be deleterious to virus infectivity. No virus could be isolated from meat taken from the carcases of experimentally infected and viremic sheep and ostriches which had been hung at 4°C for 1–7 days [51, 58].

11.3. SIGNS AND SYMPTOMS OF INFECTION IN HUMANS

Incubation periods commonly range from 1 to 3 days after infection by tick bite, but occasionally extend to 7 days, and are usually 5–6 days in people exposed to infected blood or other tissues of livestock or human patients, but may extend to 9 days or more [20, 21, 33, 42, 51, 54–57, 62].

Onset of the disease is usually very sudden, with severe headache frequently accompanied by dizziness, neck pain and stiffness, sore eyes, photophobia, fever, rigor and chills, plus myalgia with intense backache or leg pains. Malaise, nausea, sore throat, and vomiting occur early in the illness and there may be nonlocalized abdominal pain and diarrhea. Fever may be intermittent and patients often undergo sharp changes of mood over the first 2 days, with feelings of confusion and aggression. By day 2–4 of illness patients often exhibit lassitude, depression and somnolence, and have a flushed appearance with injected conjunctivae and chemosis. At this stage, tenderness may localize in the right upper quadrant of the abdomen, and hepatomegaly may be discernible. There may be tachycardia with slight hypotension, plus lymphadenopathy and exanthema and petechiae of the throat, tonsils and buccal mucosa.

A petechial rash appears on the trunk and limbs by day 3–6 of illness, often followed rapidly by the appearance of large bruises and ecchymoses, especially in the antecubital fossae, upper arms, axillae, and groin. Epistaxis, hematemesis, hematuria, melena, gingival bleeding, and bleeding from the vagina may commence on day 4–5 of illness, or occasionally even earlier. In some patients, a hemorrhagic tendency is evident only from the oozing of blood from injection or venipuncture sites. Internal bleeding, including retroperitoneal and intracranial hemorrhage, occurs in some patients. Hepatorenal and pulmonary failure occur from about day 5 onwards in severely ill patients, and they become progressively drowsy, stuporous, and comatose. Jaundice is sometimes evident during the 2nd week of illness. The mortality rate is approximately 30%, and deaths generally occur on days 5–14 of illness. Patients who recover usually manifest subjective improvement on day 9–10, but asthenia, conjunctivitis, slight confusion, and amnesia may continue for 1 month or longer.

During the first few days of illness there may be leukocytosis or leukopenia, elevated serum aspartate transaminase (AST), alanine transaminase (ALT), gamma-glutamyltransferase, lactic dehydrogenase, alkaline phosphatase and creatine kinase levels, thrombocytopenia, elevation of the prothrombin time, activated thromboplastin time (APTT), thrombin time, and fibrin degradation products, as well as depression of fibrinogen and hemoglobin values. Bilirubin, creatinine, and urea levels increase and serum protein levels decline during the 2nd week. During the first 5 days of illness any of the following clinical pathology values are highly predictive of fatal outcome: leukocyte counts $\geq 10 \times 10^9$/L; platelet counts $\leq 20 \times 10^9$/L; AST $\geq$ 200U/L; ALT $\geq$ 150U/L; APTT $\geq$ 60 s; and fibrinogen $\leq$ 110 mg/dL [57]. Leukopenia does not have the same poor prognostic connotation as leukocytosis at this early stage, and all clinical pathology values may be grossly abnormal after day 5 of illness without necessarily being indicative of a poor prognosis. Antibody response is rarely demonstrable in fatal illness, and thus detection of antibodies is generally a favorable sign [3].

Treatment of CCHF involves the use of barrier-nursing techniques for the protection of medical staff, and consists essentially of supportive and replacement therapy with blood products. Immune plasma has been used in therapy, but its efficacy has never been established, largely due to the lack of a uniform product. Promising results have been obtained with the chemotherapeutic drug ribavirin [16, 51] particularly when administered before day 5 after the onset of illness. No vaccines are currently available [21].

11.4. DIAGNOSIS

CCHF may be suspected when there is a sudden onset of febrile illness with a short incubation period, usually less than week, after exposure to tick bite, or fresh blood or other tissues of livestock or human patients, but occasionally there is merely a history of residing in or visiting a rural environment where exposure to ticks and livestock is possible. Often clinicians request clinical pathology studies before the disease is suspected, and the occurrence of leukopenia or leukocytosis, thrombocytopenia, and raised ALT and AST values early in the course of the illness provides supportive evidence for a diagnosis of CCHF.

Clotted blood is usually submitted for laboratory confirmation of the diagnosis, but if earlier samples have been taken with various preservatives for blood counts or other clinical pathology studies, these should also be forwarded to the virus laboratory. Pathologists are usually reluctant to perform autopsies on patients who have died of suspected CCHF, but clinicians should be encouraged to take liver samples from deceased patients with biopsy needles, and to submit samples in duplicate in viral transport medium and formalin fixative.

In the acute phase of illness the diagnosis can be confirmed by the detection of viral nucleic acid by real-time or conventional reverse transcription polymerase chain reaction (RT-PCR), or by demonstration of viral antigen by enzyme-linked immunosorbent assay (ELISA) in serum samples, or by isolation of virus [6, 13, 40, 41, 56]. Virus can be isolated in cell cultures, commonly Vero cells, or by intracerebral inoculation of day-old mice. Virus is detected and identified in cell cultures by immunofluorescence (IF), and isolation can be achieved in 1–5 days compared to 5–9 days in mice, but mouse inoculation is more sensitive for isolating virus present in low concentration. Viremia may be demonstrable for up to 13 days after the onset of illness, and nucleic acid detected in serum by RT-PCR for up to 16 days [6].

Both immunoglobulin G (IgG) and immunoglobulin M (IgM) antibodies become demonstrable by indirect immunofluorescence or ELISA from about day 5 of illness onwards, and are present in the sera of all survivors of the disease by day 9 at the latest. The IgM antibody activity declines to undetectable levels by the 4th month after infection, and IgG titers may decline gradually, but remain demonstrable for at least 5 years. Recent or current infection is confirmed by demonstrating seroconversion, ≥ fourfold increase in antibody activity in paired serum samples, or IgM activity in a single specimen [3, 47, 56] Patients who die frequently fail to develop a demonstrable antibody response, and the diagnosis has to be confirmed by isolation of virus or detection of viral nucleic acid in serum samples, or liver samples taken after death, or by demonstration of CCHF antigen using immunohistochemical techniques on paraffin-embedded liver sections [3, 5, 6]. Virus antigen may sometimes be demonstrated in liver impression smears by immunofluorescence, or in serum or liver homogenate by ELISA. Observation of necrotic lesions compatible with CCHF in sections of liver provides presumptive evidence in support of the diagnosis.

CCHF should be distinguished from other tick-borne infections [4], particularly tick-borne typhus (*Rickettsia conorii* infection, commonly known as tick-bite fever) which is highly prevalent in southern Africa. Tick-bite fever has a longer incubation period than CCHF, 7–10 days, and there is usually a characteristic necrotic lesion, an eschar, at the site of the bite. Severe cases can present with hemorrhagic signs similar to CCHF, but the disease can be treated with broad-spectrum antibiotics. Rift Valley fever can also be acquired from contact with the tissues of livestock in Africa, but this usually occurs in the context massive outbreaks of disease in sheep and cattle following heavy rains which favor breeding of mosquito vectors of the virus. Marburg and Ebola hemorrhagic fevers cause sporadic outbreaks of lethal disease in tropical Africa, and although the source of these viruses in nature remains unknown humans frequently acquire Ebola virus infection from contact with diseased carcases of nonhuman primates. Lassa fever is caused by a rodent-borne virus in certain countries of West Africa, and humans acquire infection through contamination of food and house dust with rodent urine. Outbreaks of infection with the mosquito-borne yellow fever and dengue viruses occur periodically in West Africa and less frequently in East Africa. Yellow fever can be highly lethal, but there is a very effective vaccine. Dengue infection is usually benign but can be lethal in the very young or aged, or where there is sequential infection with a second serotype (there are four serotypes of the virus).

11.5. PREVENTION AND CONTROL

The control of CCHF through the application of acaricides to livestock is impractical under the extensive farming conditions which prevail in the arid areas where *Hyalomma* ticks are most prevalent. Following the occurrence of an outbreak of CCHF among workers in an ostrich abattoir in South Africa in 1996, regulations were promulgated requiring ostriches to be treated for ticks with pyrethroids which are nontoxic to livestock and humans, and kept enclosed in a tick-free environment for 2 weeks prior to slaughter in order to reduce the risk of exposure of workers and consumers to infection [58]. Similar measures could be applied to other slaughter animals. Veterinarians, slaughtermen, and others involved with livestock should be aware of CCHF and take practical steps, such as the wearing of gloves and protective clothing, to limit or avoid exposure of naked skin to fresh blood and other tissues of farm animals. Pyrethroid preparations can be used to kill ticks which come into contact with treated human clothing. There is no vaccine.

REFERENCES

1. Apanaskevich DA, Horak IG (2006) The genus *Hyalomma* Koch, 1844. I. Reinstatement of *Hyalomma* (*Euhyalomma*) *glabrum* Delpy, 1949 (Acari, Ixodidae) as a valid species with a re-description of the adults, first description of the immature stages and notes on its biology. Onderstepoort J Vet Res 73:1–12
2. Burt FJ, Swanepoel R, Braack LEO (1993) Enzyme-linked immunosorbent assays for the detection of antibody to Crimean-Congo haemorrhagic fever virus in the sera of livestock and wild vertebrates. Epidemiol Infect 111:547–557
3. Burt FJ, Leman PA, Abbott JC, Swanepoel R (1994) Serodiagnosis of Crimean-Congo haemorrhagic fever. Epidemiol Infect 113:551–562
4. Burt FJ, Spencer DC, Leman PA, Patterson B, Swanepoel R (1996) Investigation of tick-borne viruses as pathogens of humans in South Africa and evidence of Dugbe virus infection in a patient with prolonged thrombocytopenia. Epidemiol Infect 116:353–361
5. Burt FJ, Swanepoel R, Shieh WJ, Smith JF, Leman PA, Greer PW, Coffield LM, Rollin PE, Ksiazek TG, Peters CJ, Zaki S (1997) Immunohistochemical and in situ localization of Crimean-Congo hemorrhagic fever (CCHF) virus in human tissues and implications for CCHF pathogenesis. Arch Pathol Lab Med 121:839–846

6. Burt FJ, Leman PA, Smith JF, Swanepoel R (1998) The use of a reverse transcription-polymerase chain reaction for the detection of viral nucleic acid in the diagnosis of Crimean-Congo haemorrhagic fever. J Virol Meth 70:129–137
7. Burt FJ, Swanepoel R (2005) Molecular epidemiology of African and Asian Crimean-Congo haemorrhagic fever isolates. Epidemiol Infect 133:659–666
8. Casals J (1969) Antigenic similarity between the virus causing Crimean hemorrhagic fever and Congo virus. Proc Soc Exp Biol Med 131:233–236
9. Casals J, Tignor GH (1980) The *Nairovirus* genus: serological relationships. Intervirol 14:144–147
10. Chumakov MP (1974) Contribution to 30 years of investigation of Crimean-Congo haemorrhagic fever. In: Chumakov MP (ed.) Medical virology. Trudy Inst Polio Virus Enstef Akad Med Nauk SSSR (In Russian). English translation: NAMRU3-T950
11. Chumakov MP, Smirnova SE, Tkachenko EA (1970) Relationship between strains of Crimean haemorrhagic fever and Congo viruses. Acta Virol 14:82–85
12. Donets MA, Chumakov MP, Korolev MB, Rubin SG (1977) Physicochemical characteristics, morphology and morphogenesis of virions of the causative agent of Crimean hemorrhagic fever. Intervirol 8:294–308
13. Drosten C, Gottig S, Schilling S, Asper M, Panning M, Schmitz H, Gunther S (2002) Rapid detection and quantification of RNA of Ebola and Marburg Viruses, Lassa Virus, dengue Virus, and yellow fever virus by real-time reverse transcription-PCR. J Clin Microbiol 40:2323–2330
14. Drosten C, Minnak D, Emmerich P, Schmitz H, Reinecke T (2002) Crimean-Congo hemorrhagic fever in Kosovo. J Clin Microbiol 40:1122–1123
15. Fisher-Hoch SP, McCormick JB, Swanepoel R, Van Middlekoop A, Harvey S, Kustner HG (1992) Risk of human infections with Crimean-Congo hemorrhagic fever virus in a South African rural community. Am J Trop Med Hyg 47:337–345
16. Fisher-Hoch SP, Khan JA, Rehman S, Mirza S, Khurshid M, McCormick JB (1995) Crimean Congo-haemorrhagic fever treated with oral ribavirin. Lancet 346:472–475
17. Fourie LJ, Horak IG, Marais L (1988). The seasonal abundance of adult ticks on Merino sheep in the south-western Orange Free State. J S Afr Vet Assoc 59:191–194
18. Fourie LJ, Horak IG (1991) The seasonal activity of adult ixodid ticks on Angora goats in the south-western Orange Free State. J S Afr Vet Assoc 62:104–106
19. Fourie LJ, Kok DJ, Heyne H (1996) Adult ixodid ticks on two cattle breeds in the south-western Free State, and their seasonal dynamics. Onderstepoort J Vet Res 63:19–23
20. Gear JH, Thomson PD, Hopp M, Andronikou S, Cohn RJ, Ledger J, Berkowitz FE (1982) Congo-Crimean haemorrhagic fever in South Africa. Report of a fatal case in the Transvaal. S Afr Med J 62:576–580
21. Hoogstraal H (1979) The epidemiology of tick-borne Crimean-Congo hemorrhagic fever in Asia, Europe, and Africa. J Med Entomol 15:307–417
22. Hoogstraal H (1981) Changing patterns of tick-borne diseases in modern society. Ann Rev Entomol 26:75–99
23. Hoogstraal H, Kaiser MN, Traylor MA, Gaber S, Guindy E (1961) Ticks (Ixodoidea) on birds migrating from Africa to Europe and Asia. Bull WHO 24:197–212
24. Hoogstraal H, Kaiser MN, Traylor MA, Gaber S, Guindy E (1961) Ticks (Ixodoidea) on birds migrating from Europe and Asia to Africa. Bull WHO 24:197–212
25. Horak IG, De Vos V, Brown MR (1983) Parasites of domestic and wild animals in South Africa. XVI. Helminth and arthropod parasites of blue and black wildebeest (*Connochaetes taurinus* and *Connochaetes gnou*). Onderstepoort J Vet Res 50:243–255
26. Horak IG, Potgieter FT, Walker JB, De Vos V, Boomker J (1983) The ixodid tick burdens of various large ruminant species in South African nature reserves. Onderstepoort J Vet Res 50:221–228
27. Horak IG, De Vos V, De Klerk BD (1984) Parasites of domestic and wild animals in South Africa. XVII. Arthropod parasites of Burchells's zebra, *Equus burchelli*, in the eastern Transvaal lowveld. Onderstepoort J Vet Res 51:145–154

28. Horak IG, Fourie LJ, Novellie PA, Williams EJ (1991) Parasites of domestic and wild animals in South Africa. XXVI. The mosaic of ixodid tick infestations on birds and mammals in the Mountain Zebra National Park. Onderstepoort J Vet Res 58:125–136
29. Horak IG, Anthonissen M, Krecek RC, Boomker J (1992) Arthropod parasites of springbok, gemsbok, kudus, giraffes and Burchell's and Hartmann's zebras in the Etosha and Hardap Nature Reserves, Namibia. Onderstepoort J Vet Res 59: 253–257
30. Horak IG, Fourie LJ (1992) Parasites of domestic and wild animals in South Africa. XXXI. Adult ticks on sheep in the Cape Province and in the Orange Free State. Onderstepoort J Vet Res 59:275–283
31. Horak IG, Swanepoel R, Gummow B (2002) The distribution of *Hyalomma* spp. and human cases of Crimean-Congo haemorrhagic fever in South Africa. In: Proceedings of the 10th Conference of the Association of Institutions for Tropical Veterinary Medicine, Copenhagen, Denmark 20–23 August 2001, pp 501–509
32. Howell CJ, Walker JB, Nevill EM (1978) Ticks, mites and insects infesting domestic animals in South Africa. Science Bulletin, No 393. Department of Agricultural Technical Services, Pretoria
33. Joubert JR, King JB, Rossouw DJ Cooper R (1985) A nosocomial outbreak of Crimean-Congo haemorrhagic fever at Tygerberg Hospital. Part III. Clinical pathology and pathogenesis. SAMJ 68:722–728
34. Khan AS, Maupin GO, Rollin PE, Noor AM, Shurie HH, Shalabi AG, Wasef S, Haddad, YM, Sadek R, Ijaz K, Peters CJ Ksiazek TG (1997) An outbreak of Crimean-Congo hemorrhagic fever in the United Arab Emirates, 1994–1995. Am J Trop Med Hyg 57:519–525
35. Marriott AC, Nuttall PA (1996) Molecular biology of nairoviruses In: Elliott RM (ed.) The Bunyaviridae. Plenum Press, New York
36. Norval RAI (1982) The ticks of Zimbabwe. IV. The genus Hyalomma. Zimbabwe Vet J 13: 2–10
37. Papa A, Ma B, Kouidou S, Tang Q, Hang C, Antoniadis A (2002a) Genetic characterization of the M RNA segment of Crimean Congo hemorrhagic fever virus strains, China. Emerg Infect Dis 8:50–53
38. Rechav Y (1986) Seasonal activity and hosts of the vectors of Crimean-Congo haemorrhagic fever in South Africa. S Afr Med J 69:364–368
39. Rechav Y, De Jager C (1991) Seasonal abundance of ticks associated with indigenous goats on a Northern Transvaal farm. J S Afr Vet Assoc 62:10–11
40. Rodriguez LL, Maupin GO, Ksiazek TG, Rollin PE, Khan AS, Schwarz TF, Lofts RS, Smith JF, Noor AM, Peters CJ, Nichol ST (1997) Molecular investigation of a multisource outbreak of Crimean-Congo hemorrhagic fever in the United Arab Emirates. Am J Trop Med Hyg 57:512–518
41. Schwarz TF, Nsanze H, Longson M, Nitschko H, Gilch S, Shurie H, Ameen A, Zahir AR, Acharya UG, Jager G (1996) Polymerase chain reaction for diagnosis and identification of distinct variants of Crimean-Congo hemorrhagic fever virus in the United Arab Emirates. Am J Trop Med Hyg 55:190–196
42. Shepherd AJ, Swanepoel R, Shepherd SP, Leman PA, Blackburn NK, Hallett AF (1985) A nosocomial outbreak of Crimean-Congo haemorrhagic fever at Tygerberg Hospital: part V. Virological and serological observations. S Afr Med J 68:733–736
43. Shepherd AJ, Swanepoel R, Leman PA, Shepherd SP (1987) Field and laboratory investigation of Crimean-Congo haemorrhagic fever virus (*Nairovirus*, family Bunyaviridae) infection in birds. Trans R Soc Trop Med Hyg 81:1004–1007
44. Shepherd AJ, Swanepoel R, Shepherd SP, McGillivray GM, Searle LA (1987) Antibody to Crimean-Congo hemorrhagic fever virus in wild mammals from southern Africa. Am J Trop Med Hyg 36:133–142
45. Shepherd AJ, Leman PA, Swanepoel R (1989) Viraemia and antibody response of small African and laboratory animals to Crimean-Congo hemorrhagic fever virus infection. Am J Trop Med Hyg 40:541–547
46. Shepherd AJ, Swanepoel R, Cornel AJ, Mathee O (1989) Experimental studies on the replication and transmission of Crimean-Congo hemorrhagic fever virus in some African tick species. Am J Trop Med Hyg 40:326–331

47. Shepherd AJ, Swanepoel R, Leman PA (1989) Antibody response in Crimean-Congo hemorrhagic fever. Rev Infect Dis 11:S801-S806
48. Shepherd AJ, Swanepoel R, Shepherd SP, Leman PA, Mathee O (1991) Viraemic transmission of Crimean-Congo haemorrhagic fever virus to ticks. Epidemiol Infect 106:373–382
49. Simpson DKH, Knight EM, Courtois MCG, Weinbren MP, Kibukamusoke JW (1967) Congo virus: a hitherto undescribed virus occurring in Africa. Part 1. Human isolations-clinical notes. E Afr Med J 44:87–92
50. Spickett AM, Horak IG, Braack LEO, Van Ark H (1991) Drag-sampling of free-living ixodid ticks in the Kruger National Park. Onderstepoort J Vet Res 58:27–32
51. Swanepoel R (1981–2006) National Institute for Communicable Diseases, Private Bag X4, Sandringham 2131, Johannesburg, South Africa (unpublished)
52. Swanepoel R (1994) Crimean-Congo haemorrhagic fever. In: Coetzer JAW, Thomson GR, Tustin RC (eds) Infectious Diseases of Livestock with Special Reference to Southern Africa. Oxford University Press, Cape Town, pp 723–729
53. Swanepoel R (1998) Crimean-Congo haemorrhagic fever. In: Palmer SR, Soulsby EJL, Simpson DIH (eds) Zoonoses. Oxford University Press, Oxford, pp 461–470
54. Swanepoel R, Struthers JK, Shepherd AJ, McGillivray GM, Nel MJ, Jupp PG (1983) Crimean-Congo hemorrhagic fever in South Africa. Am J Trop Med Hyg 32:1407–1415
55. Swanepoel R, Shepherd AJ, Leman PA, Shepherd SP (1985) Investigations following initial recognition of Crimean-Congo haemorrhagic fever in South Africa and the diagnosis of 2 further cases. S Afr Med J 68:638–641
56. Swanepoel R, Shepherd AJ, Leman PA, Shepherd SP, McGillivray GM, Erasmus MJ, Searle LA, Gill DE (1987) Epidemiologic and clinical features of Crimean-Congo hemorrhagic fever in southern Africa. Am J Trop Med Hyg 36:120–132
57. Swanepoel, R, Gill DE, Shepherd AJ, Leman PA, Mynhardt JH, Harvey S (1989) The clinical pathology of Crimean-Congo hemorrhagic fever. Rev Infect Dis 11:S794–S800
58. Swanepoel R, Leman PA, Burt FJ, Jardine J, Verwoerd DJ, Capua I, Bruckner GK, Burger WP (1998) Experimental infection of ostriches with Crimean-Congo haemorrhagic fever virus. Epidemiol Infect 121:427–432
59. Swanepoel R, Burt FJ (2004) Crimean-Congo haemorrhagic fever. In: Coetzer JAW, Tustin RC (eds) Infectious Diseases of Livestock, 2nd edn. Oxford University Press, Cape Town, pp 1077–1085
60. Theiler G (1956) Zoological survey of the Union of South Africa. Tick Survey, Part IX. The distribution of the three South African *Hyalommas* or bontpoots. Onderstepoort J Vet Res 27:239–269 + 3 maps
61. Theiler G (1962). The Ixodoidea parasites of vertebrates in Africa South of the Sahara (Ethiopian Region). Project S 9958. Report to the Director Veterinary Services, Onderstepoort
62. Van Eeden PJ, Joubert JR, Van De Wal BW, King JB, De Kock A, Groenewald JH (1985) A nosocomial outbreak of Crimean-Congo haemorrhagic fever at Tygerberg Hospital. Part I. Clinical features. S Afr Med J 68:711–717
63. Walker JB (1991) A review of the ixodid ticks (Acari, Ixodidae) occurring in southern Africa. Onderstepoort J Vet Res 58:81–105
64. Watts DM, Ksiazek TG, Linthicum KJ, Hoogstraal, H (1989) Crimean-Congo haemorrhagic fever. In: Monath TP (ed.) The Arboviruses: Epidemiology and Ecology, vol II. CRC Press, Boca Raton, FL, pp 177–222
65. Zeller HG, Cornet JP Camicas JL (1994) Crimean-Congo haemorrhagic fever virus infection in birds: field investigations in Senegal. Res Virol 145:105–109
66. Zeller HG, Cornet JP Camicas JL (1994) Experimental transmission of Crimean-Congo haemorrhagic fever virus by West African ground-feeding birds to *Hyalomma marginatum rufipes* ticks. Am J Trop Med Hyg 50:676–681

CHAPTER 12

ROLE OF TICKS IN THE TRANSMISSION OF CRIMEAN-CONGO HEMORRHAGIC FEVER VIRUS

MICHAEL J. TURELL, PH.D.*

Virology Division, United States Army Medical Institute of Infectious Diseases, Fort Detrick, Maryland, USA; Virology Division, USAMRIID, 1425 Porter Street, Fort Detrick, MD 21702-5011, USA. Tel.: +1-301-619-4921; Fax: +301-619-2290; E-mail: michael.turell@amedd.army.mil

12.1. INTRODUCTION

Ticks were suspected of transmitting Crimean-Congo hemorrhagic fever (CCHF) virus (CCHFV) shortly after the disease was formally described in the mid-1940s [6], and inoculation of tick suspensions into human volunteers confirmed that ticks contained a filterable agent that caused CCHF [5, 6, 8,]. Initial epidemiological studies indicated that cases of this newly described disease occurred during the spring and summer; cases were often sporadic with no direct contact between patients; nearly all of the patients had a history of a recent tick bite; and there was little evidence of mosquito, sand fly, or other insect bites among these patients [6]. Studies conducted since then have consistently found a relationship between various tick species and the presence of CCHFV, and ticks are believed to be the principal means of viral transmission and persistence in nature [4, 8, 17, 43]. CCHFV has been detected in or isolated from >30 species of ticks (Table 12-1). In addition, CCHFV has been isolated from *Culicoides* species flies on at least two occasions [3, 25]. However, the mere isolation of a virus from an arthropod does not mean that the arthropod is involved in the natural transmission of that virus.

For an arthropod to be incriminated as an actual vector, several criteria must be met [35]. These include repeated isolation of virus from field-caught individuals of the species and demonstration in the laboratory that the species is able to

*The views, opinions, and findings contained herein are those of the author and should not be constructed as an official Department of Army position, policy, or decision unless so designated by other documentation.

O. Ergonul and C. A. Whitehouse (eds.), Crimean-Congo Hemorrhagic Fever, 143–154.

Table 12-1. Association of ticks and CCHFV

Species[a]	Isolation[b]	Inoc-vector competence[c]	Oral-vector competence[d]	Vertical transmission[e]
Amblyomma variegatum (Fabricius)	3,49	13,30	–	13
Dermacentor daghestanicus Olenev	17,43	–	–	–
Dermacentor marginatus (Sulzer)	17,43	–	21	–
Dermacentor niveus Neumann	31,32	–	–	–
Haemaphysalis parva Neumann	37	–	–	–
Haemaphysalis punctata Canestrini and Fanzago	17,43	–	–	–
Hyalomma anatolicum anatolicum Koch	31,32,44	–	–	–
Hyalomma anatolicum excavatum Koch	3,44	–	–	–
Hyalomma asiaticum asiaticum Schulze and Schlottke	31,32	–	–	–
Hyalomma detritum detritum Schulze	17,43	–	–	–
Hyalomma dromedarii Koch	43	28	–	–
Hyalomma impeltatum Schulze and Schlottke	3	28	10	–
Hyalomma marginatum impressum Koch	17,43	–	–	–
Hyalomma marginatum marginatum Koch[f]	6, 29,42	–	21,50	50
Hyalomma marginatum rufipes Koch	3,36,47,49	13,30	39,48	13,26,48
Hyalomma marginatum turanicum Pomerantsev	17,43	–	–	–
Hyalomma nitidum Schulze	17,43	–	–	–
Hyalomma truncatum Koch	3,13	27,39	45[g]	13,15,45
Ixodes ricinus (Linnaeus)	17,43	–	–	–
Rhipicephalus annulatus (Say)[h]	37	–	–	–
Rhipicephalus appendiculatus Neumann	43	28	–	–
Rhipicephalus bursa Canestrini and Fanzago	17,43	–	–	–
Rhipicephalus decoloratus Koch[g]	3	–	–	–
Rhipicephalus evertsi Neumann	41,44,49	14	14	14
Rhipicephalus guilhoni Morel and Vassiliades	49	–	–	–
Rhipicephalus microplus Canestrini[g]	17,43	–	–	–
Rhipicephalus pulchellus Gerstäcker	17,43	–	–	–
Rhipicephalus pumilio Schulze	17,43	–	–	–
Rhipicephalus rossicus Yakimov and Kol-Yakimova	22	–	21	21
Rhipicephalus sanguineus (Latreille)	17,43	–	–	–
Rhipicephalus turanicus Pomerantsev	17,43	–	–	–
Argas persicus (Oken)	17,43	–	–	–
Argas lahorensis (Neumann)[i]	43	–	–	–

[a] Tick taxonomy follows that of Horak et al. [18]
[b] References for detection of CCHFV from ticks collected in nature
[c] References for replication and transmission of CCHFV by ticks that were inoculated intracoelomically in the laboratory
[d] References for replication and transmission of CCHFV by ticks that were orally exposed to CCHFV in the laboratory
[e] References for vertical transmission of CCHFV by ticks
[f] Often reported as *Hyalomma plumbeum plumbeum* in the Russian literature
[g] Demonstrated replication, but transmission was not attempted
[h] Formerly in the genus *Boophilus*
[i] Reported as *Ornithodoros lahorensis*

become infected and transmit the virus by bite after it ingests a viremic blood meal (i.e. vector competence). In addition, there needs to be data demonstrating that the arthropod species feeds, in nature, on a host that develops an appropriate viremia and that it is active at the time of year that viral transmission is occurring. However, the relative role that the species will play in viral transmission is determined not only by how efficient a vector is in the laboratory, but also by its relative abundance, feeding patterns, longevity, and environmental factors such as temperature and rainfall. All of these need to be taken into account when trying to determine the importance of a particular species as a potential vector.

12.2. LACK OF POTENTIAL FOR MOSQUITOES AND OTHER DIPTERANS TO TRANSMIT CCHFV

Despite the recovery of CCHFV from field-collected *Culicoides* species on two separate occasions [3, 25], there are no data to support involvement of dipteran species in the natural transmission of CCHFV. These isolations were most likely made from viremic blood ingested by the *Culicoides* from an animal infected with CCHFV shortly before the flies were captured and therefore represent isolations from vertebrate blood, rather than from an infected fly. Laboratory studies, which included feeding mosquitoes on CCHF viremic blood or inoculating them with CCHFV, failed to find any evidence of the ability of this virus to replicate in mosquito tissue [1]. Similarly, Hazara virus (a member of the CCHFV group) did not produce plaques in *Aedes albopictus* cells [46]. In addition, despite testing more than 25,000 mosquitoes from areas where CCHF was occurring in the late 1960s, there was no evidence of CCHFV in these mosquitoes [7]. Taken together, there is no evidence that insects play a role in the natural transmission of CCHFV.

12.3. LABORATORY STUDIES

12.3.1. Laboratory studies implicating ticks as vectors of CCHF

Based on the initial epidemiological studies [6], ticks were suspected as vectors of this new disease, and CCHFV was detected in *Hyalomma marginatum* ticks shortly after the disease was described [6, 8]. Since the initial studies in the 1940s, CCHFV has been isolated from ticks on numerous occasions, and to date, this virus has been detected in >30 different species of ticks (Table 12-1). However, as mentioned earlier, the mere isolation of virus from an arthropod does not incriminate it as an actual vector. Although relatively few studies have examined the ability of ticks to actually transmit CCHFV, those that have examined vector competence have consistently shown that ixodid (hard) ticks, particularly members of the genus *Hyalomma* (Fig. 12-1), are highly susceptible to infection with CCHFV and that infected ticks can transmit this virus by bite. In contrast, several studies indicate that argasid (soft) ticks are not competent CCHFV vectors. These include Shepherd et al. [38], who found that three species of argasid ticks, *Argas walkerae*, *Ornithodoros porcinus*, and

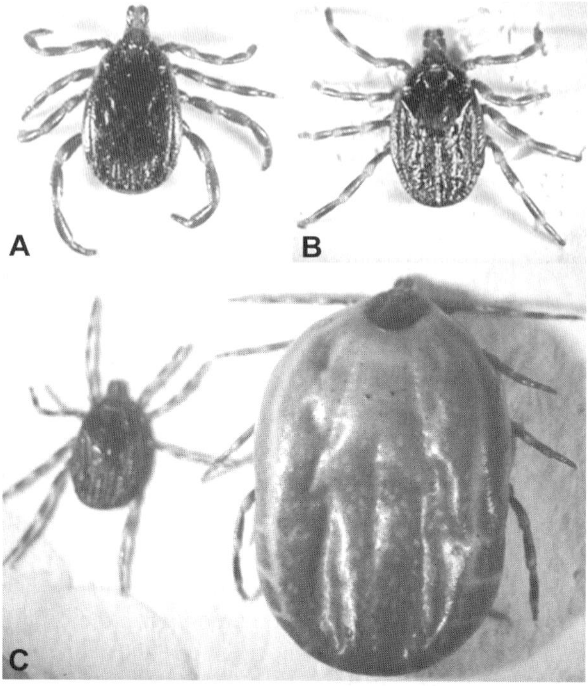

Fig. 12-1. Hyalomma spp. ticks, as illustrated here by *Hyalomma marginatum marginatum*, are the principal vectors of CCHFV. (A) Adult male, (B) adult female, (C) unfed and engorged adult females. (*See Color Plates*)

Ornithodoros savignyi, did not become infected after feeding on viremic animals. Even when these ticks were inoculated intracoelomically, CCHFV could be detected for only 24 h after inoculation. Similarly, Durden et al. [11] found that *Ornithodoros sonrai* did not become infected when they fed on viremic suckling mice, even though these mice had viremias that infected *Hyalomma truncatum* [27]. Therefore, the relatively few virus isolations that have been made from *Argas* or *Ornithodoros* species probably represent recently ingested viremic vertebrate blood, rather than an actual infection. Taken together with the epidemiological evidence (spring–summer transmission, history of tick bites among patients with CCHF, and the repeated isolation of this virus from ixodid ticks), the vector competence data clearly indicate that ixodid ticks, primarily members of the genus *Hyalomma*, serve as CCHFV vectors.

12.3.2. Replication of CCHFV in ticks

When a tick feeds on a CCHF viremic vertebrate host, it ingests virus along with the blood meal. If the appropriate viral receptors are not present, the blood meal, along with the virus, will be digested and excreted, and the tick will not become infected.

However, if the appropriate receptors are present, the gut may become infected, thus initiating the infection process in the tick. Dickson and Turell [9] examined the time course of CCHFV infection in various tissues of *H. truncatum* after intracoelomic inoculation of the ticks. They found that viral titers remained low for about 2 days and then gradually increased in various tissues. Of particular interest were those tissues involved in possible transmission of CCHFV (i.e. salivary glands for transmission by bite and reproductive tissues for venereal or vertical transmission). For the three tissues examined (salivary glands, ovaries, and testes), viral titers were significantly higher in those ticks that took a blood meal than in unfed controls matched for time since intracoelomic inoculation. In contrast, titers in other tissues, including muscle, malpighian tubules, midguts, and nervous tissue were similar in fed and unfed ticks. Therefore, viral replication in tissues associated with possible transmission of CCHFV in an infected tick may be stimulated by attachment and feeding on a susceptible host. This might reduce the stress on a tick induced by viral replication while the tick was waiting to find a vertebrate host, but increase the potential for viral transmission once a host had been acquired.

12.3.3. Time from attachment until viral transmission

A common question is how long a tick must be attached before transmission of a pathogen can occur. Studies conducted with *Ixodes scapularis* have shown that ticks infected with *Borrelia burgdorferi*, the causative agent of Lyme disease, or *Anaplasma phagocytophilum*, the causative agent of human granulocytic anaplasmosis, need to be attached to their vertebrate host for at least 24 h before efficient transmission can occur [20, 33, 34]. Although this has not been examined with CCHFV in its vectors, studies with Powassan virus in *I. scapularis* indicate that these ticks can transmit virus within 15 min of attachment [12]. Therefore, there may not be a delay between tick attachment and CCHFV transmission as is observed with nonviral agents such as *B. burgdorferi*, *Babesia microti*, and *A. phagocytophilum*. Timely tick removal is thus recommended using forceps or effective tick removal tools [40].

12.4. ROLE OF TICKS IN THE LONG-TERM MAINTENANCE AND SPREAD OF CCHFV

Hard ticks feed only once during each developmental stage (larva, nymph, and adult). Therefore, to serve as a vector, the tick must ingest virus at one stage, become infected, transmit the virus transstadially to the next stage, and then transmit the virus horizontally by bite to another vertebrate host when this stage feeds. A typical transmission cycle is illustrated in Fig. 12-2. Because most hard ticks require at least 2 years to complete their life cycle, infected ticks would provide a means for the virus to persist in nature over the winter season and provide a mechanism for reintroduction of virus the following spring. Thus, unlike most arthropods, ticks can be so long-lived that they may also serve as disease reservoirs.

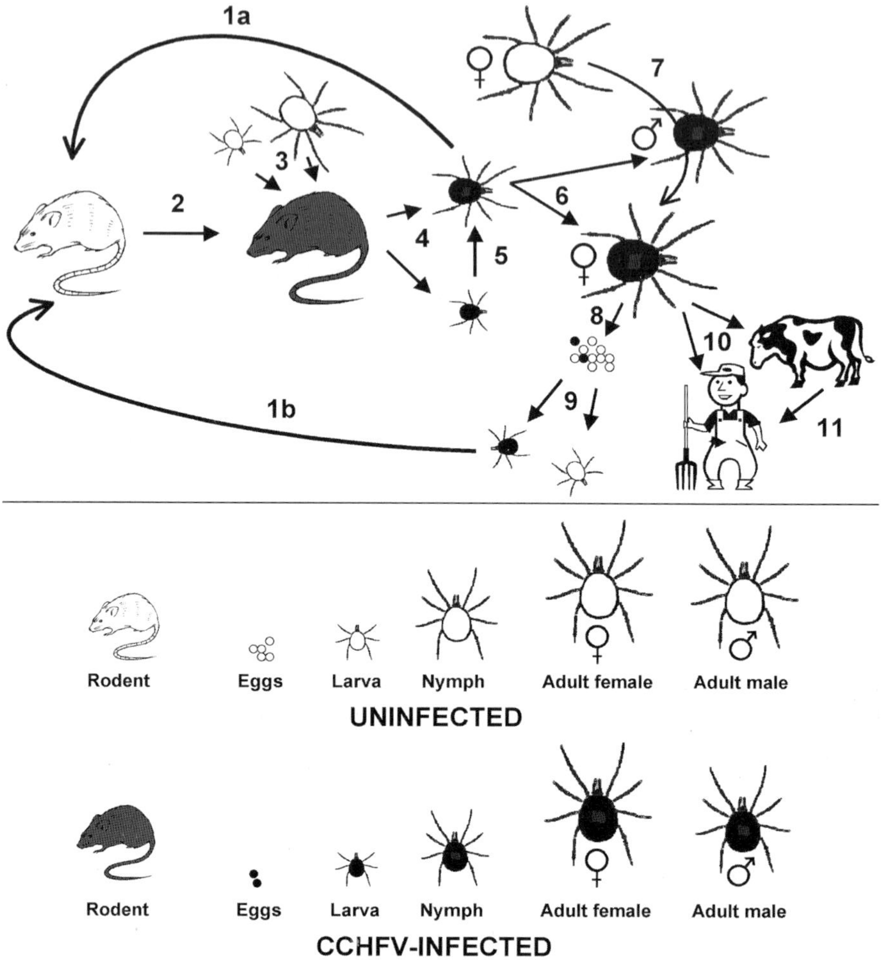

Fig. 12-2. Illustrated life cycle of CCHFV 1a: infected nymph (or 1b: infected larva) feeds on an uninfected rodent and initiation transmission cycle; 2: rodent becomes infected and infectious; 3: uninfected larvae and nymphs feed on the rodent; 4: larval and nymphal ticks become infected; 5: infected larvae transmit virus transstadially as they molt to nymphs; 6: infected nymphs transmit virus transstadially as they molt to adult females and males; 7: uninfected adult females become infected venereally by mating with infected males; 8: infected adult females transmit virus vertically to a portion of their eggs; 9: these eggs hatch to provide both infected and uninfected larvae; 10: infected nymphs and adults transmit virus horizontally to humans and other large mammals; 11: humans become infected by contact with fluids from infected large mammals.

12.4.1. Vertical transmission

In addition to transstadial transmission allowing virus to be maintained between different feeding stages, some infected female ticks may be able to transmit virus to their eggs. When these eggs hatch, the larvae are already infected and able to transmit virus by bite during their first blood meal. This is referred to as vertical transmission (i.e. transgenerational transmission to the F_1 progeny) and provides an additional means for long-term maintenance of virus in nature. Some authors refer to this as transovarial transmission, but the transgenerational transmission could be either via infected ovaries (transovarial) or the developed eggs could be infected directly (transovum). Therefore, it is better to use the more general term, vertical transmission, as it encompasses both means of transgenerational transmission. While vertical transmission was not documented by some investigators [27, 28, 38, 39], others have demonstrated that a portion of the progeny of infected female ticks is infected and able to transmit virus by bite (Table 12-1). Vertical transmission of CCHFV has been demonstrated for *H. marginatum* (reported as *H. plumbeum*), *Rhipicephalus rossicus*, and *Dermacentor marginatus* [21, 22, 50]; *H. truncatum* [15, 45]; *Hyalomma rufipes* (now *H. marginatum rufipes* [18]) [13, 26, 48]; and for *Rhipicephalus evertsi* and *Amblyomma variegatum* [13, 14]. In addition to the laboratory studies demonstrating vertical transmission, isolation of CCHFV from eggs obtained from field-collected *H. marginatum* provide evidence that vertical transmission occurs in nature [22].

12.4.2. Co-feeding studies

Ticks are able to become infected with an arbovirus when co-feeding with virus-infected ticks on the same vertebrate host, even if the vertebrate does not develop a detectable viremia [19, 23, 24]. For example, larval *H. truncatum* and *Hyalomma impeltatum* became infected with CCHFV when co-fed with infected adults of these species, despite a lack of a detectable viremia in the vertebrate host [16]. These infected larvae were able to transmit CCHFV transstadially to nymphs and adults, but transstadial transmission rates were very low. Similar studies by Gonzalez et al. [15] demonstrated transmission during co-feeding by adult *H. truncatum*.

12.4.3. Venereal transmission

In addition, CCHFV-infected male *H. truncatum* ticks transmitted virus venereally (during copulation) to female ticks, and these ticks then transmitted virus vertically to their progeny [15]. To rule out that this was due to co-feeding on the same vertebrate host, they showed that hypostomectomized male *H. truncatum* did not transmit CCHFV to co-feeding female *H. truncatum* when they used gonopore-closed female ticks (i.e. the mere presence of infected male ticks was not sufficient to infect female ticks). However, transmission did occur when intact infected males co-fed with gonopore-closed females (transmission during co-feeding, but not venereally) or when hypostomectomized males co-fed with intact females (venereal transmission, but not due to co-feeding). When allowed to oviposit, the venereally infected female *H. truncatum* vertically transmitted

CCHFV to their progeny. Vertical transmission, transmission during co-feeding on an animal without a viremia, and venereal transmission all allow for additional ticks to become infected without the need for a viremic animal and may play a role in the amplification and maintenance of CCHFV in nature.

Despite the importance of this disease, there have been only a limited number of laboratory vector competence studies with ticks and CCHFV. In part, this is due to the high-level containment (biological safety level-4) required for most strains of CCHFV. Additionally, it is a challenge to work with tick-borne viruses. Not only is it difficult to restrain ticks on a vertebrate host, compounded by working with a highly pathogenic virus such as CCHFV, but also the life cycle of ticks, often taking up to 2 years to complete a single generation, makes these studies difficult to complete in a timely fashion.

12.4.4. Persistence of CCHFV infection at reduced environmental temperature

The survival of ticks over a winter provides the opportunity for them to maintain virus in nature. However, during this period, the ticks may be exposed to reduced environmental temperatures. To examine the potential effect of these reduced temperatures on the persistence of CCHFV in ticks, Zgurskaya et al. [50] held *H. marginatum* ticks at reduced temperature (4°C) after allowing them to feed on a viremic animal. They detected CCHFV in these ticks for up to 700 days after the viremic meal, and these ticks transmitted CCHFV by bite after storage at 4°C for up to 10 months. This provides additional evidence that ticks serve as the long-term reservoir for CCHFV.

12.5. POTENTIAL ROLE OF BIRDS AND TICKS IN THE SPREAD OF CCHFV

Avian species are generally not regarded as amplifying hosts for CCHFV, and earlier studies indicated that they did not produce a viremia, even when inoculated with CCHFV [2]. However, it is believed that they may transport transovarially infected larval ticks or infected nymphal ticks along avian migration routes. Thus, avian transport of infected ticks may be important in the geographic spread of CCHFV. Zeller et al. [48] indicated that birds may play an even greater role than merely dispersing infected ticks. They reported that *H. marginatum rufipes* fed on red-beaked hornbills (*Tockus erythrorhynchus*) or a long-tailed glossy starling (*Lamprotornis caudatus*) inoculated with CCHFV became infected despite a lack of detectable viremia in the birds. When allowed to refeed, the ticks went on to transmit CCHFV to rabbits, and one group of F_1 progeny successfully transmitted virus by bite, indicating that these ticks were able to transmit CCHFV vertically. They also isolated CCHFV from *H. marginatum rufipes* nymphs collected on a hornbill [47]. Therefore, certain avian species may actually serve as amplification hosts for CCHFV in addition to dispersing virus-infected ticks.

12.6. SUMMARY

Various species of ixodid ticks, primarily in the genus *Hyalomma*, transmit CCHFV between vertebrate amplifying hosts and are the principal source of human infection. Ticks are able to become infected, transmit CCHFV transstadially, and then transmit this virus horizontally to another vertebrate host or vertically to their offspring. Therefore, ticks serve not only as the vector, but also as the long-term reservoir for this virus. The actual species involved in transmission depends on which species of ticks are present in a geographic area. Various mammals, and possibly some avian species, serve as amplifying hosts that can infect ticks feeding on them. Because of the paucity of data on vector competence for many tick species and the lack of information on the effect of environmental factors on CCHFV transmission, studies are needed to evaluate additional tick species as well as the various factors that might affect virus transmission.

Acknowledgments

I thank Glen R. Needham, Ohio State University, Columbus, OH; Robert S. Lane, University of California, Berkeley, CA; and Richard G. Robbins, Walter Reed Army Medical Center, Washington, DC, for their helpful comments.

REFERENCES

1. Ardoin P (1965) Congo group transmission experiments. Rep E Afr Virus Res Inst 14:52
2. Berezin VV, Chumakov PM, Reshetknikov IA, Zgurskaya GN (1971) Study of the role of birds in the ecology of Crimean hemorrhagic fever virus. Mater 6 Simp Izuch Virus Ekol Svyazan Ptits (Omsk, 1971):94–95 (in Russian) (in English, NAMRU3-T721)
3. Causey OR, Kemp GE, Madbouly MH, David-West TS (1970) Congo virus from domestic livestock, African hedgehog, and arthropods in Nigeria. Am J Trop Med Hyg 19:846–850
4. Camicas JL, Cornet JP, Gonzalez JP, Wilson ML, Adam F, Zeller HG (1994) Crimean-Congo hemorrhagic fever in Senegal. Latest data on the ecology of the CCHF virus. Bull Soc Pathol Exot 87:11–16 (in French)
5. Casals J (1978) Crimean-Congo hemorrhagic fever. In: Pattyn SR (ed.) Ebola virus hemorrhagic fever. Elsevier/North-Holland Biomedical Press, Amsterdam, The Netherlands, pp 301–317
6. Chumakov, MP (1947) A new viral disease – Crimean hemorrhagic fever. Nov Med 4:9–11 (in Russian) (in English, NAMRU3-T900)
7. Chumakov MP, Butenko AM, Rubin SG, Berezin VV, Karinskaya GA, Vasilenko SM, Smirnova SE, Bashkirtsev VV, Derbedeneva MP, Badalov ME, Stolbov DN (1972) Question on the ecology of Crimean hemorrhagic fever (CHF) virus. In: Cherepanov AI (ed) Transcontinental connections of migratory birds and their role in distribution of arboviruses. Mater 5 Simp Izuch Roli Pereletn Ptitsepererab Rasprostr Arbovirus, Novosibirsk, pp 222–229 (In Russian) (In English, NAMRU3-T877)
8. Chumakov, MP (1974) On 30 years of investigation of Crimean hemorrhagic fever. Tr Inst Polio Virusn Entsefalitov Akad Med Nauk SSSR 22:5–18 (in Russian) (in English, NAMRU3-T950)
9. Dickson DL, Turell MJ (1992) Replication and tissue tropisms of Crimean-Congo hemorrhagic fever virus in experimentally infected adult *Hyalomma truncatum* (Acari: Ixodidae). J Med Entomol 29:767–773

10. Dohm DJ, Logan TM, Linthicum KJ, Rossi CA, Turell MJ (1996) Transmission of Crimean-Congo hemorrhagic fever virus by *Hyalomma impeltatum* (Acari:Ixodidae) after experimental infection. J Med Entomol 33:848–851
11. Durden LA, Logan TM, Wilson ML, Linthicum KJ (1993) Experimental vector incompetence of a soft tick, *Ornithodoros sonrai* (Acari: Argasidae), for Crimean-Congo hemorrhagic fever virus. J Med Entomol 30:493–496
12. Ebel GD, Kramer LD (2004) Duration of tick attachment required for transmission of Powassan virus by deer ticks. Am J Trop Med Hyg 71:268–271
13. Faye O, Cornet JP, Camicas JL, Fontenille D, Gonzalez JP (1999) Experimental transmission of Crimean-Congo hemorrhagic fever virus: role of 3 vector species in the maintenance and transmission cycles in Senegal (Article in French). Parasite 6:27–32
14. Faye O, Fontenille D, Thonnon J, Gonzalez JP, Cornet JP, Camicas JL (1999) Experimental transmission of Crimean-Congo hemorrhagic fever virus by *Rhipicephalus evertsi evertsi* (Acarina:Ixodidae) (Article in French). Bull Soc Pathol Exot 92:143–147
15. Gonzalez JP, Camicas JL, Cornet JP, Faye O, Wilson ML (1992) Sexual and transovarian transmission of Crimean-Congo haemorrhagic fever virus in *Hyalomma truncatum* ticks. Res Virol 143:23–28
16. Gordon SW, Linthicum KJ, Moulton JR (1993) Transmission of Crimean-Congo hemorrhagic fever virus in two species of *Hyalomma* ticks from infected adults to cofeeding immature forms. Am J Trop Med Hyg 48:576–580
17. Hoogstraal H (1979) The epidemiology of tick-borne Crimean-Congo hemorrhagic fever in Asia, Europe, and Africa. J Med Entomol 15:307–417
18. Horak IG, Camicas J-L, Keirans JE (2002) The Argasidae, Ixodidae and Nuttalliellidae (Acari: Ixodida): a world list of valid tick names. Exp Appl Acarol 28:27–54
19. Jones LD, Davies CR, Steele GM, Nuttall PA (1987) A novel mode of arbovirus transmission involving a nonviremic host. Science 237:775–777
20. Katavolos P, Armstrong PM, Dawson JE, Telford SR III (1998) Duration of tick attachment required for transmission of granulocytic ehrlichiosis. J Infect Dis 177:1422–1425
21. Kondratenko VF (1976) Importance of ixodid ticks in transmitting and preserving the Crimean hemorrhagic fever agent in infection foci. Parazitologiya 10:297–302 (in Russian) (in English, NAMRU3-T1116)
22. Kondratenko VF, Blagoveshchenskaya NM, Butenko AM, Rubin SG, Berezin VV, Vishnivetskaya LK, Zarubina LV, Milyutin VN, Kuchin VV, Novikova EM, Rabinovich VD, Shevchenco SF, Chumakov MP (1970) Results of virological investigation of ixodid ticks in Crimean hemorrhagic fever focus in Rostov Oblast. In: Chumakov MP (ed.) Crimean Hemorrhagic Fever. Mater 3 Oblast Nauch Prakt Konf, Rostov-na-Donu, pp 29–35 (in Russian) (in English, NAMRU3-T524)
23. Labuda M, Jones LD, Williams T, Danielova V, Nuttall PA (1993) Efficient transmission of tick-borne encephalitis virus between cofeeding ticks. J Med Entomol 30:295–299
24. Labuda M, Kozuch O, Zuffova E, Eleckova E, Hails RS, Nuttall PA (1997) Tick-borne encephalitis virus transmission between ticks cofeeding on specific immune natural rodent hosts. Virology 235:138–143
25. Lee VH (1979) Isolation of viruses from field populations of *Culicoides* (Diptera: Ceratopogonidae) in Nigeria. J Med Entomol 16:76–79
26. Lee VH, Kemp GE (1970) Congo virus: experimental infection of *Hyalomma rufipes* and transmission to a calf. Bull Entomol Soc Nigereria 2:133–135
27. Logan TM, Linthicum KJ, Bailey CL, Watts DM, Moulton JR (1989) Experimental transmission of Crimean-Congo hemorrhagic fever virus by *Hyalomma truncatum* Koch. Am J Trop Med Hyg 40:207–212
28. Logan TM, Linthicum KJ, Bailey CL (1990) Replication of Crimean-Congo hemorrhagic fever virus in four species of ixodid ticks (Acari) infected experimentally. J Med Entomol 27:537–542
29. L'vov DN, Dzharkenov AF, Aristova VA, Kovtunov AI, Gromashevskii VL, Vyshemirskii OI, Galkina IV, Larichev VF, Butenko AM, L'vov DK (2002) The isolation of Dhori viruses

(Orthomyxoviridae, Thogotovirus) and Crimean-Congo hemorrhagic fever virus (Bunyaviridae, Nairovirus) from the hare (*Lepus europaeus*) and its ticks *Hyalomma marginatum* in the middle zone of the Volga delta, Astrakhan region, 2001 (Article in Russian). Vopr Virusol 47:32–36

30. Okorie TG (1991) Comparative studies on the vector capacity of the different stages of *Amblyomma variegatum* Fabricius and *Hyalomma rufipes* Koch for Congo virus, after intra-coelomic inoculation. Vet Parasitol 38:215–223
31. Onishchenko GG, Tumanova IIu, Vyshemirskii OI, Kuhn J, Seregin SV, Tiunnikov GI, Petrova ID, Tishkova F, Ospanov KS, Kazakov SV, Karimov SK, Esmagambetova AS, Netesov SV, Petrov VS (2005) ELISA and RT-PCR-based research of viruses in the ticks collected in the foci of Crimean-Congo fever in Kazakhstan and Tajikistan in 2001–2002 (Article in Russian). Vopr Virusol 50:23–26
32. Onishchenko GG, Tumanova IIu, Vyshemirskii OI, Kuhn J, Seregin SV, Tiunnikov GI, Petrova ID, Tishkova FKh, Ospanov KS, Kazakov SV, Karimov SK, Esmagambetova AS, Netesov SV, Petrov VS (2005) Study of virus contamination of *Ixodes* ticks in the foci of Crimean-Congo hemorrhagic fever in Kazakhstan and Tajikistan (Article in Russian). Zh Mikrobiol Epidemiol Immunobiol 1:27–31
33. Peavey CA, Lane RS (1995) Transmission of *Borrelia burgdorferi* by *Ixodes pacificus* nymphs and reservoir competence of deer mice (*Peromyscus maniculatus*) infected by tick-bite. J Parasitol 81:175–178
34. Piesman J, Mather TN, Sinsky RJ, Spielman A (1987) Duration of tick attachment and *Borrelia burgdorferi* transmission. J Clin Microbiol 25:557–558
35. Reeves WC (1957) Arthropods as vectors and reservoirs of animal pathogenic viruses. In: Hallauer C, Meyer KF (eds) Handbuch der Virusforschung, vol 4 (Suppl 3). Springer-Verlag, Vienna, pp 177–202
36. Saluzzo JF, Digoutte JP, Camicas JL, Chauvancy G (1985) Crimean-Congo haemorrhagic fever and Rift Valley fever in south-eastern Mauritania. Lancet 1:116
37. Shchelkanov MIu, Kolobukhina LV, Moskvina TM, Aushev ID, Kartoev AA, Kelli EI, Merkulova LN, Grenkova EP, Samokhvalov EI, Petriaev VG, Serobian AG, Klimova EA, Galkina IV, Malyshev NA, Aristova VA, Slavskii AA, Luk'ianova NA, Deriabin PG, Gromashevskii VL, Efremenko VI, Onishchenko GG, L'vov DK (2005) Detection of the circulation of Crimean-Congo hemorrhagic fever virus in the piedmont steppes of the North Caucasus (Article in Russian). Vopr Virusol 50:9–15
38. Shepherd AJ, Swanepoel R, Cornel AJ, Mathee O (1989) Experimental studies on the replication and transmission of Crimean-Congo hemorrhagic fever virus in some African tick species. Am J Trop Med Hyg 40:326–331
39. Shepherd AJ, Swanepoel R, Shepherd SP, Leman PA, Mathee O (1991) Viraemic transmission of Crimean-Congo haemorrhagic fever virus to ticks. Epidemiol Infect 106:373–382
40. Stewart RL, Burgdorfer W, Needham GR (1998) Evaluation of three commercial tick removal tools. Wilderness Environ Med 9:137–142
41. Swanepoel R, Struthers JK, Shepherd AJ, McGillivray GM, Nel MJ, Jupp PG (1983) Crimean-Congo hemorrhagic fever in South Africa. Am J Trop Med Hyg 32:1407–1415
42. Tsilinsky YaYa, Lebedev AD, Pak TP, Gromashevsky VL, Timofeev EM, Ershov FI, Tsirkin YuM, L'vov DK (1972) Isolation of Crimean hemorrhagic fever (CHF) virus from *Hyalomma plumbeum* ticks in Tadzhikistan. Mater Simp Itogi 6 Simp Izuch Virus Ekol Svyazan Ptits (Omsk, 1971):94–97 (in Russian) (in English, NAMRU3-T665)
43. Watts DM, Ksiazek TG, Linthicum KJ, Hoogstraal H (1988) Crimean-Congo Hemorrhagic fever. In: Monath TP (ed.) The Arboviruses: Epidemiology and Ecology, vol 2. CRC Press, Boca Raton, FL, pp 177–222
44. Williams RJ, Al-Busaidy S, Mehta FR, Maupin GO, Wagoner KD, Al-Awaidy S, Suleiman AJ, Khan AS, Peters CJ, Ksiazek TG (2000) Crimean-Congo haemorrhagic fever: a seroepidemiological and tick survey in the Sultanate of Oman. Trop Med Int Health 5:99–106

45. Wilson ML, Gonzalez JP, Cornet JP, Camicas JL (1991) Transmission of Crimean-Congo haemorrhagic fever virus from experimentally infected sheep to *Hyalomma truncatum* ticks. Res Virol 142:395–404
46. Yunker CE, Cory J (1975) Plaque production by arboviruses in Singh's *Aedes albopictus* cells. Appl Microbiol 29:81–89
47. Zeller HG, Cornet JP, Camicas JL (1994) Crimean-Congo haemorrhagic fever virus infection in birds: field investigations in Senegal. Res Virol 145:105–109
48. Zeller HG, Cornet JP, Camicas JL (1994) Experimental transmission of Crimean-Congo hemorrhagic fever virus by West African wild ground-feeding birds to *Hyalomma marginatum rufipes* ticks. Am J Trop Med Hyg 50:676–681
49. Zeller HG, Cornet JP, Diop A, Camicas JL (1997) Crimean-Congo hemorrhagic fever in ticks (Acari: Ixodidae) and ruminants: field observations of an epizootic in Bandia, Senegal (1989–1992). J Med Entomol 34:511–516
50. Zgurskaya GN, Berezin VV, Smirnova SE, Chumakov MP (1971) Investigation of the question of Crimean hemorrhagic fever virus transmission and interepidemic survival in the tick *Hyalomma plumbeum plumbeum* Panzer. In: Chumakov MP (ed.) Viral hemorrhagic fevers. Crimean hemorrhagic fever, Omsk hemorrhagic fever, and hemorrhagic fever with renal syndrome. Trudy Inst Polio Virusn Entsefalitov Akad Med Nauk SSSR 19:217–220 (in Russian) (in English, NAMRU3-T911)

CHAPTER 13

CRIMEAN-CONGO HEMORRHAGIC FEVER VIRUS INFECTION AMONG ANIMALS*

AYSEGUL NALCA, M.D., PH.D.[1] AND CHRIS A. WHITEHOUSE, PH.D.[2]

[1]*Center for Aerobiological Sciences, United States Army Institute of Infectious Diseases (USAMRIID), Fort Detrick, Frederick, MD 21702-5011, USA*

[2]*Diagnostic Systems Division, United States Army Institute of Infectious Diseases (USAMRIID), Fort Detrick, Frederick, MD 21702-5011, USA*

13.1. INTRODUCTION

Like other tick-borne zoonotic pathogens, Crimean-Congo hemorrhagic fever virus (CCHFV) generally circulates in nature unnoticed in an enzootic tick–vertebrate–tick cycle. CCHFV, similar to other zoonotic agents, appears to produce little or no disease in its natural hosts, but causes severe disease in humans. CCHFV infection has been detected in numerous domestic and wild vertebrates. Primarily these studies have been conducted by seroepidemiological surveys using a variety of techniques to detect antibodies to CCHFV. Such surveys of animal sera could be beneficial to disclose the presence of an otherwise unrecognized CCHFV source. In addition, they can be useful to determine the prevalence of infection and thus the risk of human exposure to infected tick bites. Furthermore, attempts have been made to experimentally infect some animals with the virus in the laboratory. This chapter will review the various vertebrate animals that have been infected with CCHFV, either naturally or experimentally, emphasizing what role, if any, they play in the ecology and zoogeography of the virus.

*The views, opinions, and findings contained herein are those of the authors and should not be construed as an official Department of the Army position, policy, or decision unless so designated by other documentation.

O. Ergonul and C. A. Whitehouse (eds.), Crimean-Congo Hemorrhagic Fever, 155–165.

13.2. INFECTIONS IN DOMESTIC ANIMALS

Results of seroepidemiological surveys in areas where CCHF is endemic (enzootic) reveals a number of domestic animals have been infected with the virus with high prevalence, particularly among cattle and sheep. For instance, in a survey of sera collected from 2,205 animals from three different faunal areas of Iraq, antibodies to CCHFV were detected in 443/769 (57.6%) sheep, 279/562 (49.6%) goats, 122/411 (29.3%) cattle, 148/252 (58.8%) horses, and 23/99 (23.2%) camels [32]. Likewise, in a 1975 study from neighboring Iran, 277/728 (38%) sheep, 49/135 (36%) goats, and 23/130 (18%) cattle were found to have antibodies to CCHFV [23]. Interestingly, a few years earlier Chumakov et al. [9] first suggested the presence of CCHFV in Iran when they detected antibodies to the virus in sera of 45 of 100 sheep sent to Moscow from a Tehran abattoir (for more information on CCHF in Iran, see Chapter 8). Similarly, antibodies against CCHFV were detected in livestock imported from Sudan to Saudi Arabia [17]. In another study, sera from domestic camels and imported ruminants in United Arab Emirates were tested for the presence of immunoglobulin G (IgG) antibody to CCHFV by enzyme-linked immunosorbent assay (ELISA) [20]. Serum samples were obtained and tested from 268 domestic animals (58 cattle, 74 sheep, 42 goats, and 94 camels) originating from eight countries (Somalia, Iran, Pakistan, United Arab Emirates, Sudan, Australia, India, and the Netherlands). Animals from Somalia, Iran, Pakistan, United Arab Emirates, and Sudan were positive; whereas, none of the animals imported from Australia, India, or the Netherlands were positive. In total, 19/268 (7%) animals, representing all for livestock types, were positive for CCHFV IgG antibodies.

Antibody against CCHFV has also been detected in sera of domestic animals over vast areas of South Africa and Zimbabwe [30]. Using the reversed passive hemagglutination-inhibition assay, antibody to CCHFV was found in 2,460/8,667 (28%) cattle sera (140 out of 180 herds) in South Africa. Likewise, in Zimbabwe, 347/763 (45%) cattle sera in 32 out of 34 herds was positive for CCHFV antibody. In general, antibody found in all major cattle farming areas, but was of low prevalence along the southern coast, where two of the species of *Hyalomma* tick which occur in South Africa are absent [30].

In addition to the presence of antibodies against CCHFV, virus isolation from ticks, domestic and wild vertebrates, or humans in different countries of Africa demonstrates the wide distribution of CCHFV on the continent [36].

The high infection rates among domestic animals suggest that these animals are an important part of the ecology of CCHF, if only to provide a source of blood meal to infected ticks. Domestic ruminants can develop demonstrable viremia and are capable of infecting ticks [29], but it is not known how significant a role they play in the ecology of disease. The role of wild and domestic animals as reservoirs of CCHFV depends on the level of viremia during infection, as only viremia above a certain threshold level will be sufficient to infect feeding ticks.

13.3. INFECTIONS IN WILD MAMMALS

Although CCHFV has been isolated from a few wild animal species, most field data are primarily restricted to serosurveys in areas where the disease is endemic. Seroepidemiological studies conducted in different endemic regions of Europe, Africa, and Asia have shown that large herbivores (the principal hosts of adult *Hyalomma* spp. ticks) exhibit the highest antibody prevalence.

Increased awareness of CCHF cases in southern Africa forced researchers to determine the role of wild mammals in the natural ecology of virus. In one study, a survey of sera was performed for stored sera from wild animals collected in South Africa and Zimbabwe between 1964 and 1985 [27]. Antibodies against CCHFV were detected in 100% ($n = 3$) of giraffe (*Giraffa cameleopardalis*), 54% ($n = 13$) of rhinoceros (*Ceratotherium simium* and *Diceros bicornis*), 46% ($n = 127$) of eland (*Taurotragus oryx*), 20% ($n = 287$) of buffalo (*Syncerus caffer*), 22% ($n = 78$) of kudu (*Tragelaphus strepsiceros*), and 17% ($n = 93$) of zebra (*Equus burchelli*). Similarly, sera from small mammals showed antibody to CCHFV in 14% ($n = 293$) of hares, 1.7% ($n = 1{,}305$) of rodents, and 1.4% ($n = 74$) of wild small carnivores. Interestingly, none of 522 primates tested were positive, but antibody was found in 6% ($n = 1{,}978$) of the domestic dogs tested.

In other study, sera of 29 species of wild vertebrates collected in the Kruger National Park, South Africa between 1974 and 1992 were tested for the presence of antibodies to CCHFV [6]. Antibodies to CCHFV were detected in some of the larger mammals such as buffalo (10%), white rhinoceroses (68%), and giraffe (23%). These results agree with the observation that CCHFV appears to occur most frequently in larger mammals, which are the preferred hosts of adult *Hyalomma* ticks [27]. On the other hand immature *Hyalomma* ticks feed primarily on small vertebrates such as hares, and contrary to previous findings, no antibody was detected in any of the 63 hares that were tested. Furthermore, detection of viremia in several small mammal species, such as hares in the republics of former USSR, hedgehogs in Nigeria [8], and a multimammate mouse in the Central African Republic [12], as well as antibodies against CCHFV in variety of wild animals [35] as seen in Table 13-1, demonstrates the presence of CCHFV infection in various wild animals. However, this does not necessarily mean that all of these species are natural hosts for the virus. Hares are a notable exception to this, as they have been shown to play a role as reservoir hosts for CCHFV in the former USSR and Bulgaria, as well as in southern Africa [18, 28].

13.4. EXPERIMENTAL INFECTIONS IN ANIMALS

Although over the years, several attempts have been made to establish an animal model for CCHF, currently, the only animal (other than humans) that manifests disease is the newborn mouse. Other laboratory animals, including nonhuman primates, show little or no sign of infection or disease when

Table 13-1. Summary of serological evidence of CCHFV infection in small wild vertebrates

Order	Vertebrate species	Location	% Antibody positive (*n*)	Reference
Insectivora				
	Long-eared hedgehog (*Hemiechinus auritius*)	Turkmenia (now Turkmenistan)	Positive	[18]
Lagomorpha				
	Cape hare (*Lepus capensis*)	South Africa	22.6 (62)	[27]
	Scrub hare (*Lepus saxatilis*)	South Africa	14.5 (131)	[27]
	Hare (*Lepus* spp.)	South Africa	14.3 (49)	[27]
	European hare (*Lepus europaeus*)	Bulgaria	Positive	[18]
		Russia	20 (20)	[18, 5]
Rodentia				
	Ground squirrel (*Xerus inauris*)	South Africa	2.7 (37)	[27]
	Springhare (*Pedetes capensis*)	South Africa	12.1 (33)	[27]
	Highveld gerbil (*Tatera brantsii*)	South Africa	2.2 (224)	[27]
	Bushveld gerbil (*Tater l eucogaster*)	South Africa	9.8 (61)	[27]
	Naked-soled gerbil (*Tatera indica*)	Pakistan	6 (157)	[11]
	Namaqua rock mouse (*Aethomys namaquensis*)	South Africa	1.1 (95)	[27]
	Striped mouse (*Rhabdomys pumilio*)	South Africa	0.6 (344)	[27]
	Porcupine (*Hystrix africaeaustralis*)	South Africa	12.5 (8)	[27]
	House mouse (*Mus musculus*)	Iran	Positive	[23]
	Black rat (*Rattus rattus*)	Egypt	4 (72)	[10]
	Field rat (*Arvicanthis niloticus*)	Egypt	5 (113)	[10]
		Mauritania	16 (43)	[24]
	Norway rat (*Rattus norvegicus*)	Egypt	7 (176)	[10]
	Multimammate rat (*Mastomys natalensis*)	Senegal	Positive	[18]
	Common field mouse (*Apodemus sylvaticus*)	Russia	Positive	[18]
	Long-clawed ground squirrel (*Spermophilopsis leptodactylus*)	Russia	Positive	[18]
	Williams' jerboa (*Allactaga euphrata williamsi*)	Iran	Positive	[23]
	Swinhoe's jird (*Meriones crassus swinhoei*)	Iran	Positive	[23]
Carnivora				
	Domestic dog (*Canis familiaris*)	South Africa	6.0 (1,978)	[27]
	Common red fox (*Vulpes vulpes*)	Turkmenia (now Turkmenistan)	Positive	[18]
	Pallas' cat (*Otocolobus manul*)	Turkmenia (now Turkmenistan)	Positive	[18]
	Genet (Genetta genetta senegalensis)	Senegal	Positive	[18]

infected with CCHFV [7, 14]. In one study using infant mice infected with CCHFV intraperitoneally, mice succumbed to disease around day 8 postinfection [33]. Serial sacrifice studies showed that virus replication was first detected in the liver, with subsequent spread to the blood (serum). Virus was detected very late after infection in other tissues including the heart (day 6) and the brain (day 7).

Horses and donkeys have been used for experimental CCHFV infections. Experimentally infected donkeys develop a low-level viremia [22]. Experimentally infected horses developed little or no viremia, but developed high levels of virus-neutralizing antibodies, which remained stable for at least 3 months. These studies suggest that horses may be useful in the laboratory to obtain serum for diagnostic and possible therapeutic purposes [4, 21].

Calves were another species of domestic animals used for experimental infections by several investigators. Causey et al. [8] observed mild illness, characterized by dullness, lassitude, and decreased appetite, when they infected two calves with varying amounts of CCHFV. Later, in 1976, Zarubinsky et al. [38] reported that the age of the calves at the time of infection was important for the outcome. Although there was no clinical signs, CCHFV was recovered from the blood of a 2-month-old calf on days 3 and 7 postinfection. Six-month-old calves did not develop viremia. Both calves had high titers of antibody against CCHFV [38]. The same researchers also experimentally infected lambs with CCHFV. Viremia was present in the blood of lambs up to day 8 of the 10-day postinfection period. These authors concluded that calves, as well as lambs, may participate in the circulation of CCHFV in nature [38]. More recently, Gonzalez et al. [15] experimentally infected 17 West African sheep either by intraperitoneal inoculation or by infestation with virus-infected ticks. A moderate, but constant fever, which correlated with viremia, was observed. Virus was reisolated from blood samples taken from day 3 to 9 postinfection. Interestingly, five of the infected sheep showed hepatic dysfunction with a moderate serum aspartate transferase increase, and two showed an abnormal blood cell count, with marked neutrophilia lasting for 2 weeks.

Shepherd et al. [28] infected 11 species of small African wild mammals and laboratory rabbits, guinea pigs, and Syrian hamsters with CCHFV. Animals were bled daily and viremia was measured by intracranial inoculation of heparinized whole blood samples into litters of 1-day-old mice. The mice were observed for 14 days and presence of CCHFV was evaluated on the brains. While CCHF viremia was detected in scrub hares (*Lepus saxatilis*), cape ground squirrels (*Xerus inauris*), red veld rats (*Aethomys chrysophilus*), white-tailed rats (*Mystromys pumilio*), and guinea pigs; South African hedgehogs (*Atelerix frontalis*), highveld gerbils (*Tatera brantsii*), Namaqua gerbils (*Desmodillus auricularis*), two species of multimammate mouse (*Mastomys natalensis* and *Mastomys coucha*) and Syrian hamsters were negative. All species regardless of viremia levels developed antibody response against CCHFV.

Table 13-2. Animal species experimentally infected with CCHFV

Animal			
Common name	Taxonomic name	Outcome	References
Mouse (newborn)	*Mus musculus*	Viremia, mortality	[33]
Horse	*Equus caballus*	Low-level viremia, high levels of neutralizing antibody	[4, 21]
Donkey	*Equus asinus*	Low-level viremia	[22]
Cow	*Bos taurus*	Mild illness, dullness, lassitude, decreased appetite	[8]
Cow (calf)	*B. taurus*	2-month-old: no clinical signs, viremia. 6-month-old: no clinical signs, no viremia. Both groups, high titers of antibody	[38]
Sheep (lamb)	*Ovis aries*	Viremia	[38]
Sheep	*O. aries*	Fever, viremia, hepatic dysfunction, abnormal blood cell count	[15]
Scrub hare	*Lepus saxatilis*	Low-level viremia, neutralizing antibodies	[28]
Cape ground squirrel	*Xerus inauris*	Low-level viremia, neutralizing antibodies	[28]
Red veld rat	*Aethomys chrysophilus*	Low-level viremia, neutralizing antibodies	[28]
White-tailed rat	*Mystromys albicaudatus*	Low-level viremia, neutralizing antibodies	[28]
Bushveld gerbil	*Tatera leucogaster*	Low-level viremia, neutralizing antibodies	[28]
Striped mouse	*Rhabdomys pumilio*	Low-level viremia, neutralizing antibodies	[28]
Guinea pig	*Cavia porcellus*	Low-level viremia, neutralizing antibodies, fever	[28]
South African hedgehog	*Atelerix frontalis*	No clinical signs, neutralizing antibodies	[28]
Highveld gerbil	*Tatera brantsii*	No clinical signs, neutralizing antibodies	[28]
Namaqua gerbil	*Desmodillus auricularis*	No clinical signs, neutralizing antibodies	[28]
Multimammate mouse	*Mastomys natalensis* and *Mcoucha coucha*	No clinical signs, neutralizing antibodies	[28]
Syrian hamster	*Mesocricetus auratus*	No clinical signs, neutralizing antibodies	[28]
Rabbit	*Oryctolagus cuniculi*	No clinical signs, neutralizing antibodies	[28]
African green monkey	*Cercopithecus aethiops*	No clinical signs, neutralizing antibodies	[7]
Patas monkey	*Erythrocebus patas*	Low-level viremia	[13]
Guinea baboon	*Papio papio*	Low-level viremia, neutralizing antibody	[13]

All guinea pigs had low-level viremia without clinical signs except elevated temperatures within the first 7 days after infection. Interestingly, there was a correlation with the route of infection and onset of viremia. Intravenous and intracranially infected animals showed onset of viremia earlier than those infected by the subcutenous or intraperitoneal routes. With the exception of a single rabbit, which had low-level viremia, none of the other rabbits or the Syrian hamsters had viremia or exhibited any clinical signs of disease.

Butenko et al. [7] used African green monkeys (*Cercopithecus aethiops*) for experimental CCHFV infections. The animals did not show signs of disease, except one monkey had a 40.3°C fever on day 4 postinfection [7]. Antibodies to the virus were detected in three out of five monkeys, including the one with fever. In 1975, Fagbami et al. [13] infected two Patas monkeys (*Cercopithecus (Erythrocebus) patas*) and one Guinea baboon (*Papio papio*) with CCHFV. While all three animals had low-level viremia between days 1 and 5 after inoculation, only the baboon serum had neutralizing antibody activity on day 137 postinfection.

Susceptibility to CCHFV is can often be species-dependent. For example, studies have shown that South African hedgehogs were not susceptible to CCHF, but nevertheless, developed neutralizing antibody against CCHFV. Furthermore, while CCHFV-infected long-eared hedgehogs can serve as a source of infectious blood meal for *Hyalomma marginatum marginatum*, European hedgehogs are apparently refractory to infection [5, 40]. Table 13-2 shows data on the various animals that have been experimentally infected with CCHFV.

13.5. INFECTIONS IN BIRDS

Although many domestic and wild vertebrates are infected with CCHFV, as mentioned above, birds, in general, appear to be refractory to infection with CCHFV. For instance, early experiments by Berezin et al. [2, 3] showed that after experimental inoculation of birds (rooks and rock doves) with CCHFV, they remained healthy, and evidence of viremia or an antibody response could not be demonstrated. Furthermore, work by the same group [1], showed that even though CCHFV could be isolated from nymphal ticks collected from over 600 birds, the birds remained serologically negative for antibody to CCHFV. Attempts to isolate virus from the blood and organs of 360 of those birds were uniformly negative. Several additional examples are known from the 1970s in which CCHFV has been isolated from ticks infecting numerous species of birds, which remain serologically negative for the virus [18]. Taken together, these studies suggest that birds appear to be refractory to CCHF viremia even though they can support large numbers of CCHFV-infected ticks. However some exceptions do exist; Semashko et al [25] detected antibodies to CCHFV in 1 of 428 sera tested from chickens and ducks in Kazakhstan and Zarubinsky et al. [37] also detected CCHFV antibodies in the serum of a magpie (*Pica pica*). However, in more recent pathogenicity studies, domestic chickens proved to be refractory to CCHFV infection [26]. Also, in the same study, low CCHF viremia was detected in a blue-helmeted guinea fowl (*Numidia meleagris*) after experimental infection with CCHFV. In another more recent study, antibodies were detected after CCHFV inoculation in a red-beaked hornbill and a glossy starling (but not in two laughing doves or six domestic chickens); however, none of the birds showed detectable viremia [39].

Another interesting exception has been the disease's apparent association with the commercial ostrich meat industry in South Africa. In 1984, a case of CCHF occurred in a worker who became ill after slaughtering ostriches (*Struthio camelus*) on a farm in South Africa [34]. Antibody to CCHFV was detected in 24% of ostriches from surrounding farms, including six of nine ostriches from the farm where the patient worked. Interestingly, none of 460 birds of 37 other species tested during that study had detectable antibodies to CCHFV [26]. Also, in 1996, there was an outbreak of 17 cases of CCHF among workers at an ostrich abattoir [31]. In both instances it was suspected that infection was acquired either by contact with ostrich blood or by inadvertently crushing infected ticks while skinning the ostriches. Ostriches have also been experimentally infected with CCHFV [31]. Ostriches, raised under tick-free conditions, were infected with CCHFV subcutaneously then bled daily for viral titers and antibody production. Unlike, most other birds, the ostriches developed viremia 1–4 days postinfection, and virus was detectable in the visceral organs up to 5 days postinfection. Although the ostriches did not show signs of illness, they developed antibodies against CCHFV beginning on day 5 postinfection and by day 13 all experimentally infected ostriches had antibodies. It was concluded from these studies that infection in ostriches at abattoirs could be prevented by keeping the birds free of ticks for a certain period of time before slaughter. This led to the standard 14-day tick-free preslaughter quarantine period currently enforced in South African ostrich export facilities.

Clearly, ground-feeding birds may play an important role in the ecology and epizoology of CCHF by transporting virus-infected ticks (even though the birds themselves may remain nonviremic). In addition, they may play a role in the ecology of disease by being involved in other modes of virus transmission, such as nonviremic transmission or cofeeding [16, 19]. The role, if any, for the birds themselves are not clear and additional work needs to be done in this area to resolve these issues.

In summary, vertebrates are essential as a source of blood for vector ticks and the number of species of vertebrates implicated in the natural history of CCHF is extensive, the exact role, if any, of vertebrates in the maintenance and transmission of the virus remains to be determined.

REFERENCES

1. Berezin VV, Chumakov MP, Rubin SG, Stolbov DN, Butenko AM, Bashkirtsev VA (1969) Contribution to the ecology of Crimean Hemorrhagic fever virus in the lower Volga River. Mater 16 Nauchn Sess Inst Polio Virusn Entsefalitov (Moscow, October 1969) 2:120 (in Russian; in English, NAMRU3-T912)

2. Berezin VV, Chumakov MP, Reshetnikov IA, Zgurskaya GN (1971a) Study of the role of birds in the ecology of Crimean hemorrhagic fever virus. Mater 6 Simp Izuch Virus Ekol Svyazan Ptits (Omsk, 1971):94–95 (in Russian; in English, NAMRU3-T721)
3. Berezin VV, Chumakov MP, Stolbov, DN, Butenko AM (1971b) On the problem of natural hosts of Crimean hemorrhagic fever virus in Astrakhan Region. Tr Inst Polio Virusn Entsefalitov Akad Med Nauk SSSR 19:210–216 (in Russian; in English, NAMRU3-T912)
4. Blagoveshchenskaya NM, Butenko AM, Vyshnivetskaya LK, Zavodova, TI, Zarubina LV, Karinskaya, GA, Kuchin VV, Milyutin VN, Novikova EM, Rubin, SG, Chumakov MP (1969) Experimental infection of horses with Crimean hemorrhagic fever virus. Report 2. Virological and serological observations. Mater 16 Nauchn Sess Inst Polio Virusn Entsefalitov (Moscow, October 1969) 2:126–127 (in Russian; in English, NAMRU3-T840)
5. Blagoveshchenskaya NM, Donets MA, Zarubina LV, Kondratenko VF, Kuchin VV (1975) Study of susceptibility to Crimean hemorrhagic fever (CHF) virus in European and long-eared hedgehogs. Tezisy Konf Vop med Virus:269–270
6. Burt FJ, Swanepoel R, Braack LE (1993) Enzyme-linked immunosorbent assays for the detection of antibody to Crimean-Congo haemorrhagic fever virus in the sera of livestock and wild vertebrates. Epidemiol Infect 111:547–557
7. Butenko AM, Chumakov MP, Smirnova SE, Vasilenko SM, Zavodova TI, Tkachenko EA, Zarubina LV, Bashkirtsev VN, Zgurskaya GN, Vyshnivetskaya LK (1970) Isolation of Crimean hemorrhagic fever virus from blood of patients and corpse material (from 1968–1969 investigation data) in Rostov, Astrakhan Oblast, and Bulgaria. Mater 3 Oblast Nauchn Prakt Konf (Rostov-on-Don, May 1970):6–25. (in Russian; in English, NAMRU3-T522)
8. Causey OR, Kemp GE, Madbouly MH, David-West TS (1970) Congo virus from domestic livestock, African hedgehogs, and arthropods in Nigeria. Am J Trop Med Hyg 19:846–850
9. Chumakov, MP, Ismailova, ST, Rubin, SG, Smirnova, SE, Zgurskaya, GN, Khankishiev, Ash, Berezin, VV, Solovei, AE (1970) Detection of Crimean hemorrhagic fever foci in Azerbaijan SSR from results from serological investigations of domestic animals. Tr Inst Polio Virusn Entsefalitov Akad Med Nauk SSSR 18:120–122 (in Russian; in English, NAMRU3-T941)
10. Darwish MA, Imam IZ, Omar FM, Hoogstraal H. (1978) Results of a preliminary seroepidemiological survey for Crimean-Congo hemorrhagic fever virus in Egypt. Acta Virol 22:77
11. Darwish MA, Hoogstraal H, Roberts TJ, Ghazi R, Amer T (1983) A sero-epidemiological survey for Bunyaviridae and certain other arboviruses in Pakistan. Trans R Soc Trop Med Hyg 77:446–450
12. Digoutte JP, Heme G (1985) Rapport annuel, 1984. Rapport Institut Pasteur, Dakar
13. Fagbami AH, Tomori O, Fabiyi A, Isoun TT (1975) Experimental Congo virus (IB-AN 7620) infection in primates. Rev Roum Med Virol 26:33–37
14. Gear GHS (1988) Crimean-Congo hemorrhagic fever. In: Gear GHS (ed.) CRC Handbook of Viral and Rickettsial Hemorrhagic Fevers. CRC Press, Boca Raton, FL, pp 121–129
15. Gonzalez JP, Camicas JL, Cornet JP, Wilson ML (1988) Biological and clinical responses of West African sheep to Crimean-Congo haemorrhagic fever virus experimental infection. Res Virol 149:445–455
16. Gordon SW, Linthicum KJ, Moulton JR (1993) Transmission of Crimean-Congo hemorrhagic fever virus in two species of *Hyalomma* ticks from infected adults to cofeeding immature forms. Am J Trop Med Hyg 48:576–580
17. Hassanein KM, El-Azazy OM, Yousef HM (1997) Detection of Crimean-Congo haemorrhagic fever virus antibodies in humans and imported livestock in Saudia Arabia. Trans R Soc Trop Med Hyg 91:356–537
18. Hoogstraal H (1979) The epidemiology of tick-borne Crimean-Congo hemorrhagic fever in Asia, Europe, and Africa. J Med Entomol 15:307–417
19. Jones LD, Davies CR, Steele GM, Nuttall PA (1987) A novel mode of arbovirus transmission involving a nonviremic host. Science 237:775–777
20. Khan AS, Maupin GO, Rollin PE, Noor AM, Shurie HH, Shalabi AG, Wasef S, Haddad YM, Sadek R, Ijaz K, Peters CJ, Ksiazek TG (1997) An outbreak of Crimean-Congo hemorrhagic fever in the United Arab Emirates, 1994–1995. Am J Trop Med Hyg 57:519–525

21. Milyutin VN, Butenko AM, Artyushenko AA, Bliznichenko AG, Zavodova TI, Zarubina LV, Novikova, EM Rubin SG, Cherynshev NI, Chumakov MP (1969) Experimental infection of horses with Crimean hemorrhagic fever virus. Report I. Clinical observations. Mater 16 Nauchn Sess Inst Polio Virusn Entsefalitov (Moscow) 2:145–146 (in Russian; in English, NAMRU3-T851)
22. Rabinovich VD, Milyutin VN, Artyushenko AA, Buryakov BG, Chumakov MP (1972) Possibility of extracting hyperimmune gammaglobulin against CHF from donkey blood sera. Tezisy 17 Nauchn Sess Inst Posvyashch Aktual Probl Virus Profilakt Virus Zabolev (Moscow, October 1972):350–351 (in Russian; in English, NAMRU3-T1177)
23. Saidi S, Casals J, Faghih, MA (1975) Crimean hemorrhagic fever-Congo (CHF-C) virus antibodies in man, and in domestic and small mammals, in Iran. Am J Trop Med Hyg 24:353–357
24. Saluzzo JF, Digoutte JP, Camicas JL, Chauvancy G (1985) Crimean-Congo haemorrhagic fever and Rift Valley fever in south-eastern Mauritania. Lancet 1:116
25. Semashko IV, Dobritsa PG, Bashkirtsev VN, Chumakov MP (1975) Results from investigating blood sera from healthy persons, animals, and birds collected in southern Kazakhstan for antibodies to CHF-Congo virus. Mater 9 Simp Ekol Virus (Dushanbe, October 1975):43–44 (in Russian; in English, NAMRU3-T1128)
26. Shepherd AJ, Swanepoel R, Leman PA, Shepherd SP (1987) Field and laboratory investigation of Crimean-Congo haemorrhagic fever virus (*Nairovirus*, family Bunyaviridae) infection in birds. Trans R Soc Trop Med Hyg 81:1004–1007
27. Shepherd AJ, Swanepoel R, Shepherd SP, McGillivray GM, Searle LA (1987) Antibody to Crimean-Congo hemorrhagic fever virus in wild mammals from southern Africa. Am J Trop Med Hyg 36:133–142
28. Shepherd AJ, Leman, PA, Swanepoel R (1989) Viremia and antibody response of small African and laboratory animals to Crimean-Congo hemorrhagic fever virus infection. Am J Trop Med Hyg 40:541–547
29. Shepherd AJ, Swanepoel R, Cornel AJ, Mathee O (1989) Experimental studies on the replication and transmission of Crimean-Congo hemorrhagic fever virus in some African tick species. Am J Trop Med Hyg 40:326–331
30. Swanepoel R, Shepherd AJ, Leman PA, Shepherd SP, McGillivray GM, Erasmus MJ, Searle LA, Gill DE (1987) Epidemiologic and clinical features of Crimean-Congo hemorrhagic fever in southern Africa. Am J Trop Med Hyg 36:120–132
31. Swanepoel R, Leman PA, Burt FJ, Jardine J, Verwoerd DJ, Capua I, Bruckner GK, Burger WP (1998) Experimental infection of ostriches with Crimean-Congo haemorrhagic fever virus. Epidemiol Infect 121:427–432
32. Tantawi HH, Shony MO, Al-Tikriti SK (1981) Antibodies to Crimean-Congo haemorrhagic fever virus in domestic animals in Iraq: a seroepidemiological survey. Int J Zoonoses 8:115–120
33. Tignor GH, Hanham, CA (1993) Ribavirin efficacy in an in vivo model of Crimean-Congo hemorrhagic fever virus (CCHF) infection. Antiviral Res 22:309–325
34. Van Eeden PJ, Joubert JR, van de Wal BW, King JB, de Kock A, Groenewald JH (1985) A nosocomial outbreak of Crimean-Congo haemorrhagic fever at Tygerberg Hospital. Part I. Clinical features. S Afr Med J 68:711–717
35. Watts DM, Ksiazek TG, Linthicum KJ, Hoogstraal H (1989) Crimean-Congo hemorrhagic fever. In: Monath TP (ed.) The Arboviruses: Epidemiology and Ecology, vol II. CRC Press, Boca Raton, FL, pp 177–222
36. Wilson ML, LeGuenno B, Guillaud M, Desoutter D, Gonzalez JP, Camicas JL (1990) Distribution of Crimean-Congo hemorrhagic fever viral antibody in Senegal: environmental and vectorial correlates. Am J Trop Med Hyg 43:557–566
37. Zarubinsky VYa, Klisenko GA, Kuchin VV, Timchenko VV, Shanoyan NK (1975) Application of the indirect hemagglutination inhibition test for serological investigation of Crimean Hemorrhagic fever focus in Rostov Oblast. Sb Tr Inst Virusn Imeni DI Ivanovsky Akad Med Nauk SSSR 2:73–77 (in Russian; in English, NAMRU3-T1145)

38. Zarubinsky VYa, Kondratenko VF, Blagoveshchenskaya NM, Zarubina LV, Kuchin VV (1976) Susceptibility of calves and lambs to Crimean hemorrhagic fever virus. Tezisy Dokl 9 Vses Konf Prirod Ochag Bolez Chelov Zhivot (Omsk, May 1976):130–131 (in Russian; in English, NAMRU3-T1178)
39. Zeller HG, Cornet JP, Camicas JL (1994) Experimental transmission of Crimean-Congo hemorrhagic fever virus by West African wild ground-feeding birds to *Hyalomma marginatum rufipes* ticks. Am J Trop Med Hyg 50:676–681
40. Zgurskaya, BN, Berezin, SE, Smirnova, SE, Chumakov, MP (1971) Investigation of the question of Crimean hemorrhagic fever virus transmission and interepidemic survival in the tick Hyalomma plumbeum plumbeum. Panzer Tr Inst Polio Virusn Entsefalitov Akad Med Nauk SSSR 19:217–220

CHAPTER 14

ECOLOGY OF TICK-BORNE DISEASE AND THE ROLE OF CLIMATE

SARAH E. RANDOLPH, PH.D. AND DAVID J. ROGERS, PH.D.
Department of Zoology, University of Oxford, South Parks Road, Oxford OX1 3PS, UK.
E-mail: sarah.randolph@zoo.ox.ac.uk

14.1. INTRODUCTION: ECOLOGY UNDERPINS EPIDEMIOLOGY

Crimean-Congo hemorrhagic Fever (CCHF) is unusual among vector-borne zoonoses in that, as well as the virus being transmitted to humans if they accidentally intrude on the natural transmission cycle by being bitten by ticks, the virus is also commonly transmitted directly to humans from its natural wildlife and livestock hosts (and even human patients) via contact or contamination with infected tissue or blood. The epidemiology of human disease therefore follows a pattern that is greatly influenced by employment practices that bring humans into close contact with livestock, whether alive or, even more riskily, dead. In tanneries, for example, ticks that detach from hides may reattach to humans, potentially transferring the infection much more rapidly than via the conventional transmission route that involves a long delay between the engorged infected tick of one stage and the feeding tick of the next stage. Nevertheless, like all other vector-borne diseases, the presence and persistence of zoonotic foci of infection depend on biological and ecological relationships between three very different kinds of organisms: virus, ticks, and vertebrates. These three must interact not only physically and biologically to permit each complete act of transmission, but also ecologically to permit continuing cycles of transmission. Although the focus of epidemiological interest is on the pattern of human cases, explanations for the described distribution and abundance of infections must come from understanding the underlying ecological processes.

O. Ergonul and C. A. Whitehouse (eds.), Crimean-Congo Hemorrhagic Fever, 167–186.

14.2. ECOLOGICAL HURDLES TO PERSISTENT TRANSMISSION CYCLES OF TICK-BORNE DISEASES

14.2.1. Biotic specificity

14.2.1.1. Vector competence

The first task is to identify the key players. In the CCHF disease system, the main unresolved question concerns the identity of any significant vectors in addition to *Hyalomma* spp., particularly *Hyalomma marginatum, Hyalomma truncatum*, and *Hyalomma anatolicum*, the three most widely recognized vector tick species [3, 15, 56] (See also Chapter 12). It is not sufficient to base conclusions on the coincidence in distributions of CCHF and certain tick genera or species, because other ticks may be competent but not yet "active" vectors, ready to play a more significant role if abiotic or biotic environmental conditions change. We need to identify the factors that limit the competence of each potential vector, and assess their flexibility relative to likely environmental changes.

In addition to the genus *Hyalomma*, ticks of other genera (*Dermacentor*, *Rhipicephalus*, *Ixodes*, *Amblyomma*, *Boophilus*) are biologically competent in the laboratory and can even become infected under natural conditions [3]. The prevalence of virus in unfed ticks questing for hosts reveals whether or not these additional species can play a significant role in nature. Such ticks can only have become infected from an infected host during feeding by previous tick stages as long as they maintain sufficient viral load during interstadial development and molting. They must then be able to develop mature infections in the salivary glands for onward transmission to the next host. Each of these steps depends on the ability of a microbe to overcome the many intrinsic biological (molecular, cellular, physiological, and physical) barriers during its passage from host to host via the vector (Fig. 14-1). Even then, although biologically possible, this cycle may not proceed with sufficient force to support persistent transmission cycles. That depends on the quantitative balance of the rates of all the processes involved in each complete transmission cycle. The majority of these rates are affected by extrinsic environmental factors, both biotic and abiotic (Fig. 14-1), although host-acquired immunity also plays a significant role. Wherever the basic reproductive number, R_0 (defined as the number of new infections that arise over each transmission cycle from a single index case introduced into a wholly susceptible population [1]) falls below the absolute threshold value of 1, the infection is not sustainable. This may be due to a shortfall in any one of the followingfactors: (1) tick abundance; (2) tick contact rates with competent hosts; (3) tick survival rates; (4) tick interstadial development rates; and (5) transmission coefficients from host to tick, interstadially between ticks, or from tick to host.

This concept of ecological specificity, highlighting the distinction between biological and ecological competence, is well illustrated by tick-borne encephalitis (TBE) virus (TBEV). As with CCHFV, many tick species have been shown to be competent vectors in the laboratory, but in the field only *Ixodes ricinus* seems

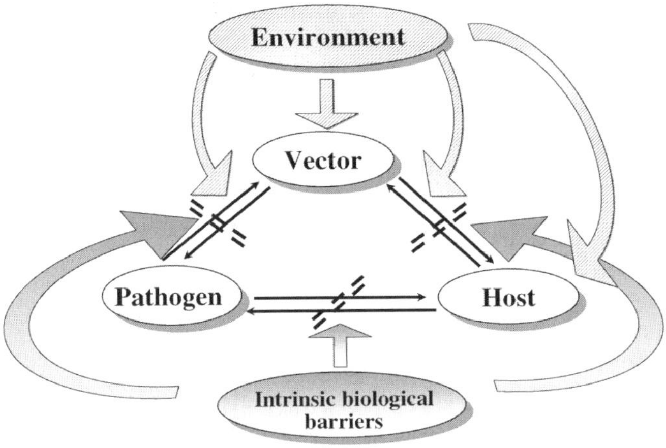

Fig. 14-1. The triangle of host–vector–pathogen interactions, showing the points of action of the intrinsic biological barriers to transmission and the extrinsic environmental factors.

to play a significant role in Europe [22]. Within TBE foci in Slovakia, a second biologically competent tick species, *Dermacentor reticulatus*, feeds on rodents alongside *I. ricinus*, but *D. reticulatus* develops very rapidly so that larvae molt and give rise to feeding nymphs within a month. Each immature stage therefore feeds sequentially within the summer on the competent rodent hosts, with very little overlap and therefore little co-feeding by larvae and nymphs on the same host [37]. Co-feeding is a necessary condition for the transmission of virus from infected nymphs to infectible larvae for any tick-borne virus that remains infective within the host for only a few days (as is true of CCHFV) [32]. Our understanding of the ecology of TBEV transmission tells us that *D. reticulatus* is unlikely to assume a significant role as vector of this virus under changing environmental conditions, because the timing of its developmental cycle is entirely wrong. Furthermore, in this region at least, *D. reticulatus* feeds to a significantly lesser extent on mice, *Apodemus flavicollis*, than on voles, *Clethrionomys glareolus*; the former rodents are quantitatively superior as transmission hosts because they support nonviremic transmission better [21], and also do not acquire immunological resistance to feeding ticks [5].

Not enough is known quantitatively about the necessary biotic conditions for the maintenance of CCHFV cycles in nature to allow a firm conclusion on the relative roles, potential, or realized, of the various biologically competent tick species. This is a significant gap, limiting our ability to assess the potential for epidemiological change in the future (see below).

14.2.1.2. Vertebrate amplification hosts

The edge of the distribution of any tick-borne disease system (i.e. where $R_0 < 1$) is defined by the product of all the factors within the R_0 equation [32]. The closer the value is to 1, the more fragile are the transmission cycles and the greater the likelihood that the distribution will be focal, occurring only in well-defined places where all the conditions are satisfied quantitatively. Conversely, robust systems will be more widespread. Many of the most widespread tick-borne pathogens utilize a wide range of ubiquitous hosts; *Borrelia burgdorferi* sensu lato is an obvious example, whose long period of infectivity within its hosts makes this system even more robust. In the case of CCHF virus, both wild and domesticated ruminants appear to be commonly infected hosts, with a variety of small mammals (e.g. susliks, hares, hedgehogs, rodents) [3, 15, 50, 52, 56] and possibly some birds (e.g. ostriches and guinea fowl) [3, 51, 60] also warranting full experimental investigation of their exact transmission potential.

Just as with ticks, however, infection or even transmission competence per se are not necessarily sufficient for persistent cycles if the pattern of attachment by different tick stages does not allow sufficient amplification. In order to overcome the inevitable high mortality between each tick stage, transmission must occur "backwards" through the tick's life cycle, from one infected nymph to many larvae, or from one infected adult to many larvae and/or nymphs. Only if transovarial transmission is efficient, from one infected female to a large proportion of her progeny, does pathogen amplification run in the same direction as the tick's life cycle. For this reason, for CCHFV, for which transovarial transmission does occur but is not very efficient [59], at least two tick stages must feed on any one host. Immature stages of *Hyalomma* species, or indeed most other tick species from which CCHFV has been isolated, are much more commonly recorded on small mammals and birds than on ruminants [41, 54, 57]. In a national park in Cape Province, South Africa, there was a complete separation between immature stages of *H. marginatum turanicum* and *H. truncatum* on hares (the most highly infested host species), ground-feeding birds or small rodents, and adult ticks on zebra and eland [16]. This pattern of tick biology raises the distinct possibility that, while ruminants may be vital in feeding adult ticks and therefore supporting tick populations, and domestic livestock infected by these adult ticks may be instrumental in bringing CCHFV to humans, smaller vertebrates may be the principle maintenance hosts. This "division of labor" among the vertebrate components of tick-borne disease systems is not uncommon and has been well documented for Lyme borreliosis, TBE and Louping ill [11, 19, 22, 34], all transmitted by *I. ricinus*, a tick with an exceptionally catholic host range. An additional complication for CCHF, unquantified in nature, is the variability shown by *H. truncatum* and *H. marginatum rufipes* between a two-host feeding pattern (i.e. both larvae and nymphs feeding from the same host and adults from a second) and three-host feeding pattern (each stage feeding on a different host), apparently depending on the species of larval hosts [23, 41]. For two-hosts ticks showing distinct host relations as immature stages and adults (above), the sustainability of natural virus transmission cycles depends on the efficiency of complete vertical transmission, including transovarial transmission, and the degree of amplification amongst immature stages feeding together on the same host individual. This latter factor is augmented by the commonly observed aggregated distribution of ticks amongst their hosts, with the few hosts that carry many ticks making an essential, disproportionately large, contribution [37]. All these factors need to be quantified before natural transmission cycles can be fully understood.

14.2.2. Abiotic specificity

Because of the ubiquity of the putative vertebrate hosts of CCHFV, and their common infestation by ticks of many genera including *Hyalomma*, host availability is unlikely to be a limiting ecological factor in the distribution, or even the prevalence, of this virus. Moreover, a nonviremic route has been indicated by reports of transmission among co-feeding ticks via hosts that do not develop viremia above a notional threshold level [60]. This may add both qualitatively (more species than conventionally recognized acting as hosts) and quantitatively to the transmission potential [35].

Rather, abiotic factors are more likely to determine patterns of epidemiological risk, because they affect the rates of many of the tick population processes critical to the dynamics of tick-borne disease systems. The most obvious outcomes that vary with climatic conditions are tick abundance and patterns of seasonal tick population dynamics; these are the product of interstadial development rates and the timing of questing activity. Microclimate may also have a direct impact on tick behavior, in particular the height on the vegetation at which they quest for hosts, affecting tick–host contact rates and therefore pathogen transmission potential [40]. Adult ticks, which typically quest at higher levels in the vegetation than do immature stages, will be unavailable to smaller hosts species moving about at ground level. At the same time, host dynamics and behavior vary independently, also determining contacts rates. Any seasonal variation in the differential availability of each host species results in different tick–host ratios in different places. Smaller hosts show stronger seasonal cycles of abundance wherever the climate dictates seasonal breeding, but even among larger hosts exposure to ticks may vary with, for example, livestock husbandry practices or changes in behavior (e.g. squirrels shifting from ground- to tree-based foraging in the summer [4]).

It is one thing to argue in principal for the likely importance of abiotic limiting factors, and quite another to identify those factors precisely and define the necessary conditions for CCHF presence. Yet without such knowledge we can neither explain the present situation nor predict any future changes.

14.3. METHODS FOR IDENTIFYING THE LIMITING CONDITIONS FOR CCHF PRESENCE

14.3.1. The biological approach

The fuller and more quantitative the description of any system, the better our understanding of the causes of the observed patterns and the more reliable our predictions of possible change will be. Patterns are simply the product of the underlying processes. Even in the case of complex vector-borne disease systems these include the same small handful of principal events, namely vector birth, death, and feeding, pathogen developmental cycles within the vector, and vertebrate host recovery from infection. The myriad of patterns, different for each system and varying in both time and space for any single system, is simply the outcome of interactions of the variable rates of these processes. These rates are typically driven by environmental factors. If we could quantify the relationship between each process rate and its environmental drivers, we could build full biological, process-driven models (Fig. 14-2). Unfortunately this exceeds our current knowledge; there is no single fully functional biological model for any vector-borne disease system.

The most variable term in such a model is vector abundance, which commonly varies by up to several orders of magnitude seasonally, annually, or geographically. Yet our ability to model vector population dynamics is the biggest single gap in our biological toolbox. For ticks, this exercise is particularly demanding as the changing abundance of all three tick stages must be predicted simultaneously, because at least two stages are required for pathogen transmission (see above). Indeed, for *Hyalomma* ticks, it seems that we do not even have good descriptions of vector abundance, as these ticks are typically counted on cattle, upon which typically only the adults feed [56], leaving the all-important immature stages out of the equation (but see the exemplary work of Ivan Horak as an exception to this, e.g. [16]). While the framework necessary to direct new empirical measures of input data is relatively simple [33], and a variety of tick population models has been proposed [26, 38, 49], there has been limited success in using these models to predict tick abundance at locations independent of the input data [33, 38].

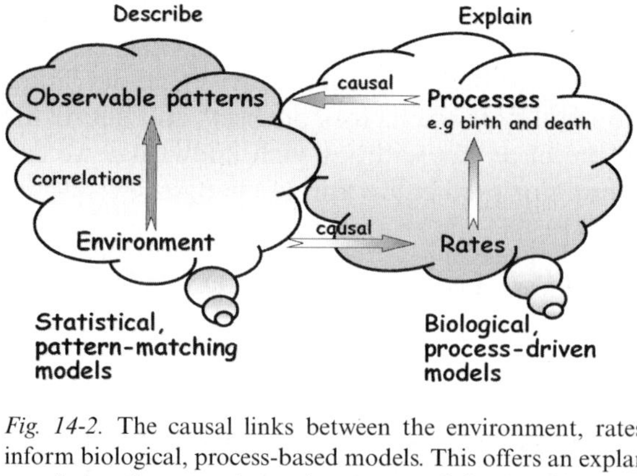

Fig. 14-2. The causal links between the environment, rates of processes and resultant patterns inform biological, process-based models. This offers an explanation of the system. Alternatively, the correlations between the environment and the described patterns allow the creation of statistical, pattern-matching models.

14.3.2. The statistical approach

14.3.2.1. Basic principles

Nil desperandum: because of the causal link between the environment, process rates, and resultant patterns, the shortcut correlational link between environmental conditions and patterns can be exploited (Fig. 14-2). By identifying the best environmental correlates of the known presence or absence of any organism (whether an elephant, a cactus, or a tick-borne virus), the relative significance of each environmental variable can be quantified by determining its distribution. The multivariate correlation can then be extrapolated beyond existing observations, to fill in the gaps in our knowledge and so create predictive risk maps for the full potential global distribution of the organism. This has all been made much easier by the development of geographical information systems (GIS), software programs that collate, analyze, and display spatially explicit data. The reliability of such predictive risk maps depends, of course, on the quality of the input data used to train the correlations, the quality and spatial resolution of the environmental data, the statistical handling of these data to take account of multivariate nonnormality across wide heterogeneous areas, and the particular statistical pattern-matching program chosen. Detailed considerations of these points can be found in several reviews [12, 13, 31, 47]. Moreover, the reliability of these maps, expressed as percentage correct matches between the predictions and the test data, can appear misleadingly good if, for example, the latter are inflated by a disproportionately large sample of absence points taken from far outside the observed distributional limits. Predicting the absence of tsetse flies outside sub-Saharan Africa, for example, or of *Amblyomma hebraeum* outside southern Africa, carries very little merit. It is not the production of such maps that is now the challenge, but the intelligent use to which they are put.

14.3.2.2. Data sources: the environment

There is now a major disparity between the availability of environmental data and of useful spatially explicit epidemiological data. The era of satellites ushered in an abundance of information about conditions on the earth's surface, albeit measured remotely as surrogates of ground-based conditions. Data in 3–7 wavebands (channels) within 0.3–14 μm of the electromagnetic spectrum are usually processed to produce indices related to the following ground-based variables: (1) thermal conditions (land surface temperature – Price LST, near-surface air temperature – TvX, and middle infrared band – MIR); (2) moisture conditions (normalized difference vegetation index – NDVI, actually an index of plant photosynthetic activity); and (3) rainfall (cold cloud duration – CCD) [14, 47].

Any approximation in absolute measures of temperature and moisture conditions is offset by the benefit of the near-global coverage. Whereas ground meteorological stations are unevenly, and commonly sparsely, distributed around the globe, satellites provide blanket-cover information on continental scales at spatial and temporal resolutions to match our epidemiological questions. There are, however, inevitable trade-offs between the spectral, spatial, and temporal resolutions, with different satellites designed for different purposes and suitable for different epidemiological uses [13, 47].

At one extreme, earth-observing satellites produce data with high spatial resolutions (i.e. pixel sizes) of between 1 and 4 m (Ikonos-2) and between 30 and 120 m (Landsats 1–5), but low repeat frequencies of only 2–3 weeks. As clouds commonly obscure images, however, these low-frequency images provide only occasional snapshots. In contrast, orbiting oceanographic, or geostationary meteorological satellites have lower spatial resolutions, down to 1.1 km (National Oceanographic and Atmospheric Administration Advanced Very High Resolution Radiometer [NOAA AVHRR]), but they produce two or more images per day of the entire earth's surface. These high-frequency images can be combined to produce relatively cloud-free monthly "maximum value composites" (MVCs). Because many of the dynamical processes within vector-borne disease systems follow seasonal ecological patterns, the high-frequency images are of the much greater use. The more recent Terra (EOS AM-1) and Aqua (EOS PM-2) spacecraft include a moderate resolution imaging spectroradiometer (MODIS) with 36 spectral channels and spatial resolutions of between 250 and 1,000 m. This instrument has a 2-day repeat frequency and the images it produces have much greater geolocational accuracy than those of the AVHRR instrument, giving better quality MVCs of seasonal processes.

The problem now is one of a superabundance of environmental data, with a high degree of inherent redundancy in serial monthly imagery. Temporal Fourier processing offers a method of overcoming this, without losing the biologically meaningful signals [42, 43]. The French mathematician Joseph Fourier (1768–1830) showed that any complex time series can always be expressed as the sum of a series of sine and cosine curves with different amplitudes, frequencies, and phases (i.e. timings) around a characteristic mean. Multitemporal satellite data can be processed to yield information about the annual, biannual, and triannual cycles of rainfall, temperature, etc. that characterize the natural environments of diseases (Fig. 14-3). In temperate and tropical environments with a dominant annual cycle, the biannual and triannual Fourier cycles modulate the simple annual Fourier cycle (i.e. a sine curve with a period of 1 year) to characterize the particular shape of the rising and falling curves between the annual minima and maxima. The output of temporal Fourier analysis is a set of orthogonal (i.e. uncorrelated) variables that capture the seasonality that is of vital interest in epidemiology [42, 43, 45], and these variables may therefore be used to classify habitats in ways that are relevant to arthropod vectors and transmitted pathogens [44]. For disease systems, Fourier variables are the environmental equivalent of the genes of individual pathogens, and whole Fourier-processed images [48], that capture all the interactive space–time features of a habitat, may be likened to the organismal genome.

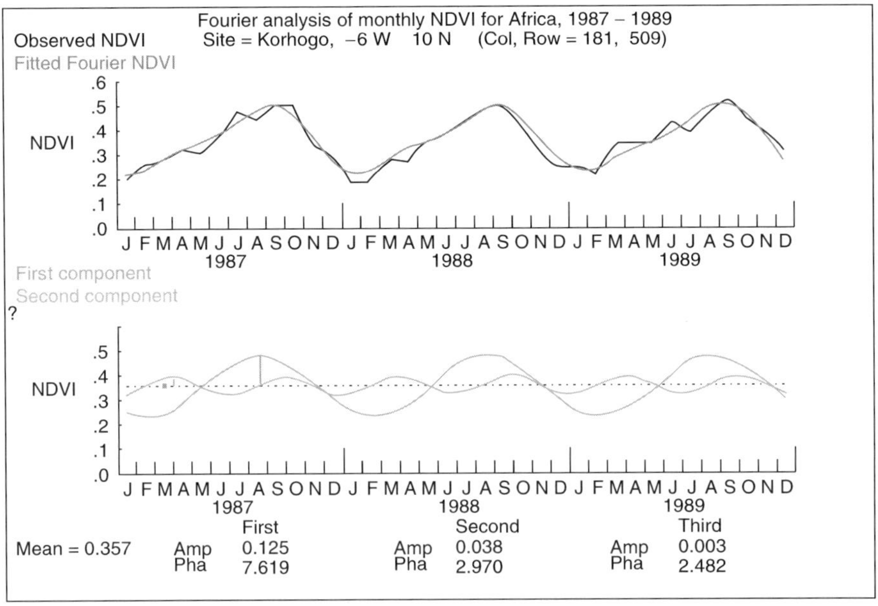

Fig. 14-3. Any regular seasonal cycle (black line), in this case of the remotely sensed normalized difference vegetation index (NDVI) at Khorogo, Côte d'Ivoire, Africa may be deconstructed by temporal Fourier analysis into its component harmonic sine waves with annual (red line), biannual (blue line), and triannual (green line) periodicities. Each sine wave (rescaled here to fluctuate around zero) has its characteristic amplitude (variation around the mean) and phase (timing of peak). The component cycles may be reconstructed to the fitted Fourier mean (purple line) to give the smoothed seasonal signal characteristic of each location over the period of observation.

14.3.2.3. Data sources: CCHF

The statistical prediction exercise must always start with some baseline knowledge that is expected to be incomplete (hence the need for the exercise). CCHF has the largest geographic distribution of any tick-borne viral disease, so it is not surprising that its full range is described in broad brush-strokes with little subnational precision. In addition to much of sub-Saharan Africa and a belt from the Balkans and Ukraine throughout southern central Asia including Kazakhstan, the map based on human cases [2] shows foci in Portugal and southern France. An alternative map based on both human disease and viral isolates [58] shows a much more extensive range to the east across all of China and Mongolia, and includes the Arabian land mass but not India. Recently, more spatially explicit data were acquired by extracting 622 unique case records of CCHF from the literature (1920–2006), as part of a major project on risk mapping for a variety of infectious diseases. This database was reduced to 378 positive records after the elimination of duplicates and insufficiently exact locations (such as when cases are reported only to the level of large administrative regions). This database does not, however, include the hitherto unpublished records of several hundred cases in Turkey each year since 2002. Records of definite absence were too scarce to be useful in defining the disease's geographical limits. Instead, 5,000 locations, randomly selected between 0.5° and 10° away from the nearest site for CCHF presence, were taken as negative locations. Satellite data were then extracted for each positive and negative location. Details of the precise procedure and the model building methods are given in Hay et al. [12]. In brief, 100 bootstrap samples (400 each of presence and absence points) were selected at random (with replacement) from the training data set. Nonlinear, maximum likelihood discriminant analysis was used to select the best set of descriptor variables for each bootstrap sample, using the information-theoretic approach of stepwise selection of variables that minimized Akaike's information criterion (AIC), corrected for the number of variables in the model (i.e. the AICc). Each resulting model was then applied to the full extent of the satellite image data to make global predictions of the posterior probabilities of habitat suitability for CCHF.

14.4. USES AND INTERPRETATIONS OF PREDICTIVE RISK MAPS FOR CCHF

14.4.1. Risk maps as blueprints

The most immediate use of a risk map is as a blueprint to inform people about the full potential extent of the risk of infection. If the training data comprise human case records, as here, the predicted distribution will also refer to human risk, the tip of the zoonotic iceberg, but almost certainly not the full range of natural enzootic cycles, the whole iceberg. A variety of factors may cause the iceberg to emerge further above the surface without necessarily any change in enzootic transmission potential, i.e. greater human exposure to risk without any change in that intrinsic risk (see below).

The risk map for CCHF (Fig. 14-4) shows a greater than 0.5 probability of human disease not only where cases have indeed been recorded, but also extensively in other parts of the world: notably in large parts of North America, south, western and eastern parts of South America (Argentina, Chile, Bolivia, Peru, Brazil), eastern China, Australia, Spain and Italy, and the northern edge of Morocco, Tunisia, and Libya. At the same time, only the southeastern edge of Kazakhstan is predicted as highly suitable, even though a few positive records come from central and northern Kazakhstan. Most of India and central Africa have little or no predicted habitat suitability, despite recent records from these areas. While the very few false negative predictions obviously arise from errors or inadequacies in this modeling procedure, the much more extensive false positive predictions are far more interesting. They could also be dismissed as errors, as any prediction of West Nile virus in the Americas would have been prior to 1999 (when this virus arrived in the New World for the first time). Alternatively,

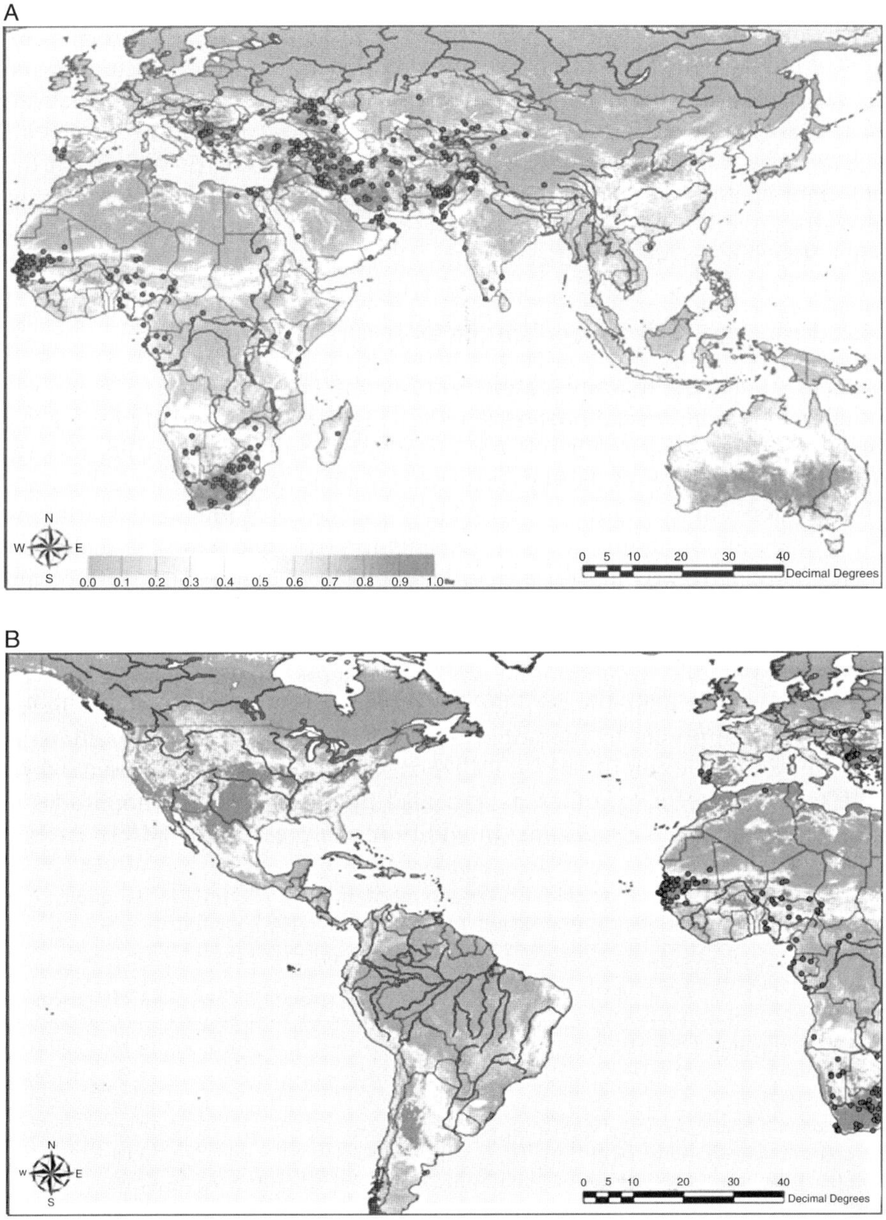

Fig. 14-4. Predictive risk map for CCHF (A) Old World and (B) New World. The data source and methods of creation are summarily described in the text (further methodological details in [12]). The mean posterior probabilities of environmental suitability for CCHF from 100 bootstrapped models are here shown on a color scale from green (low probability) to red (high probability). The original literature records of CCHF presence, on which the models were based, are shown as blue points on the map(s). *(See Color Plates)*

they can be taken as a wake-up call, alerting public health authorities to regions potentially at risk of invasion if the virus were to be introduced across the current oceanic barriers. Such an invasion would depend on the virus finding permissive biotic conditions (i.e. competent vector species and vertebrate transmission hosts) within the apparently abiotically hospitable regions. Within the previously well-recognized broad-brush range of CCHF, predictions at this much finer spatial scale (here down to 0.10° longitude/latitude, ~10 km^2 pixel size at the equator) allow intervention strategies to be targeted much more precisely, achieving greater cost-effectiveness.

14.4.2. Risk maps to throw spotlights on the underlying biology

For scientists, an equally valuable and far more interesting use of risk maps is as spotlights on the biological processes underlying the observed patterns, seeking explanations upon which to base more versatile, robust predictions. For this, the critical information lies in the key predictor variables: which abiotic factors best define the necessary conditions for CCHF presence, what are these conditions, what do they tell us about the virus transmission dynamics, and are these conditions changing to account for past epidemiological changes and warn of any future changes?

Table 14-1 shows the top ten variables in ranked order from the best of the 100 bootstrap models (i.e. minimum AICc value across all models). Many of the other models had similar, but not identical, combinations of predictor variables. All the evidence points to the importance of seasonal thermal conditions in determining suitable conditions for CCHF risk: it appears that the specific form of the seasonal thermal profiles, as indicated by the phases and amplitudes of the harmonic cycles, is significant. In addition, the presence of CCHF cases is

Table 14-1. The key predictor variables (shown in ranked order of selection) for the presence of CCHF identified by stepwise forward selection for the best of the 100 bootstrap models

Variable	Mean values CCHF present	Mean values CCHF absent
Price LST minimum (°C)	18.91	18.16
MIR variance	78.47	78.79
MIR mean (°C)	36.42	34.89
NDVI triannual phase, decimal months	1.80	2.05
Price LST annual amplitude (°C)	13.49	13.04
Price LST biannual amplitude (°C)	2.60	2.58
NDVI maximum	0.36	0.39
Digital elevation model	7.85	7.08
MIR biannual amplitude (°C)	2.41	2.23
Price LST maximum (°C)	46.91	45.00

LST, land surface temperature; MIR, middle infrared; NDVI, normalized difference vegetation index. Not all models selected the same top ten variables.

consistently associated with warmer temperatures. In particular, minimum LST was selected as the top ranked variable by the best individual model, and had by far the highest mean rank across the various bootstrap models, while mean middle infrared was also highly ranked.

Until a great deal more is known about the specific biotic interactions involved, such as the necessary tick infestation patterns (both qualitative and quantitative) on the key transmission hosts, it is impossible to progress from identifying the key predictor variables to interpreting their relevance for CCHF virus transmission. Even for TBE virus, whose biotic interactions are well established and whose ecology has therefore been quantitatively analyzed in detail by an integrated biological and statistical approach, there remains a major gap in the story: there is a clear biological link between the cellular transmission route amongst co-feeding ticks via rodent hosts [20] and the climate-driven seasonal dynamics of the ticks to ensure a high degree of co-feeding between larval and nymphal *I. ricinus* [22]; there is a strong statistical correlation between a particular feature of the remotely sensed autumnal thermal conditions (a higher than average rate of cooling) and both the seasonal synchrony of these tick stage and the presence of TBE cases [36]; but biologically it seems more likely that rates of increasing temperature in the spring would determine the degree of seasonal synchrony as larvae have a higher temperature threshold for activity than do nymphs [33]. Nevertheless, identifying the key predictor variables for TBE presence, by creating a risk map, was an important step in the (hitherto incomplete) analytical process, and so it could be for CCHF.

14.5. POTENTIAL FOR EPIDEMIOLOGICAL CHANGE OF CCHF

14.5.1. Predicting future change

It is clear that we do not yet have sufficient quantitative knowledge of either the biotic or the abiotic factors limiting the distribution of CCHF foci to predict whether, as environmental factors change, the disease will emerge more widely in human populations. For any tick-borne disease whose existing foci depend on a delicate balance of factors and rates, the system may equally well be disrupted by environmental change, leading to retrenchment rather than emergence. Predictions must rest on system-specific analyses, taking into account the particular ecology of the major tick vectors and wildlife hosts. The wide geographical distribution of CCHF, wider than for any other tick-borne disease, indicates an ecological robustness in the system, which may therefore confer considerable potential for epidemiological expansion in the future. This potential may be enhanced by the wide range of biologically competent tick species if those that are not yet ecologically competent are rendered so by changing environmental conditions. There is no evidence that *I. ricinus*, for example, currently plays a major role in the transmission of CCHF virus; its biological competence in the laboratory and widespread distribution throughout Europe, however, are sufficient reasons to place CCHF on the UK Health Protection Agency's "at risk" list of potential invaders (http://www.hpa.org.uk).

Just as there are two complementary methods for predicting the present distribution of any organism, the biological process-based approach and the statistical pattern-matching approach, so these two methods may be applied to predictions of future change. Ideally, if we had a full biological model for the CCHF system driven by environmental variables, we could apply the environmental conditions forecast under the scenarios of climate change to predict new process rates and therefore distribution patterns at specified points in the future. Failing that, for with an incomplete, not to say nonexistent, biological model this approach is doomed to fail, we can extrapolate the identified correlations between present environmental and epidemiological patterns into the future according to forecast climate scenarios. This latter approach depends very much on the assumption that all other nonclimatic factors will stay the same so that the shape and structure of the underpinning correlations do not change. In reality, climate change will have an impact on all sorts of human behavioral factors, which in turn may alter livestock and wildlife host availability for ticks and CCHF virus. Past exercises of this sort, taking into account changes in both temperature and moisture conditions, have suggested that vector-borne diseases may appear in some new places but disappear from parts of their current range, mostly around the distributional edges, with little net change for malaria [46] but considerable decrease in range for TBE [39]. This is entirely consistent with the complexity of these disease systems and the greater quantitative fragility of TBE. Nevertheless, such conclusions, however plausible, are not scientific because they are inherently untestable until the future arrives.

14.5.2. Explaining past changes

A more rigorous approach is to seek explanations for the past. Specifically, we should search for key changes in abiotic, biotic, or socioeconomic factors associated with specific epidemiological events. We may then warn of potential future changes if conditions continue to change in the same way. The danger is that correlations between recent epidemiological phenomena and recent climate changes are almost inevitable, given that climate change is the general backdrop to all recent and current events, but this does not necessarily indicate causality. Only if disease emergence in new places or increases in incidence, for example, are strictly consistent in time and space with environmental changes may we impute causality. This is analogous to Koch's postulates, which demand absolutely consistent associations between the presence of a pathogen and evidence of infection. The present task, however, is beset not by the technical microbiological hurdles of the 19th century, but the analytical problems of dealing with a complex network of environmental and human factors, operating over large heterogeneous regions and each changing independently yet indirectly linked, which may act synergistically or antagonistically on the enzootic transmission cycles and/or degree of human exposure.

The last 5 years have witnessed notable increases in reports of CCHF, including first reports of cases in Kenya [6], Iran [24], Albania [29], Kosovo [30], and Turkey [9]. As ticks are obviously subject to seasonal climate effects, it is reasonable to suggest that recent climate change may have promoted these outbreaks of CCHF. As is typical of arthropods, tick development rates and activity increase with temperature, while *Hyalomma* ticks are better adapted to surviving in dry conditions than are many other ticks. This is consistent with the mean values of the key predictor variables for the presence of CCHF (Table 14-1), which indicate that CCHF occurs where average conditions are warmer. In the northern hemisphere, *H. marginatum marginatum* is usually activated by increasing temperature in the spring, with adults appearing as average daily temperatures reach 5–9°C in April or May, and the immature stages appearing somewhat later from May onwards; tick activity continues throughout the summer to early autumn [7, 15, 55]. Hoogstraal (1979) attributes the reduction in case numbers of CCHF in the Astrakhan Oblast of Russia in the late 1960s to a reduction in tick densities due to the severe winter of 1968–1969 in that region.

Conversely, Ergönül et al. [8] examined trends in April temperatures and annual rainfall to test for a climatic cause for the emergence of CCHF in Turkey. CCHF infection was first recognized in 2002 in Turkey [9, 18], with marked increases in the numbers and spread of cases recorded through to 2006. Cases were mainly from three provinces of central Turkey, namely Tokat, Sivas, and Yozgat, for which meteorological records were available over the last 40, 70, and 60 years, respectively. Significant increasing trends in monthly mean and minimum temperatures, and in the number of days in April with temperatures of >5°C, were detected only for Sivas, but not for Tokat or Yozgat. Likewise, a significant increase in annual rainfall was detected only in Tokat [8]. Changing climate does not, therefore, provide a sufficiently consistent explanation for this CCHF emergence in Turkey. Although shorter-term variation in climate, not detectable as statistically significant trends, might still indicate an abiotic contribution, other possible human-induced biotic changes have been suggested [7]: because of terrorist activities in the region, hunting and pastoral activities had been abandoned between 1995 and 2001, allowing an increase in populations of mammals such as hares and wild boar. After 2001, hunting resumed and the fields became available again, exposing cattle and sheep to increased populations of virus-carrying ticks. Similar environmental changes due to wartime events have been held responsible for an outbreak of CCHF in Crimea in 1944–1945 [15]: after the occupation of Crimea during World War II, disruption of normal agricultural activities and the abandonment of hare hunting allowed fields to become overgrown with weeds and occupied by high densities of hares, thereby improving both the abiotic and biotic conditions for ticks. When susceptible Soviet troops reoccupied the region in 1944, a major epidemic occurred (200 cases, 10% mortality [7]).

Such considerations confirm the conclusion, obvious from first biological principles, that CCHF, like all other vector-borne disease systems, is potentially highly dynamic in response to a wide range of disparate factors, of which climate change per se is only one and not necessarily the most influential. Changes in human factors can be very abrupt, perhaps matching the abrupt outbreaks of large numbers of cases of CCHF over the last 60 years [7].

14.5.3. Vehicles of invasion

The predictive risk map (Fig. 14-4) shows far-flung areas of apparent habitat suitability for CCHF, many at considerable distance, even on different continents, from any records of presence to date. Are we to consider these as areas at real, rather than merely theoretical, risk of invasion? Such an invasion depends primarily on the introduction of the virus. The source of any newly arrived pathogen can now be identified by molecular epidemiology, allowing appropriate suspected vehicles to be investigated. The movement of livestock is now more carefully regulated than in the past, although illegal trade is still common (and the centralization of abattoirs in the UK following the bovine spongiform encephalopathy [BSE] crisis has resulted in increased animal traffic). The movement of humans and undetected arthropod stowaways in planes and ships, however, continues to grow inexorably and has been associated with the introduction of exotic vector-borne microbes [*inter alia* 10, 53].

Migrating birds are largely outside legislative control and have been held responsible for the dispersal of a large range of infectious agents (and are assumed to be so now and in the future [25] even if there is evidence against their involvement [28]). These events are either well established and account for the disjunct global distribution of closely related microbes [27], or occur rarely through exceptional random events [17], or may become newly established through changing patterns of migration in response to climate change (virtually all reports of the impact of climate change on bird migration refer to the timing, not direction, of migration – http://wok.mimas.ac.uk – which could be equally critical for any seasonal disease system). With tick-borne diseases, ticks will only be introduced if they do not complete their feeding and drop off before the birds complete their migration (although transport over part of the migration route is still possible if birds feed on the ground *en route*). Birds most commonly feed immature tick stages, which have feeding periods of a few days, and which must be introduced in sufficiently large numbers to offset the normal 80–90% interstadial tick mortality if breeding populations are to become established. Any introduced ticks would only be infected if the birds were competent to permit pathogen transmission to ticks, or allowed the maintenance of pre-existing infections in ticks as they fed on birds. Alternatively, the birds themselves must transport the pathogens, retaining the infection during the course of their migration. For CCHF virus, none of these alternatives has been demonstrated, and none seems likely on the basis of current knowledge, even though CCHF virus has been transmitted experimentally from birds to *H. marginatum rufipes* ticks [60].

The predictive risk map at least helps to direct future investigations into the potential for long-range introductions of CCHF virus, by focusing attention on transport networks between known foci and areas identified as environmentally suitable for continuing transmission once all the key biotic elements are in place.

14.6. CONCLUSIONS

The principles behind the ecology of tick-borne disease systems are increasingly well understood and the likely significant impact of climate on CCHF enzootic cycles may be argued from first principles. The devil of any system, however, lies in the detail of each specific tripartite network of hosts, vectors, and pathogen. While increasing interest in CCHF is evident (see the front cover illustration on the April 2006 issue of *The Lancet Infectious Diseases*), greater emphasis has so far been on clinical and epidemiological features rather than on the basic biology and ecology of the system. Until the biotic interactions have been fully identified and subjected to detailed quantitative analysis, the impact of geographically, not to mention temporally, variable climate on the risk of infection must remain speculative. The creation of the first predictive risk map for CCHF offers a step towards identifying critical environmental factors, and provides a framework for intensive field work to validate the predicted fine-scale distribution. With such a widespread recorded distribution, CCHF is an ideal subject for a major international research initiative lest it turns into a major story of international emergence.

REFERENCES

1. Anderson RM, May RM (1992) Infectious Diseases of Humans: Dynamics and Control. Oxford University Press, Oxford
2. Brown RN, Lane RS, Dennis DT (2005) Geographic distribution of tick-borne diseases and their vectors. In: Goodman JL, Dennis DT, Sonenshine DE (eds) Tick-borne Diseases of Humans. ASM Press, Washington, DC, pp 363–391
3. Burt FJ, Swanepoel R (2005) Crimean-Congo hemorrhagic fever. In: Goodman JL, Dennis D, Sonenshine DE (eds) Tick-borne Diseases of Humans. ASM Press, Washington, DC, pp 164–175
4. Craine NG, Randolph SE, Nuttall PA (1995) Seasonal variation in the rôle of grey squirrels as hosts of *Ixodes ricinus*, the tick vector of the Lyme disease spirochaete, in a British woodland. Folia Parasitol 42:73–80
5. Dizij A, Kurtenbach K (1995) *Clethrionomys glareolus*, but not *Apodemus flavicollis*, acquires resistance to *Ixodes ricinus* L., the main European vector of *Borrelia burgdorferi*. Parasite Immunol 17:177–183
6. Dunster L, Dunster M, Ofula V, Beti D, Kazooba-Voskamp F, Burt FJ, Swanepoel R, DeCock KM (2002) First documentation of human Crimean-Congo hemorrhagic fever, Kenya. Emerg Infect Dis 8:1005–1006
7. Ergönül O (2006) Crimean-Congo haemorrhagic fever. Lancet Infect Dis 6:203–214

8. Ergönül O, Akgunduz S, Kocaman I, Vatansever Z, Korten V (2005) Changes in temperature and the Crimean Congo haemorrhagic fever outbreak in Turkey. Clin Microbiol Infect 11 (Suppl 2):360
9. Ergönül O, Celikbas A, Dokuzoguz B, Eren S, Baykam N, Esener H (2004) The characteristics of Crimean-Congo hemorrhagic fever in a recent outbreak in Turkey and the impact of oral ribavirin therapy. Clin Infect Dis 39:285–289
10. Giladi M, Metzkor-Cotter E, Martin D, Siegman-Igra Y, Korczyn A, Rosso R, Berger S, Campbell G, Lanciotti R (2001) West Nile Encephalitis in Israel, 1999: the New York connection. Emerg Infect Dis 7:659–661
11. Gilbert L, Norman R, Laurenson MK, Reid HW, Hudson PJ (2001) Disease persistence and apparent competition in a three-host community: an empirical and analytical study of large-scale, wild populations. J Anim Ecol 70:1053–1061
12. Hay SI, Graham AJ, Rogers DJ (2006) Global Mapping of Infectious Diseases. Academic Press, London
13. Hay SI, Randolph SE, Rogers DJ (2000) Remote Sensing and Geographical Information Systems in Epidemiology. Academic Press, London
14. Hay SI, Tucker CJ, Rogers DJ, Packer MJ (1996) Remotely sensed surrogates of meteorological data for the study of the distribution and abundance of arthropod vectors of disease. Ann Trop Med Parasitol 90:1–19
15. Hoogstraal H (1979) The epidemiology of tick borne Crimean-Congo hemorrhagic fever in Asia, Europe and Africa. J Med Entomol 15:307–417
16. Horak IG, Fourie LJ, Novelle PA, Williams EJ (1991) Parasites of domestic and wild animals in South Africa. XXVI. The mosaic of Ixodid tick infestations on birds and mammals in the Mountain Zebra National Park. Onderstepoort J Vet Res 58:125–136
17. Jääskeläinen J, Tikkakoski T, Uzcategui NY, Alekseev AN, Vaheri A, Vapalhti O (2006) Siberian sub-type tick-borne encephalitis virus, Finland. Emerg Infect Dis 12:1568–1571
18. Karti SS, Odabasi Z, Korten V (2004) Crimean-Congo hemorrhagic fever in Turkey. Emerg Infect Dis 10:1379–1384
19. Kurtenbach K, Peacey MF, Rijpkema SGT, Hoodless AN, Nuttall PA, Randolph SE (1998) Differential transmission of the genospecies of *Borrelia burgdorferi* sensu lato by game birds and small rodents in England. Appl Environ Microbiol 64:1169–1174
20. Labuda M, Austyn JM, Zuffova E, Kozuch O, Fuchsberger N, Lysy J, Nuttall PA (1996) Importance of localized skin infection in tick-borne encephalitis virus transmission. Virology 219:357–366
21. Labuda M, Nuttall PA, Kozuch O, Eleckova E, Williams T, Zuffova E, Sabo A (1993) Non-viremic transmission of tick-borne encephalitis virus: a mechanism for arbovirus survival in nature. Experientia 49:802–805
22. Labuda M, Randolph SE (1999) Survival of tick-borne encephalitis virus: cellular basis and environmental determinants. Zentralbl Bakteriol 288:513–524
23. Magano SR, Els DA, Chown SL (2000) Feeding patterns of immature stages of *Hyalomma truncatum* and *Hyalomma marginatum rufipes* on different hosts. Exp Appl Acarol 24:301–313
24. Mardani M, Jahromi MK, Naieni KH, Zeinali M (2003) The efficacy of oral ribavirin in the treatment of Crimean-Congo hemorrhagic fever in Iran. Trans R Soc Trop Med Hyg 36:1613–1618
25. McConnell J (2006) Avian influenza goes global, but don't blame the birds. Lancet Infect Dis 6:185
26. Mount GA, Haile DG, Daniels E (1997) Simulation of blacklegged tick (Acari: Ixodidae) population dynamics and transmission of *Borrelia burgdorferi*. J Med Entomol 34:461–484
27. Olsen B, Duffy DC, Jaenson TGT, Gylfe A, Bonnedahl J, Bergström SA (1995) Trans hemispheric exchange of Lyme disease spirochaetes by seabirds. J Clin Microbiol 33:3270–3274
28. Oncul O, Turham V, Cavuslu S (2006) H5N1 avian influenza: the Turkish dimension. Lancet Infect Dis 6:186–187

29. Papa A, Bino S, Llagami A, Brahimaj B, Papadimitriou E, Pavlidou V, Velo E, Cahani G, Hajdini M, Pilaca A, Harxhi A, Antoniadis A (2002) Crimean-Congo hemorrhagic fever in Albania. Eur J Clin Microbiol Infect Dis 21:603–606
30. Papa A, Bozovic B, Pavlidou V, Papadimitriou E, Pelemis M, Antoniadis A (2002) Genetic detection and isolation of Crimean-Congo hemorrhagic fever virus, Kosovo, Yugoslavia. Emerg Infect Dis 8:852–854
31. Phillips SJ, Anderson RP, Schapire RE (2006) Maximum entropy modeling of species geographic distributions. Ecol Model 190:231–259
32. Randolph SE (1998) Ticks are not insects: consequences of contrasting vector biology for transmission potential. Parasitol Today 14:186–192
33. Randolph SE (2004) Tick ecology: processes and patterns behind the epidemiological risk posed by ixodid ticks as vectors. Parasitology 129 (Suppl):S37–S66
34. Randolph SE, Craine NG (1995) General framework for comparative quantitative studies on transmission of tick-borne diseases using Lyme borreliosis Europe as an example. J Med Entomol 32:765–777
35. Randolph SE, Gern L, Nuttall PA (1996) Co-feeding ticks: epidemiological significance for tick-borne pathogen transmission. Parasitol Today 12:472–479
36. Randolph SE, Green RM, Peacey MF, Rogers DJ (2000) Seasonal synchrony: the key to tick-borne encephalitis foci identified by satellite data. Parasitology 121:15–23
37. Randolph SE, Miklisová D, Lysy J, Rogers DJ, Labuda M (1999) Incidence from coincidence: patterns of tick infestations on rodents facilitate transmission of tick-borne encephalitis virus. Parasitology 118:177–186
38. Randolph SE, Rogers DJ (1997) A generic population model for the African tick *Rhipicephalus appendiculatus*. Parasitology 115:265–279
39. Randolph SE, Rogers DJ (2000) Fragile transmission cycles of tick-borne encephalitis virus may be disrupted by predicted climate change. Proc R Soc Lond B 267:1741–744
40. Randolph SE, Storey K (1999) Impact of microclimate on immature tick-rodent interactions (Acari: Ixodidae): implications for parasite transmission. J Med Entomol 36:741–748
41. Rechav Y, Fielden LJ (1997) The effect of various host species on the feeding performance of immature stages of the tick *Hyalomma truncatum* (Acari: Ixodidae). Exp Appl Acarol 21:551–559
42. Rogers DJ (2000) Satellites, space, time and the African Trypanosomiases. Adv Parasitol 47:130–173
43. Rogers DJ, Hay SI, Packer MJ (1996) Predicting the distribution of tsetse flies in West Africa using temporal Fourier processed meteorological satellite data. Ann Trop Med Parasitol 90:225–241
44. Rogers DJ, Hay SI, Packer MJ, Wint GRW (1997) Mapping land-cover over large-areas using multispectral data derived from the NOAA-AVHRR: a case study of Nigeria. Int J Remote Sens 18:3297–3303
45. Rogers DJ, Packer MJ (1993) Vector-borne diseases, models and global change. Lancet 342:1282–1284
46. Rogers DJ, Randolph SE (2000) The global spread of malaria in a future, warmer world. Science 289:1763–1766
47. Rogers DJ, Randolph SE (2003) Studying the global distribution of infectious diseases using GIS and RS. Nat Rev Microbiol 1:231–236
48. Rogers DJ, Williams BG (1994) Tsetse distribution in Africa: seeing the wood and the trees. In: Edwards PJ, May RM, Webb NR (eds) Large-scale Ecology and Conservation Biology. Blackwell Scientific, Oxford, pp 249–273
49. Sandberg S, Awerbuch TE, Spielman A (1992) A comprehensive multiple matrix model representing the life cycle of the tick that transmits the agent of Lyme disease. J Theor Biol 157:203–220
50. Shepherd AJ, Leman PA, Swanepoel R (1989) Viremia and antibody response of small African and laboratory animals to Crimean-Congo hemorrhagic fever virus infection. Am J Trop Med Hyg 40:541–547

51. Shepherd AJ, Swanepoel R, Leman PA, Shepherd SP (1987) Field and laboratory investigation of Crimean-Congo hemorrhagic fever virus (*Nairovirus*, family Bunyaviridae) infection in birds. Trans R Soc Trop Med Hyg 81:1004–1007
52. Shepherd AJ, Swanepoel R, Shepherd SP, Lenan PA, Mathee O (1991) Viremic transmission of Crimean-Congo hemorrhagic fever virus to ticks. Epidemiol Infect 106:373–382
53. Tatem AJ, Hay SI, Rogers DJ (2006) Global traffic and disease vector dispersal. PNAS 103:6242–6247
54. Tomassone L, Camicas JL, Pagani P, Tanat Diallo O, Mannelli A, de Meneghi D (2004) Monthly dynamics of ticks (Acari: Ixodidae) infesting N'Dama cattle in the Republic of Guinea. Exp Appl Acarol 32:209–218
55. Walker AR, Bouttaour A, Camicas JL, Estrada-Peña A, Horak IG, Latif AA, Pegram RG, Preston PM (2003) Ticks of domestic animals in Africa: a guide to identification of species. Bioscience Reports, University of Edinburgh, Edinburgh
56. Watts DM, Ksiazek TG, Linthicum KJ, Hoogstraal H (1988) Crimean-Congo hemorrhagic fever. In: Monath TP (ed.) The Arboviruses: Epidemiology and Ecology. CRC Press, Boca Raton, FL, pp 177–222
57. Watts DM, Ksiazek TG, Linthicum KJ, Hoogstraal H (1989) Crimean-Congo hemorrhagic fever. In: Monath TP (ed.) The Arboviruses: Epidemiology and Ecology. CRC Press, Boca Raton, FL, pp 177–222
58. Whitehouse CA (2004) Crimean-Congo hemorrhagic fever. Antiviral Res 64:145–160
59. Wilson ML, Gonzalez J-P, Cornet J-P, Camicas J-L (1991) Transmission of Crimean-Congo haemorrhagic fever virus from experimentally infected sheep to *Hyalomma truncatum* ticks. Res Virol 142:395–404
60. Zeller HG, Cornet JP, Camicas JL (1994) Experimental transmission of Crimean-Congo hemorrhagic fever virus by West African wild ground-feeding birds to *Hyalomma marginatum rufipes* ticks. Am J Trop Med Hyg 50:676–681

CHAPTER 15

MATHEMATICAL MODELING OF CRIMEAN-CONGO HEMORRHAGIC FEVER TRANSMISSION

BEN S. COOPER, PH.D.

Statistics, Modelling and Bioinformatics Department, Centre for Infections, Health Protection Agency, London, UK. E-mail: ben.cooper@hpa.org.uk

Sensibly used, mathematical models are no more, and no less, than tools for thinking about things in a precise way. (Anderson and May 1991) [1]

This chapter is divided into five sections. Section 15.1 discusses the rationale for using mathematical models. Section 15.2 considers the specific areas where models may be useful in studying Crimean-Congo hemorrhagic fever (CCHF). Section 15.3 reviews work on modeling the dynamics of tick-borne diseases and considers the relevance of this work for CCHF. Section 15.4 considers the problem of modeling the nosocomial transmission of CCHF. Section 15.5, lastly, suggests future directions for CCHF modeling work.

15.1. WHY USE MATHEMATICAL MODELS?

Mathematical models of infectious diseases represent simplified representations of known processes and their interactions. Typically these processes are transmission, disease progression, birth, death and recovery, acquisition and loss of immunity, and immigration or emigration. Other processes appropriate for certain applications include boosting of immunity, vector dynamics, vaccination, and other control measures. By using these models we hope to capture important aspects of the behavior of the whole system and gain a full understanding of the role of the individual processes and their interactions in determining this behavior. The properties of these models and their predicted behavior under different scenarios are usually investigated either by solving the equations (in the case of deterministic models, where chance is assumed to play no part, and which can often be considered to model the average behavior of the system), or by simulating many epidemics from stochastic models (an approach known as Monte Carlo simulation because the role of chance in determining the course of

O. Ergonul and C. A. Whitehouse (eds.), Crimean-Congo Hemorrhagic Fever, 187–203.

the epidemic is explicitly accounted for). In both cases, the analysis usually makes use of a computer (though the original work with both types of models was done without one). This *in silico* modeling approach is directly comparable with the use of in vitro or in vivo models in the laboratory; all three approaches work by choosing model systems that are, it is hoped, in some important sense analogous to the system we really want to find out about. We may hope that the model is sufficiently similar to the real system in the ways that matter to give the same answers to the questions we ask, but we can seldom be sure. A cautious and critical approach to interpreting results from all models is therefore needed. Mathematical models are no exception to this rule. Nonetheless, as with other models, their use has the potential to lead to important advances in our understanding.

At the simplest level, mathematical models are just translations of reasoning from natural language into a precise mathematical formulation. The translation process has the virtue of forcing (or at least encouraging) us to think clearly, and tends to makes hidden assumptions explicit. Flaws in the reasoning of simple verbal arguments often become apparent once the models have been constructed. Models also allow the intuition that the verbal arguments rest on to be thoroughly tested. In this sense, models can be thought as formal encapsulations of hypotheses. Sometimes the resulting models support the verbal arguments and we may be encouraged that our intuition was correct, but frequently models show that our loose verbal reasoning is wrong. These models produce counterintuitive results. However, even when model results confirm our intuition, by allowing a fuller and more rigorous analysis of the consequences of the assumptions that make up the model they can broaden our understanding of the system. Often models will make new predictions that can be directly tested. If they pass the test, our confidence in the truth of the hypothesized mechanism the model represents will be strengthened. There is also a danger that new hidden assumptions are made in constructing the model. Careful modeling work should aim to highlight these assumptions and ideally to explore the sensitivity of the results to structural uncertainties in the model as well as to uncertainty in parameter values.

Models are not, as has been suggested, simply substitutes for experiments [13]. Instead, a primary use of models is to broaden our understanding, synthesize information, and to show how diverse outcomes can be understood as the result of similar underlying processes. In this way mathematical models are frequently used to help interpret experimental findings. Models themselves often suggest certain experiments or observations, and such experiments may in turn lead to revisions to the models. Nonetheless, it is true that in all the sciences where experimental manipulation is either difficult or impossible – astrophysics, economics, climatology, geology, ecology (of which infectious disease epidemiology can be considered to be one branch) – mathematical models play a prominent and sometimes central role.

The modern use of models in infectious disease epidemiology dates to the pioneering work of Ronald Ross (though the mathematical study of epidemics can in fact be traced back to work on smallpox by the great French mathematician Daniel Bernoulli [2]). Ross' work on the biomathematics of malaria originally led him to the conclusion that control of the vector would be the most efficient means of fighting malaria [18]. By the 1970s the use of mathematical models to study infectious diseases was increasing faster than exponentially, and substantial growth continues today. The range of analytical tools available to modelers has also increased considerably. In particular, the ready availability of ever-faster computers have made new types of models possible, and in the last few years new developments in statistics coupled with increased computational power have enabled more detailed and more accurate model-based analysis of epidemiological data.

Such approaches have only been applied to hemorrhagic fevers very recently [4, 9], however, and these methods have not so far been applied to CCHF. The aim of this chapter is to review the potential value of mathematical models for studying CCHF, describe the basic theory and key predictions from simple models, and to highlight some of the most important analytical techniques likely to be of value in studying CCHF. Here, we are exclusively concerned with understanding the system at the population level. Mathematical models also have a central role to play in understanding within-host progression of infectious diseases [15], but this is beyond the scope of this chapter.

15.2. THE USE OF MODELING APPROACHES FOR CCHF

There are at least four reasons why we might want to use mathematical models to study CCHF. First and foremost, they can help us to gain a qualitative understanding of the dynamics of the disease and, in this way, help us to improve our intuition. This is likely to be particularly important for studying nosocomial outbreaks where stochastic (chance) effects will be important. As casino owners know well, most people have rather poor intuition about chance, and the importance of such effects in epidemics in small populations comes as a surprise to many. Second, by highlighting key uncertainties and gaps in our knowledge models may suggest observational or experimental studies that would improve our understanding of key aspects of the whole system. This is likely to be particular important in the area of understanding the vector dynamics where major uncertainties exist for CCHF. Since models can also be considered to be hypotheses about the systems, by confronting models with data we can effectively choose between competing hypotheses. Third, models can help in the selection and evaluation of control policies. This can be done by employing models as statistical tools used to estimate the effect of interventions that have been made (and, equally importantly, to quantify the uncertainty in these estimates). Having quantified the effect of individual interventions, we can then go on to use models to ask "what-if" questions, using models predictively to determine the expected effect of hypothetical combinations of interventions. More generally, by enabling us to identify the most critical parameters affecting the behavior of the system, models can help us in setting priorities and identifying the most cost-effective control policies. Indeed, the use of dynamic models is essential for accurate economic analyses of control measures for infectious diseases [3]. Fourth, there is the potential to use models for forecasting the future of epidemics. Though popularly imagined as one of the main uses of models (perhaps by analogy to the models used to derive weather forecasts) this is one of the least developed areas in the infectious disease modeling literature, although there is increasing interest in this application.

15.3. MATHEMATICAL MODELS OF TICK-BORNE DISEASE TRANSMISSION

In this section we describe a basic framework for modeling tick-borne infections. This is adapted from the seminal work on the transmission dynamics of tick-borne infections by Medley et al. [12]. We show how this approach enables us to assess the magnitude of interventions needed to control CCHF in different regions and how interventions aimed at controlling the tick population could increase as well as decrease the risk to workers exposed to potentially infected animals. We also describe how this basic framework can be extended to address a wider class of questions.

The model presented by Medley et al. related specifically to the tick-borne transmission of *Theileria parva* in eastern Africa. The modeling framework described was, however, quite general, and with only minor modifications we can apply it to the study of CCHF. We begin by showing how seroprevalence data can be used to estimate the rate at which animals become infected with the CCHF virus. We assume that the rate at which an animal of age a is infected with the virus is $\beta(a)$ (mathematically this means that the chance an uninfected animal of age a is infected in a short time interval δt is approximately $\beta(a)\delta t$, the approximation becoming exact in the limit as δt approaches zero. The (a) following the β indicates that this rate is a function of age, and not necessarily constant). From this we can immediately write down a differential equation describing how the number of susceptible animals (those which have never been infected) changes with the age of the animals:

$$\frac{\mathrm{d}X(a)}{\mathrm{d}a} = -\beta(a)X(a) \quad (1)$$

Here we use $X(a)$ to represent the proportion of animals of age a who have not been infected with the virus and who are therefore seronegative. $\mathrm{d}X(a)/\mathrm{d}a$ is the rate at which $X(a)$ changes with age, so the equation specifies the slope of the graph plotting numbers seronegative ($X(a)$) against age. To obtain the model predictions for the actual relationship between $X(a)$ and a we solve Equation (1) by integration. This gives

$$X(a_i) = \exp\left(-\int_0^{a_i} \beta(a)\,\mathrm{d}a\right). \quad (2)$$

Two possible functional forms for $\beta(a)$ are $\beta(a) = b$, i.e. the infection rate is constant with age; and $\beta(a) = ba + c$, i.e. the infection rate starts from some baseline c and increases linearly with age. Substituting these into Equation (2) and solving gives $X(a_i) = \exp(-ba_i)$ and $X(a_i) = \exp(-ca_i - b^2\, a_i / 2)$, respectively. Equivalently these models predict that the numbers seropositive by a given age, which we call $S(a)$, are given by $S(a) = 1-\exp(-ba)$ and $S(a) = 1-\exp(-ca - ba^2/2)$, respectively. Many other functional forms for $\beta(a)$ are possible (e.g. the infection rate might saturate with increasing age), but these two are the simplest and have found to be adequate for explaining many age-seroprevalence profiles. Using these expressions for $S(a)$ we can estimate the infection rate by fitting the curves to age-seroprevalence profiles. Such profiles could be obtained by a longitudinal study, repeatedly sampling from the same animals over time, or – providing it was reasonable to assume the system was in equilibrium – by using data from a cross-sectional, age-stratified survey of animals.

Figure 15-1 below illustrates how the seroprevalence would be expected to change with age for both functional forms of $\beta(a)$. It also shows simulated data from a hypothetical cross-sectional study. In practice $\beta(a)$ would be estimated from such data by fitting curves for $S(a)$ corresponding to different functional

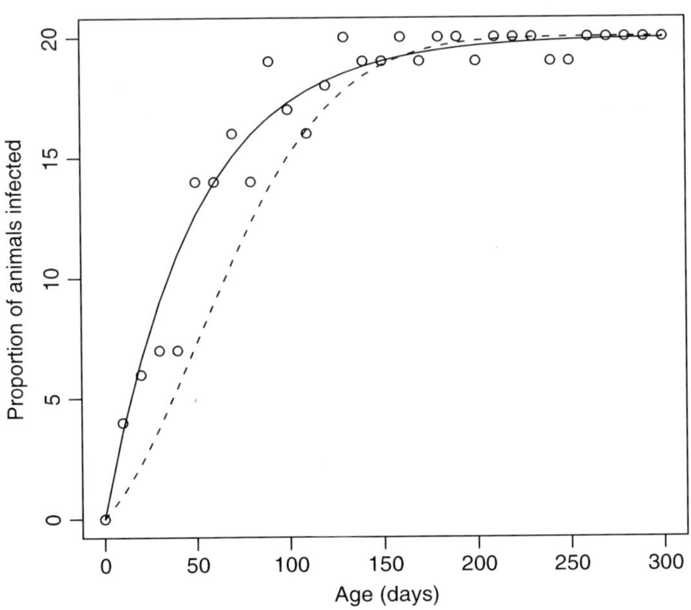

Fig. 15-1. Seroprevalence with age from a hypothetical cross-sectional study with 20 animals in each age group (0, 10, 20, . . ., 300 days). Solid line shows expected seroprevalence assuming $S(\text{age}) = 1 - \exp(-b \times \text{age})$ where $b = 0.02$. Dashed line shows expected seroprevalence when $S(\text{age}) = 1 - \exp(-c \times \text{age} - b \times \text{age}^2 / 2)$, with $b = 0.0002$ and $c = 0.004$. Dots illustrate how typical data from such a study might look when seroprevalence increases with age according to the first functional form (solid line) and is generated from this model assuming binomially distributed errors.

forms of $\beta(a)$ to the data. This can be done using standard maximum likelihood methods assuming binomially distributed data, allowing model parameters and their confidence intervals to be estimated. The best-fitting functional form of $\beta(a)$ can be selected with a likelihood ratio test.

The rate of infection, $\beta(a)$, is equal to the product of the rate of tick attachment as a function of age, $T(a)$, the probability of a tick being infected with the virus in an endemically stable environment, r^*, and the probability of transfer of infection from an infected tick to a host it is attached to, q. Rearranging this gives $T(a) = \beta(a)/(qr^*)$, so once estimates of $\beta(a)$ have been obtained using the method described above, if we also have estimates of q and r^* (which should be relatively easy to obtain), it becomes possible to derive estimates of the tick attachment rate as a function of host age.

When we also know the latent and infectious periods of the virus in a particular animal host, we can construct a dynamic transmission model of the course of infection in the population. Figure 15-2 gives a schematic illustration of the structure of such a model, which forms the basis of many disease transmission models. Each host is assumed to belong to one of four compartments: susceptible to infection (S), latently infected with the organisms (i.e. exposed) but not yet infectious (E), infectious (I), and recovered and immune (R).

As discussed above, the rate at which hosts become infected, $\beta(a)$, can be assumed to increase linearly with the proportion of ticks that are infected, and that proportion in turn would be expected to increase as the proportion of infected hosts increased. A full dynamic model of the system is needed to account for this feedback and a model incorporating both the tick and host dynamics would therefore allow $\beta(a)$ to change over time. However, when the system is in equilibrium (i.e. when the size of the host population and the amount of infection in that population is neither increasing nor decreasing with time, apart from small chance fluctuations) the proportion of ticks infected will not change over time and $\beta(a)$ will also not change over time. Under these circumstances it is possible to write down a system of ordinary differential equations that describes how the proportion of hosts in each compartment changes with the age of the hosts:

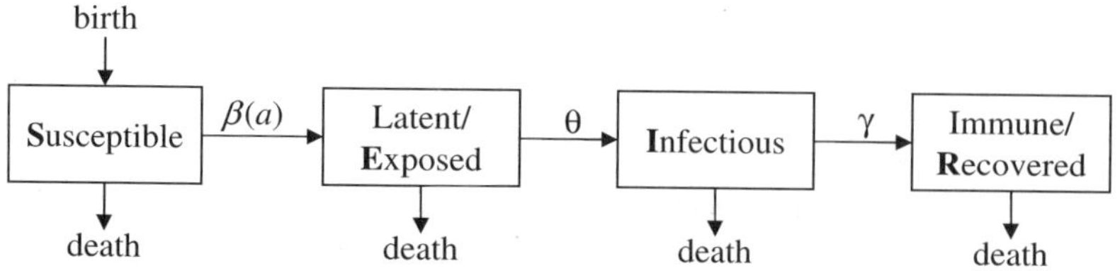

Fig. 15-2. Schematic diagram of a dynamic compartmental SEIR model.

$$
\begin{aligned}
\frac{\mathrm{d}S(a)}{\mathrm{d}a} &= -\beta(a)S(a) - \mu S(a) \\
\frac{\mathrm{d}E(a)}{\mathrm{d}a} &= \beta(a)S(a) - \theta E(a) - \mu E(a) \\
\frac{\mathrm{d}I(a)}{\mathrm{d}a} &= \theta E(a) - \gamma I(a) - \mu I(a) \\
\frac{\mathrm{d}R(a)}{\mathrm{d}a} &= \gamma I(a) - \mu R(a)
\end{aligned}
\tag{3}
$$

Here θ is that rate of progression from the latent to infectious compartment ($1/\theta$ gives the mean latent period) and γ is the rate of recovery from infection (the mean infectious period is then $1/\gamma$), and μ is the all-cause mortality rate (in general, this will be a function of age, but for simplicity we take it as a constant here implying that life expectancy is $1/\mu$ days). Essentially, these equations describe the flows between the compartments in Fig. 15-2: hosts leave the S compartment at rate $\beta(a)S(a)$ due to infection, and at rate $\mu S(a)$ due to death. Those becoming infected first flow into the E compartment at the same rate they leave the S compartment. They can leave the E compartment due to death (at rate $\mu E(a)$) or by progressing from the latently infected stage to the infectious stage (at rate $\theta E(a)$). Other terms in the above equations can be explained in a similar manner. Typical output for such a model when $\beta(a)$ is taken as a constant b is illustrated in Fig. 15-3.

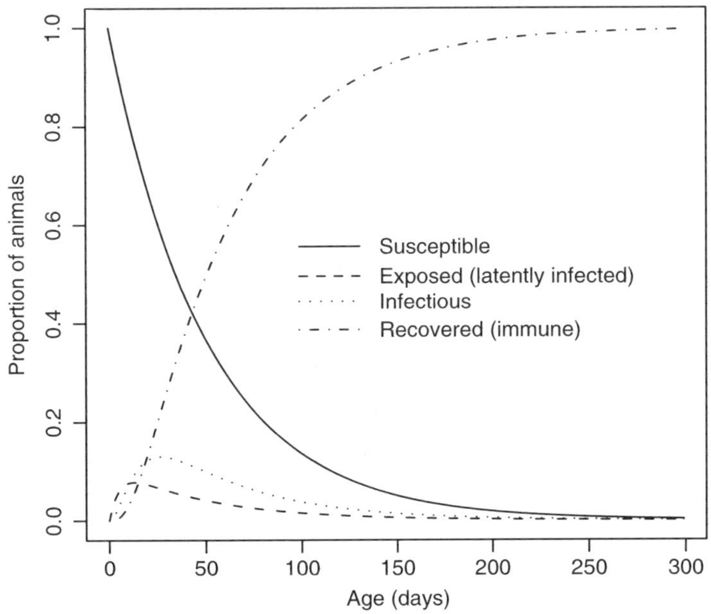

Fig. 15-3. Output from the SEIR model when mean infectious period ($1/\gamma$) is 10 days, mean latent period ($1/\theta$) is 5 days, $\mu = 0.0005$ and $b = 0.02$.

This shows the proportion of animals that are expected to be susceptible, latently infected, infectious, and immune under representative parameter values. It is noteworthy that with the relatively short infectious period assumed here (10 days), only a very small proportion of animals in most age groups are infected. This is despite the fact that infection rate, b, is high enough in this scenario to ensure that almost all animals surviving for a year or more will have been infected. Thus, despite the fact that the infection is highly endemic and would be hard to eradicate, the very low prevalence in all but the youngest age groups implies that the risk to workers handling potentially infected animals will be small provided that contacts are with animals more than about 150 days old. In general, the age profile of infectiousness will depend on all the model parameters (those used here are entirely arbitrary). Knowledge of the pattern of infection with age in different settings, however, could inform risk assessments and control policies aimed at minimizing exposures to infected animals.

A corollary of this observation is that control policies that aim to reduce the total tick population and hence the infection rate $\beta(a)$ (taken to be a constant, b, here) could, under some circumstances, increase the risk to humans. Such a perverse outcome could arise if people in high-risk occupations (veterinarians, slaughterhouse workers, etc.) were preferentially exposed to animals above the age at which most animals became infected in the absence of interventions. If the infection rate, b, was initially high, even large reductions in the tick population could have little effect on the total number of animals escaping infection, but could dramatically affect the ages at which animals became infected (Fig. 15-4). Progressive reductions in b have the effect of substantially increasing the likelihood that older animals are infected. Thus, for the highest value of b in this scenario, there are almost no infected animals which are older than 200 days. As b is reduced there becomes an appreciable chance that animals in these age groups will be infectious, putting those who have contact with them at risk. Such risks should be considered when evaluating the likely benefits of any control measure that aims to reduce infection but that is unlikely to lead to overall control of the disease.

The above equations provide a simple but general description of the infection process in an endemically stable environment (i.e. when disease incidence is neither increasing nor decreasing, and the infection rate $\beta(a)$ does not change with time). If we are interested in studying the temporal evolution of a system that is changing over time (perhaps due to the implementation of a control measure) we need to modify the approach. Most importantly, we expect the probability that a tick is infected, r, to change with the number of animal hosts infected. A simple and biologically plausible assumption is that r will increase linearly with the prevalence of infected hosts. This can be represented mathematically as $r(t) = cY(t)/N$ where c represents the probability that infection passes from an infectious hosts to an uninfected tick and N is the total number of hosts. The infection rate can then be expressed as a function of time (rather than age) as $\beta(t) = qTr(t - \varphi)$, where φ is the time a newly infected tick takes to develop and become infectious. If we

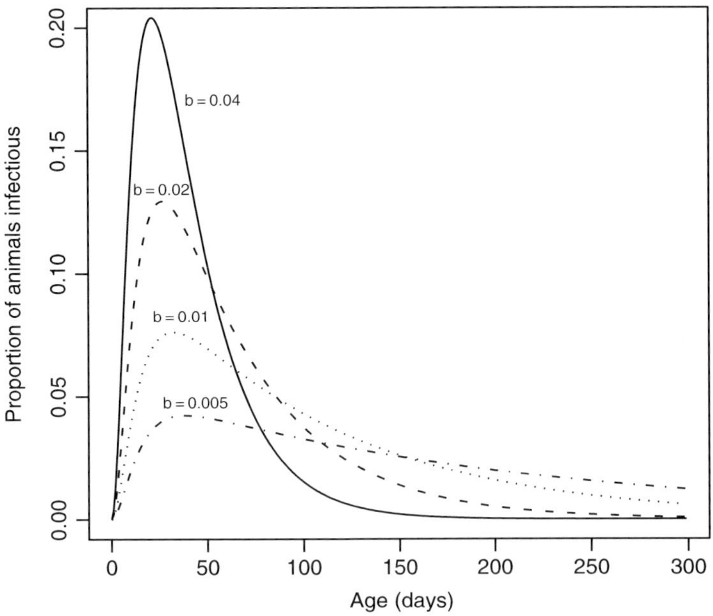

Fig. 15-4. Proportion of animals which are infectious by age of animal. Results from the SEIR model under the same assumptions as Fig. 15-3 are shown, except that the infection rate, $\beta(a) = b$, is progressively reduced by factors of 2.

replace $\beta(a)$ in Equations (3) above with this expression for $\beta(t)$, add a term μN to right-hand side of the Equation (1) to represent births (assumed to balance deaths from all compartments), and replacing a with t, we will have a model suitable for studying the nonequilibrium situation.

One of the most useful concepts in infectious disease epidemiology is the case reproduction number: the average number of secondary cases caused by a primary case. It is useful to distinguish between two reproduction numbers: the basic reproduction number (R_0) and the effective reproduction number (R_t). R_0 is defined as the average number of secondary cases produced by a typical primary case in an otherwise fully susceptible population in the absence of control measures. R_t is defined similarly, except that the population is not required to be fully susceptible and there may be control measures in place. In the absence of control measures R_t can be calculated as the product of R_0 and the proportion of the population that is susceptible. The value of R_t is therefore always less than or equal to R_0.

R_0 is of central importance because its value determines whether or not an epidemic is possible; only when it takes a value greater than one, so that each primary cases generates on average more than one secondary case, can the chain reaction that constitutes an epidemic proceed. When R_0 is less than 1 there may still be chains of disease transmission, and occasionally these may even be quite long, but these will be self-limiting and have no chances of leading to the self-sustaining chain reaction of a full-blown epidemic that affects thousands.

R_t is important since it determines the rate of epidemic growth at a given point in time. During the course of an epidemic R_t will decline from its initial value of R_0 as the number of susceptibles decreases. It may also decline as result of interventions designed to control the disease. When R_t is equal to 1 the epidemic is neither growing nor falling. If there is no supply of new susceptibles, R_t will proceed to fall below 1 as the pool of susceptibles decreases further. If there is a sufficient supply of new susceptibles (through birth, loss of immunity, or immigration) it is possible to reach an endemic equilibrium state where the generation of new susceptibles matches the loss of susceptibles due to infection. In this case R_t is maintained at a value of 1, and the amount of infection neither increases nor decreases over time. Cyclic behavior can occur when the epidemic causes R_t to fall well below 1, and a longer-term increase in susceptibles due to births eventually restores R_t to a value above 1 permitting another epidemic. This is the cause of the cyclic pattern of childhood diseases such as measles [8].

For a tick-borne disease, we can define R_0 as the average number of secondary infections in hosts from one primary infected host (when all hosts and ticks are susceptible) or, equivalently, as the average number of infectious ticks that arise from a single infected tick in a susceptible population. This definition immediately leads to an expression for R_0 in terms of parameters we have already introduced:

$$R_0 = \frac{qTc\theta}{(\gamma + \mu)(\theta + \mu)} \quad (4)$$

This can be derived by multiplying the probability that the first tick infects its host, q, the probability that the host survives to the infectious state, $\theta/(\theta + \mu)$, the mean number of ticks attaching to the infectious host (which is equal to product of the average duration of the host's infectious period, $1/(\gamma + \mu)$, and the mean rate of tick attachment, T), and the probability that each tick that attaches becomes infected, c. Using this formula (and estimates for the parameters) the impact of interventions on R_0 can be derived in terms of expected impacts of intervention on different aspects of the system. In particular, it is possible to calculate the reduction of the tick attachment rate, T, that would be needed to reduce R_0 to below 1, resulting in local eradication of the disease. An alternative formulation expresses R_0 in terms of the equilibrium infection rate β^*: $R_0 = 1 + \beta^*/\mu$. This holds only when β^* is greater than zero and, therefore, when R_0 is greater than 1. It is useful because an estimate of β^* is relatively easily obtained from age-stratified seroprevalence data as described above, and enables a simple assessment of the likely effort needed to eradicate the virus in a given population.

The above notes provide only a very broad outline of a framework for modeling tick-borne infections, but one that could easily be applied to CCHF if supported by appropriate field research. Such an approach can, and – in the context of other pathogens – has been extended in a variety of ways; for example, to address the particular biological details of different pathogens, to more fully describe the dynamics of the tick population [16], and to provide economic analyses of control policies [14]. In particular, when modeling interventions that are likely to affect the tick population, models that describe details of the tick life cycle (and how these are affected by the intervention) will usually be required. One interesting recent example of this is due to Ogden et al. [17], who developed a dynamic population model of the tick *Ixodes scapularis*, dividing the tick population into 12 developmental stages, with the aim of investigating the effects of climate (and predicting the effects of climate change) on the range and seasonality of the tick.

Though such complex models are becoming quite common and are, in many cases, entirely appropriate, the degree of detail that should be included in a model will vary according to the application. The general question of how complex models should be was succinctly addressed by Albert Einstein, who said that models should be as simple as possible, but no simpler. This applies as well to infectious disease epidemiology as it does to astrophysics: models should be only as complex as is required to address the questions at hand. Many unfamiliar with mathematical models naively believe that more complex models will provide better answers to most questions. In fact, the reverse is usually true, and for most purposes, including prediction, surprisingly simple models tend to perform better.

15.4. MODELING THE NOSOCOMIAL TRANSMISSION OF CCHF

The approach described above for modeling the tick-borne transmission of CCHF virus in animal hosts used a deterministic formulation. When the populations under investigation are large, this is a reasonable approach: the time evolution of the system is likely to be quite predictable and individual chance events (e.g. whether or not one animal gets infected, how long it takes another to recover) are unimportant. Just as casinos may lose on individual bets but are sure to win in the long run, in a large population when many are infectious there are so many individual unpredictable events that the eventual outcome becomes highly predictable. In small populations, such as hospital units, and at the beginning of epidemics in both large and small populations when only small numbers are infected, this deterministic approach fails badly. The details of the random events that make up an epidemic become important. For example, if the first infected person happens to die before he has a chance to infect anyone an epidemic will not occur, even if it had the potential to (i.e. even if R_0 was greater than 1). A deterministic model would predict a major epidemic every time, which is clearly unrealistic.

The importance of such chance events is illustrated in Figs. 15-5–15-7. These show results from Monte Carlo simulations from a stochastic susceptible-exposed-infectious-recovered (SEIR) model which has a structure similar to that

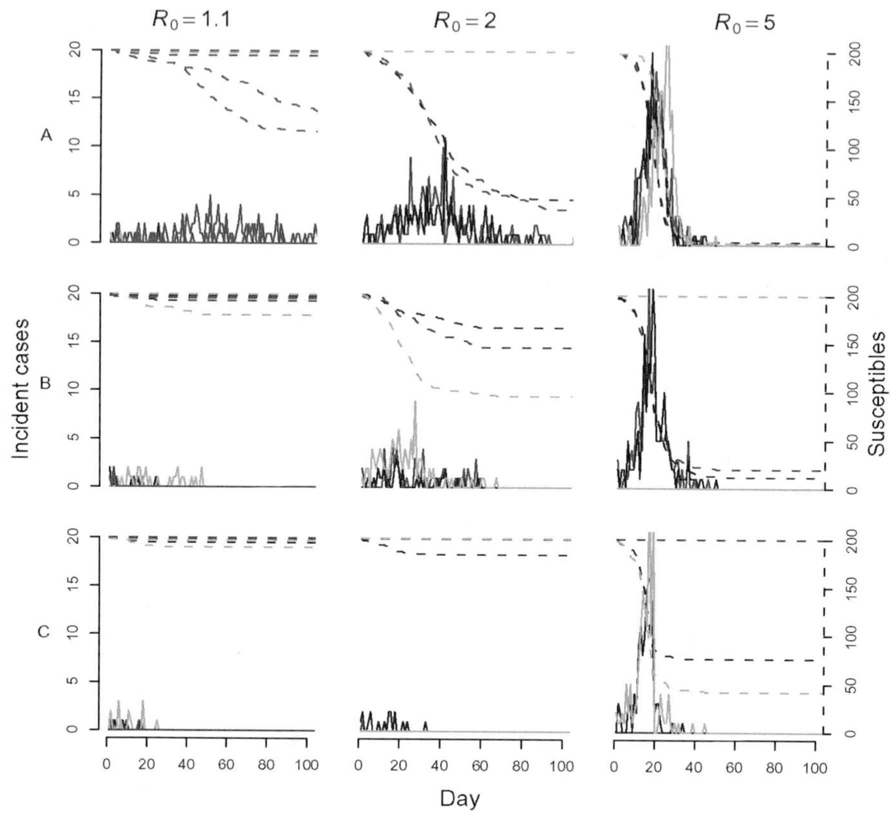

Fig. 15-5. Incident cases (*solid lines*) and number of susceptibles (*broken lines*) from stochastic simulations of an SEIR epidemic model when there is no intervention (row A), an intervention to reduce the probability of transmission from each case by 50% after 20 days (row B), and an intervention to reduce this probability by 90% (row C). Columns (*left* to *right*) show results for initial R_0 values of 1.1, 2, and 5. Three runs from each scenario are shown, except when R_0 is 1.1, when 10 runs are shown (since in this case in most simulations the epidemic does not take off). A mean incubation period of 5 days is assumed (daily probability of progressing from latent to infectious is 0.2) and a mean infectious period of 3.3 days (daily probability of ceasing to be infectious is 0.3), and at day 0 there are 200 susceptibles and one latent case.

shown in Fig. 15-2 (except that births and nondisease-related deaths can be ignored because we are interested in short timescales over which it is reasonable to assume a fixed population size). This simple model could be considered to provide an approximate description of transmission in a small population such as a hospital unit that has stopped admitted new patients (when new patients continue to be admitted the constant supply of susceptibles leads to rather different dynamics [5]). The model assumes that each day each infectious person has some fixed chance of infecting each susceptible person and some fixed probability, γ, of ceasing to be infectious. When infected, individuals enter a latent compartment, with a daily probability, θ, of progressing to the infectious compartment. The basic reproduction number, R_0, in this model is equal to the mean infectious period ($1/\gamma$) multiplied by the probability that a given susceptible person is infected by one infectious case on a given day, multiplied by the initial number of susceptibles. Figure 15-5 (top row) indicates typical model outcomes in a population of 200 initially susceptible people when no intervention to control the epidemic is made. When R_0 is greater than 1 there is a chance of an epidemic. This chance increases with R_0 (Fig. 15-6). In contrast to the deterministic model, there is also a real chance that that the epidemic dies out almost immediately. This can be seen in Fig. 15-5, where even though R_0 is greater than 1 in some of the simulation runs the epidemic fails to take off and the susceptible population stays near to its initial value of 200. Figure 15-6 shows that even when $R_0 = 2$, there is no secondary transmission at all in about one third of the simulations. As R_0 increases above 1 there is an increasing chance that if there is any secondary transmission a large number of infections will result. This gives rise to bimodal distribution when R_0 is above 1 (Fig. 15-6). It is also noticeable that as R_0 increases, when the epidemic does take off it tends to peak earlier and affects more people, though the precise course of the epidemic is not predictable (Fig. 15-5).

If an intervention is able to reduce the effective reproduction number, R_t, below 1 by reducing the probability of transmission from each case then the

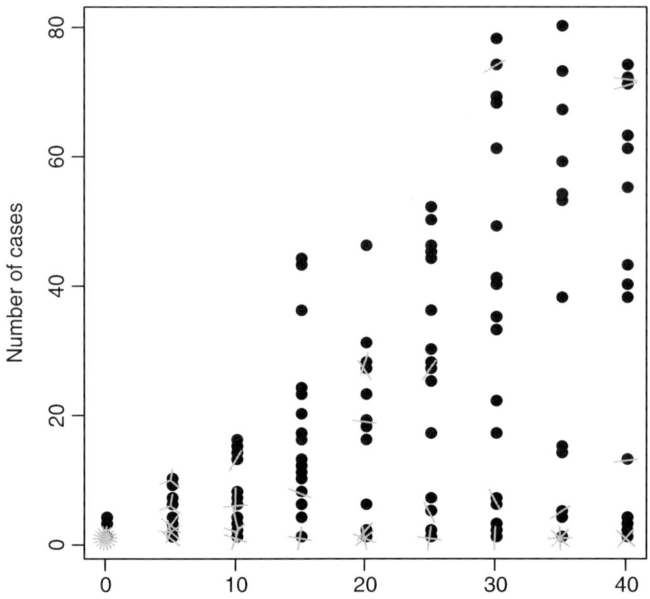

Fig. 15-6. Distributions of outbreak sizes for different values of the basic reproduction number, R_0. Results from 1,000 simulation runs are shown for each value of R_0. Model parameters as in Fig. 15-5.

epidemic will be controlled. This is shown in rows B and C of Fig. 15-5. When transmission probability is reduced by 50% after 20 days (row B) R_t is reduced to approximately half the initial R_0 value. In the first column this is sufficient to control the epidemic (because $R_t \approx 0.55$ at day 20), and though some transmission persists the epidemic comes to an end soon. When R_0 is 2 a 50% reduction in transmission gives an $R_t \approx 1$ at day 20, which is enough to permit prolonged transmission at a low level, but not to allow a large epidemic. In contrast, when R_0 is 5 a 50% transmission reduction brings R_t down to about 2.5. This is not enough to control the epidemic and most of the susceptibles go on to be infected (though fewer than would have been without the intervention). In contrast, the 90% reduction in transmission in row C is sufficient to reduce R_t below 1 in all cases, and in all simulations runs the epidemics are quickly brought under control. Nevertheless, the large number of infected cases by day 20 when R_0 is 5 is sufficient to ensure continued transmission for some time after the intervention.

Figure 15-7 shows the effect of delays in an intervention that is able to bring about control of the epidemic but unable to prevent all transmission. In this

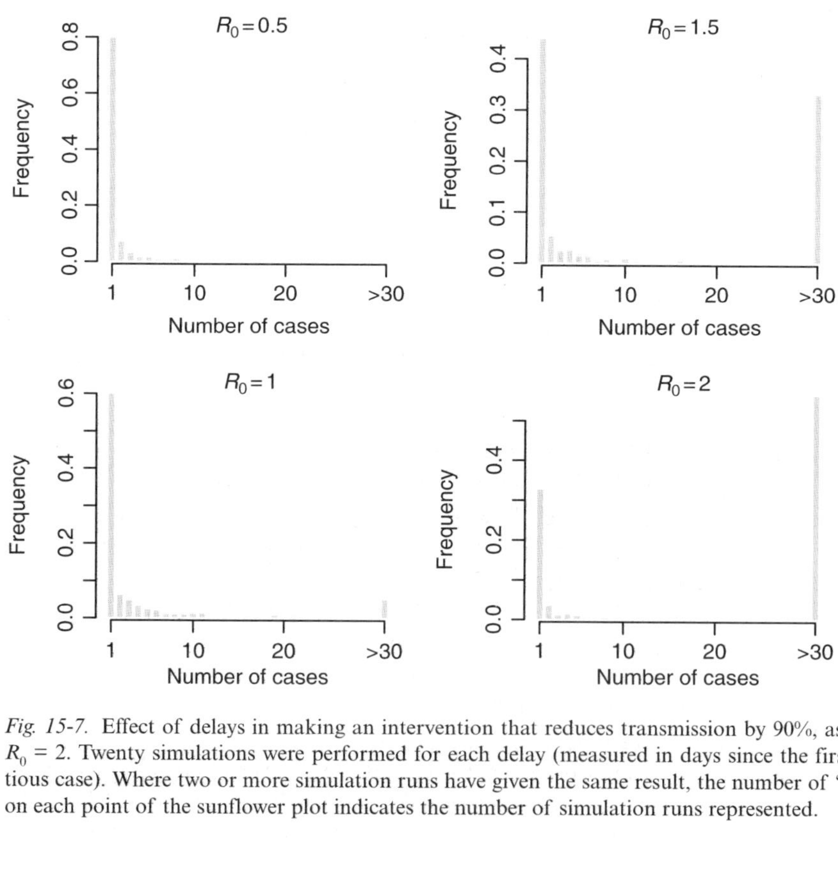

Fig. 15-7. Effect of delays in making an intervention that reduces transmission by 90%, assuming $R_0 = 2$. Twenty simulations were performed for each delay (measured in days since the first infectious case). Where two or more simulation runs have given the same result, the number of "petals" on each point of the sunflower plot indicates the number of simulation runs represented.

scenario if an intervention able to reduce the probability of transmission by 90% is made as soon as the first case is detected there is a high probability that there almost all secondary spread can be eliminated. Delays of 5 days or more lead to significant chances that much larger numbers will be affected, though there remains a substantial chance that the epidemic will die out of its own accord and affect few people. This accounts for the bimodal distribution in the number of cases that starts to become apparent when interventions are delayed by 20 days or more.

15.5. FUTURE DIRECTIONS FOR USING MODELS TO HELP UNDERSTAND CCHF

Future work modeling the spread of CCHF, whether tick-borne or nosocomial, will depend on good quality field data to enable necessary parameter estimates to be obtained and an appropriate model structure selected. Precise details of models will depend to a large extent on the questions that are being asked.

The simple SEIR modeling framework illustrated here can readily be extended to account for the fact that not all individuals infected will themselves become infectious. This can be done by constructing models that allow people to move with some probability straight from the exposed to the recovered compartment. It is also relatively easy to further modify this basic model structure to better capture observed distributions of the latent and infectious periods and to account for variable infectivity with time since infection. Appropriate modifications can also be made to account for more complex mixing patterns (e.g. probabilities of patients infecting health-care workers and other patients may differ).

Models have also proved to be useful for evaluating the role of contact tracing, quarantine, and isolation in the control of infectious diseases. This work has shown how the epidemiological characteristics of different diseases (the basic reproduction number, the length of the latent and infectious periods, and the amount of transmission that occurs prior to the onset of symptoms) largely determine the likely success of control using these measures [7]. As one might expect, the properties of CCHF put it well within the region where these measures can be expected to be effective. The severe acute respiratory syndrome (SARS) epidemic also highlighted the high degree of variability in the number of secondary cases produced by each primary case [10]. When detailed contact tracing data are examined it turns out that this pattern is seen in many infectious diseases [11]. This variability may greatly exceed that assumed in the simple stochastic models presented here. It is, nevertheless, a simple matter to account for such variability in models. Its main impact would be to make major epidemics rather less likely for a given R_0, and to make the role of chance even more dominant.

In many cases it is useful to estimate R_0. The estimate will tell us how close we are to a risking a major epidemic (if R_0 is currently below 1) and allow precautionary measure to be taken. If R_0 is greater than 1 the estimate tells us how much an intervention would have to do to bring about control or eliminate the chance of a major epidemic. An approach that has proved useful for other diseases is to estimate R_0 from the distribution of the number of cases from clusters of transmission [6]. This is possible because the probability of 0, 1, 2, . . . secondary cases varies with the value R_0 (Fig. 15-6), so the likelihood of different R_0 values can be derived from the distribution of the number of cases.

When more detailed surveillance data are available other approaches can be used to provide much better estimates of R_0 and to assess the impact of interventions. The best of these uses computationally intensive Markov chain Monte Carlo algorithms to estimate the basic reproduction number and other model parameters and quantify the uncertainty in these estimates. Such an approach has recently been used to estimate the basic reproduction number for Ebola and to evaluate the role of interventions in reducing transmission [9]. This method could be adapted relatively easily to study CCHF transmission, and detailed data from outbreaks used to evaluate the evidence of effectiveness for different interventions. Much simpler approaches based on deterministic approximations are also possible [4], but these methods appear not to accurately characterize the uncertainty and may therefore be inappropriate for assessing the evidence of effectiveness of different interventions [9].

REFERENCES

1. Anderson RM, May RM (1991) Infectious Diseases of Humans: Dynamics and Control. Oxford University Press, Oxford
2. Bernoulli D (1766) Essai d'une nouvelle analyse de la mortalité causée par la petite vérole. Mém Math Phys Acad R Sci Paris 1–45; English translation by Bradley, L. in Smallpox Inoculation: An Eighteenth-Century Mathematical Controversy, Adult Education Department, Nottingham, 1971
3. Brisson M, Edmunds WJ (2003) Economic evaluation of vaccination programs: the impact of herd-immunity. Med Decis Making 23:76–82
4. Chowell G, Hengartner NW, Castillo-Chavez C, Fenimore PW, Hyman JM (2004) The basic reproductive number of Ebola and the effects of public health measures: the cases of Congo and Uganda. J Theor Biol 229:119–126
5. Cooper BS, Medley GF, Scott GM (1999) Preliminary analysis of the transmission dynamics of nosocomial infections: stochastic and management effects. J Hosp Infect 43:131–147
6. De Serres G, Gay NJ, Farrington CP (2000) Epidemiology of transmissible diseases after elimination. Am J Epidemiol 151:1039–1048
7. Fraser C, Riley S, Anderson RM, Ferguson NM (2004) Factors that make an infectious disease outbreak controllable. Proc Natl Acad Sci USA 101:6146–6151
8. Keeling MJ (1997) Modelling the persistence of measles. Trends Microbiol 5:513–518
9. Lekone PE, Finkenstädt BF (2007) Statistical inference in a stochastic epidemic SEIR model with control intervention: Ebola as a case study. To appear in Biometrics. Available online at: http://www.blackwell-synergy.com/doi/abs/10.1111/j.1541–0420.2006.00609.x
10. Lipsitch M, Cohen T, Cooper B, Robins JM, Ma S, James L, Gopalakrishna G, Chew SK, Tan CC, Samore MH, Fisman D, Murray M (2003) Transmission dynamics and control of severe acute respiratory syndrome. Science. 300:1966–1970
11. Lloyd-Smith JO, Schreiber SJ, Kopp PE, Getz WM (2005) Superspreading and the effect of individual variation on disease emergence. Nature 438:355–359

12. Medley GF, Perry BD, Young AS (1993) Preliminary analysis of the transmission dynamics of *Theileria parva* in eastern Africa. Parasitology 106:251–264
13. Meltzer MI, Norval RAI (1993) Mathematical models of tick-borne disease transmission. Parasitology Today 9:277–278
14. Mukhebi AW, Chamboko T, O'Callaghan CJ, Peter TF, Kruska RL, Medley GF, Mahan SM, Perry BD (1999) An assessment of the economic impact of heartwater (*Cowdria ruminantium* infection) and its control in Zimbabwe. Prev Vet Med 39:173–189
15. Nowak M, May RM (2000) Virus Dynamics. Oxford University Press, Oxford
16. O'Callaghan CJ, Medley GF, Peter TF, Perry BD (1998) Investigating the epidemiology of heartwater (*Cowdria ruminantium* infection) by means of a transmission dynamics model. Parasitology 117:49–61
17. Ogden NH, Bigras-Poulin M, O'Callaghan CJ, Barker IK, Lindsay LR, Maarouf A, Smoyer-Tomic KE, Waltner-Toews D, Charron D (2005) A dynamic population model to investigate effects of climate on geographic range and seasonality of the tick *Ixodes scapularis*. Int J Parasitol 35:375–389
18. Ross R (1911) The Prevention of Malaria. Murray, London

SECTION IV

CLINICAL FEATURES, PATHOGENESIS, AND TREATMENT

CHAPTER 16

CLINICAL AND PATHOLOGIC FEATURES OF CRIMEAN-CONGO HEMORRHAGIC FEVER

ONDER ERGONUL, M.D., M.P.H.

Marmara University, School of Medicine, Infectious Diseases and Clinical Microbiology Department, Istanbul, Turkey. E-mail: ondererergonul@yahoo.com

16.1. INTRODUCTION

This chapter describes the story of the clinical process starting from the entrance of the virus to the human body to the death, if happens. Humans are the only known host of Crimean-Congo hemorrhagic fever virus (CCHFV) in which disease is manifested [18]. In general, the people living in endemic areas are at risk. But some of the people have higher risk, which were described in Section V. Among the people, who are at risk, some can get the infection, and then some get the disease (Box 16-1). In a Russian study, the probability of getting the disease for subjects who had been infected was found to be 0.215; in other words, the estimated ratio of apparent to inapparent infections suggests that one of every five persons infected develops Crimean-Congo hemorrhagic fever (CCHF) [15]. This rate was reported to be much higher than other diseases such as western equine encephalitis, St. Louis encephalitis, and tick-borne encephalitis [15]. In another study from Turkey, these rates were found to be similar [14]. In that study, the infection rate was 15/55 (27%) in the endemic region, and the attack rate was 11/55 (20%). The infection rate was significantly higher 8/19 (42%) among the individuals, who had the history of tick bite [14]. The gender and the age distribution among the cases are closely related to the exposure risk to the ticks in endemic regions. Accordingly, the rate of the diseased female is 25% in Iran [1], whereas 50% in Turkey [11].

16.2. CASE DEFINITION

Defining a case is a fundamental step in the development of a surveillance system. Surveillance definitions must balance competing needs for sensitivity, specificity, and feasibility (Box 16-2). Because of the need for simplicity, surveillance

O. Ergonul and C. A. Whitehouse (eds.), Crimean-Congo Hemorrhagic Fever, 207–220.

Box 16-1. Basic epidemiologic definitions used in the text
Infection rate: The proportion of individuals who are exposed to CCHFV who become infected, namely antibody-positive. *Attack rate*: The proportion of individuals who are exposed to an infectious agent who become clinically ill. *Index cases*: Clinically ill patients who had been admitted to the hospital because of overt CCHF infection. *Household members*: The individuals who live in the same house with the index cases. *Case fatality rate*: The proportion of the people, among those who develop a disease, who then proceed to die from the disease [24].

Box 16-2. Attributes of surveillance in CCHF infections
Sensitivity: To what extent does the system identify all of the cases in the target population? In CCHF infections, high sensitivity is required. *Timeliness*: This attribute refers to the entire cycle of information flow, ranging from information collection to dissemination. All the CCHF infections need timeliness. *Representativeness*: To what extent do events detected through the surveillance system represent persons with the condition of interest in the target population? The geographic locations should be specified for the tick bites. *Predictive value*: To what extent are reported cases really cases? Differential diagnosis for both clinical and laboratory measures is necessary. *Accuracy and completeness of descriptive information*: Forms for reporting health events often include descriptive personal information, such as demographic characteristics, clinical pattern of disease, or potential exposures. To what extent are these sections of forms completed? Is the information sufficiently reliable? *Simplicity*: Are forms easy to complete? Is data collection kept to a necessary minimum? *Flexibility*: Can the system change to address new questions? Can it adapt to evolving standards of diagnosis or medical care? [4]

case definitions are typically brief. When the definitions apply to the diseases they generally combine laboratory criteria with clinical manifestations [6]. For an individual infection, disease, or health problem, no single definition is ideal. Rather, appropriate definitions vary widely in different settings depending on information needs, methods of reporting or data collection, and staff training [4]. In CCHF infections, the precision of the case definition depends on the diagnostic facilities in a country. If polymerase chain reaction (PCR) is readily available, the diagnosis is likely within 4 h, which is usually not the case. For the provision of appropriate therapy, the cases should be classified according to their evidence of infection.

The suspected case:

- Individuals, who had fever, myalgia, malaise, diarrhea
- History of being in endemic area
  - Tick exposure history
  - Residency or travel to CCHF endemic region.

The probable case:

- Patients who had leukopenia, thrombocytopenia, elevated aspartate aminotransferase (AST), alanine aminotransferase (ALT), and lactate dehydrogenase (LDH) levels.

Confirmed case:

- CCHF immunoglobulin M (IgM) of PCR positivity in the blood or body fluids of the patient.

16.3. DISEASE COURSE

The typical course of CCHF infection was defined in four distinct phases, as incubation, prehemorrhagic, hemorrhagic, and convalescence periods [18] (Fig. 16-1). *The incubation period*, after a tick bite is usually considered to be short, 3–7 days, but it is difficult to obtain precise data [11, 29]. The incubation period could differ depending on several factors including viral dose and route of exposure, for instance, it could be shorter among bloodborne transmissions. In South Africa, the time to onset of disease after exposure to tick bite was 3.2 days, to blood or tissue of livestock was 5 days, and to blood of human cases was 5.6 days [28]. The mean duration of the disease course before the hospital was reported to be 5.5 days in Turkey [10], and 3.5 days in United Arab Emirates [26].

Almost all the patients had a *prehemorrhagic period*, which is characterized by sudden onset of fever, headache, myalgia, dizziness [11, 18, 33] (Table 16-1). Fever lasted in average for 4–5 days [18]. A typical fever pattern is not seen, and

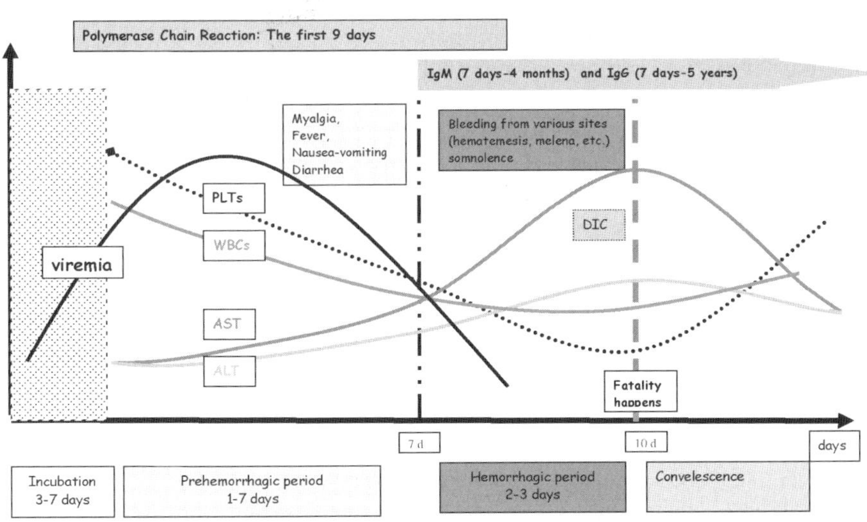

Fig. 16-1. The infection course. The starting point is the entrance of the CCHFV to the human through tick bite or a contact with infected material such as body fluids. [11] (*See Color Plates*)

Table 16-1. Symptoms of the patients [1, 3, 11, 12, 19, 22, 29]

Prehemorrhagic symptoms	%
Malaise	95–100
Nausea-vomiting	75–90
Myalgia	65–100
Fever	75–90
Headache	75
Diarrhea	30–40
Hemorrhagic symptoms	
Fever (temperature > 38°C)	21 (42)
Bleeding	
Hematemesis	7–35
Melena	1–15
Epistaxis	17–50
Hemoptysis	10
Hematuria	10–20
Vaginal	10
Gingival	8
Intraabdominal	2
Intracerebral	1
Multiple sites	3–25

the body temperature rarely exceeds 40°C. Additional symptoms of diarrhea, nausea, and vomiting could be seen [26, 29, 33]. Hyperemia of the face, neck, and chest, congested sclera, and conjunctivitis are also noted. This period lasts for 1–5 days with the average of 3 days [11]. In some of the patients, the disease course is limited to the prehemorrhagic period, and these patients may not be reported, since they possibly would not seek for further medical support. In fact, the case reports or the case series usually record the patients, whose disease extended to hemorrhagic period.

Hemorrhagic period is short, rapidly develops, and usually begins at the 3rd–5th days of disease. There is no relationship between degree of temperature and the onset of hemorrhages [11, 18]. The hemorrhagic manifestations range from petechiae to large hematomas appears on the mucous membranes and skin (Fig. 16-2). Bleeding from other sites including vagina, gingival bleeding, and cerebral hemorrhage was reported [11]. The most common bleeding sites are the nose, gastrointestinal system (hematemesis, melena, and intraabdominal), menometrorrhagia, and urinary tract (hematuria), and the respiratory tract (hemoptysis) [12, 33] (Table 16-1).

Atypical cases of CCHF are usually related to the bleeding from unexpected sites. In a patient with stubborn abdominal pain, acute appendicitis was suspected, but the hemorrhage and bleeding at the internal and external oblique

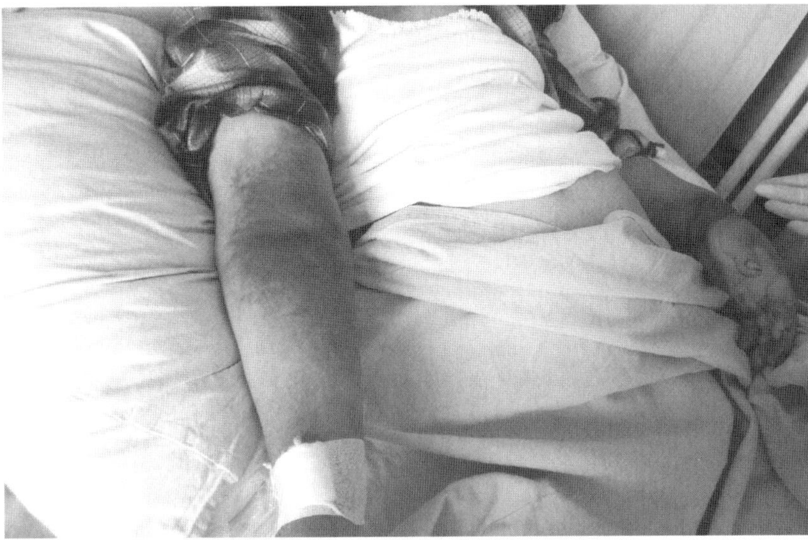

Fig. 16-2. CCHF patients, archives of Ankara Numune Education and Research Hospital. (*See Color Plates*)

muscles and cecum were detected with no pathology of appendix [7]. Bleeding to any space in the body is possible, and development of hemothorax or cerebrovascular hemorrhages could occur.

Hepatomegaly and splenomegaly were reported to be in one third of the patients, previously [18]. Hepatomegaly was detected in 20–40% of the cases [3, 10, 19, 22]. Hepatorenal insufficiency was reported from South Africa [29]. However, only one hepatorenal insufficiency was reported in the recent clinical studies [22]. Jaundice and hyperbilirubinemia are not seen in CCHF, as a differential feature from the acute viral hepatitis. However, jaundice was reported in a confirmed CCHF patient recently [22]. Splenomegaly was reported as 14–20% [3, 22] (Table 16-2).

The convalescence period among the survivors begins about 10–20 days after the onset of illness. The hospital stay is around 9–10 days [10, 26]. In convalescence period, labile pulse, tachycardia, temporary complete loss of hair, polyneuritis, difficulty in breathing, xerostomia, poor vision, loss of hearing, and loss of memory were reported in the Soviet literature [18], although none of these findings were noted in the recent reports.

Cardiovascular changes, such as bradycardia and low blood pressure were reported in an earlier review in 1979 [18], and noted only in two confirmed patients in a recent study [22] (Table 16-2). The complications because of the interventions in the hospital could be seen such as nosocomial infections such as device-related infections or bacteremia.

Table 16-2. Signs of the patients [1, 3, 11, 12, 19, 22, 29]

	%
Fever	45–85
Hepatomegaly	30–40
Lymphadenopathy	15–40
Bleeding	30–50
Maculopapular rash	30–60
Petechia and ecchymosis	30–45
Splenomegaly	15–20
Conjunctivitis	10–50
Cardiac involvement	1–10
Jaundice	1–10
Somnolence	10–20

16.4. HISTOPATHOLOGIC STUDIES

Histopathologic studies have been limited to a small number of cases [5]. In retrospective analysis of 12 cases in South Africa, the clinicopathologic features of CCHF and the diagnostic role of virus isolation as compared with serology, immunohistochemistry, and in situ hybridization were studied. The histopathologic features of CCHF were found to be resembling to the other viral hemorrhagic fevers. Immunohistochemistry and in situ hybridization analyses showed that the mononuclear phagocytes, endothelial cells, and hepatocytes are main targets of infection. Association of parenchymal necrosis in liver with viral infection suggested that cell damage might be mediated by a direct viral cytopathic effect [5]. In bone marrow biopsy of the patients, reactive hemophagocytosis was detected in 7 (50%) of 14 CCHFV-infected patients, which suggested that hemophagocytosis can play a role [19] (See also Chapter 17).

16.5. BLOOD COUNT AND BIOCHEMICAL TESTS

Blood count and biochemical tests are crucial in the early diagnosis of the disease. The rapid diagnostic microbiologic methods to detect the virus might not be available in all the settings. Therefore, blood count and biochemical tests get significance for the diagnosis and start of appropriate therapy.

Thrombocytopenia appears to be a consistent feature of CCHF infection [11, 29, 34]. The patients had leukopenia, and elevated levels of AST (serum glutamic-oxaloacetic transaminase [SGOT]), alanine transferase (ALT; serum glutamic-pyruvic transaminase [SGPT]), LDH, creatinine phosphokinase (CPK). The central tendency measures of the laboratory tests for favorable and fatal cases were presented in Table 16-3. The rate of AST to ALT is normally 0.7–1.4. However, this

Table 16-3. Pathologic laboratory findings for 75 patients (unpublished study)

	Favorable cases Mean (min–max) $N = 68$	Fatal cases Mean (min–max) $N = 7$	Significant difference
Longest prothrombin time (s)	15 (10–42)	27 (18–39)	Yes
Longest activated partial thromboplastin time (s)	42 (23–74)	69 (50–92)	Yes
Lowest platelet count (platelets/mm^3)	28,000 (3,500–108,000)	10,000 (6,000–15,000)	Yes
Highest alanine transferase level (U/L)	307 (38–1,443)	975 (219–1,785)	Yes
Highest aspartate transferase level (U/L)	820 (65–7,150)	2,600 (773–7,700)	Yes
Lowest fibrinogen level	363 (165–780)	160 (86–140)	Yes
Highest lactic dehydrogenase level (U/L)	2,220 (283–26,000)	3,870 (1,980–9,480)	No
Highest creatinine phosphokinase level (U/L)	1,095 (10–21,189)	1,618 (1,388–2,164)	No
Lowest WBC (WBCs/mm^3)	1,900 (150–13,000)	4,700 (1,200–12,000)	No

rate is increased in some diseases [32], and the level of the AST is almost always higher than the level of ALT among CCHF patients. This observation indicates the involvement of the organs besides the liver. In CCHF infection, AST or ALT is increased, and was suggested as a prognostic factor among CCHF cases [10]. The higher AST level disproportionate to ALT level was also described for Marburg fever in 1969 [20], and Lassa fever in 1987 [21].

Coagulation tests such as prothrombin time (PT), activated partial thromboplastin time (aPTT) are prolonged. Since there could be problems in standardization of PT and aPTT, the use of international normalized ratio (INR) is suggested. Fibrinogen level might be decreased, and fibrin degradation products could be increased. Laboratory tests including complete blood count and biochemical tests returned to normal levels within approximately 5–9 days [12] among surviving patients (Fig. 16-1).

16.6. PREDICTORS OF FATALITY

Swanepoel et al. [29] in South Africa described clinical laboratory criteria that could be measured early in the course of disease (first 5 days) and that predicted a fatal outcome in 90% of patients with any of these findings (see also Chapter 11);

1. White blood cell (WBC) count $\geq$ 10,000/mm^3
2. Platelet count (PLT) $\leq$ 20,000/mm^3

3. AST ≥ 200 U/L or ALT ≥ 150 U/L
4. aPTT ≥ 60 s
5. Fibrinogen ≤ 110 mg/dL

Other case series have confirmed that levels of AST, ALT, and LDH were significantly higher among severe cases [10, 12]. In a study from Turkey, higher AST (> 700) and ALT (> 900) levels were found to have higher sensitivity for severe cases, and these criteria were revisited for the patients in Turkey [12]. Although leukocytosis was one of the criteria defined by Swanepoel et al., it was not observed frequently among the fatal cases in the recent reports [11]. In a multivariate analysis, increase in ALT was found to be associated with the fatality (OR; 1.003, confidence interval; 1.001–1.005, $p = 0.007$). Increase in ALT as the disease progresses could be the indicator of liver failure [12]. According to the criteria, defined by Ergonul et al. [12], the patients, who have one of the following laboratory results were defined as severe;

1. PLT ≤ 20,000/mm^3
2. AST ≥ 700 U/L or ALT ≥ 900 U/L
3. aPTT ≥ 60 s
4. Fibrinogen ≤ 110 mg/dL

Besides the laboratory findings, melena, hematemesis, and somnolence were also defined as the parameters for the severity [12] (Table 16-4). In addition to these findings, splenomegaly was also indicated as a parameter for fatality in one study [3].

The severity criteria defined for CCHF cases have similarities with the score for disseminated intravascular coagulation (DIC) defined by International Society of Thrombosis and Hemostasis [2]. The DIC scoring system consists of four components:

1. PLT (> 100, 0; < 100, 1; < 50, 2)
2. Elevated fibrin-related marker (D-dimer was used, no increase, 0; moderate increase, 2; strong increase)

Table 16-4. Predictors of fatality

Criteria	Swanepoel [29]	Ergonul [12]
Increased WBC count ≥10,000/mm^3	Yes	No
Decreased platelet count	Yes	Yes
Elevated AST	Yes	Yes
Elevated ALT	Yes	Yes
Prolonged activated partial thromboplastin time	Yes	Yes
Prolonged thromboplastin time	No	Yes
Decreased fibrinogen ≤110 mg/dL	Yes	Yes
Melena	No	Yes
Hematemesis	No	Yes
Somnolence	No	Yes

3. Prolonged PT (< 3 s, 0; > 3 s but < 6 s, 1; > 6 s
4. Fibrinogen level (> 1.0 g/L, 0; < 1.0 g/L, 1)

The calculated DIC score ≥ 5 was defined as compatible with overt DIC. In a study performed among CCHF patients, all the fatal patients had overt DIC, and the fatal patients were found to have significantly higher DIC score than the nonfatal cases [13]. Uncontrolled release of cytokines (cytokine storm) are important in inducing a procoagulant effect [23]. The levels of proinflammatory cytokines were found to be higher among fatal cases [13].

Hematemesis, melena, and somnolence are significantly more common among patients with a fatal outcome [3, 12]. Of particular significance it is the fact that in fatal cases there is little evidence of an antibody response [12, 31], and the disease is milder among the children.

The case fatality rate differs among the countries. This could be because of several reasons:

1. The virulences of different strains might be different. However, there is no reported correlation between the severity and different CCHFV strains yet.
2. Access to the health system may differ between the countries.
3. The sensitivity threshold for the symptoms might differ. Some patients may apply to the physician with milder symptoms, and inclusion of the milder cases inflates the denominator.
4. The quality of the health system in tertiary care may differ. This is especially important for the hematologic support. Some countries have the ability to give apheresis, and some not.
5. There could be coexistent infection.

16.7. CRIMEAN-CONGO HEMORRHAGIC FEVER AMONG CHILDREN

The CCHF infection among 13 [30] and 9 [17] children were reported from Turkey. The majority of the cases had the history of tick bite. There was only one fatal case, and this fatality was because of a hospital complication. All the patients had the history of fever. Rash, myalgia, and abdominal pain were common. The AST, ALT, and the LDH levels on admission were high. The platelet and WBC counts were lower than the normal. The clinical course of the infection among children was milder and the duration of illness was shorter compared to adult cases.

16.8. PREGNANCY AND CRIMEAN-CONGO HEMORRHAGIC FEVER

Two pregnant women infected with CCHF were reported [8]. The first woman acquired the infection at the 37th week of her gestation. Her baby died because of massive bleeding, and his PCR for CCHF was positive. The baby of the mother probably acquired the infection by either intrauterine or perinatal route. In endemic areas, CCHF infection should be differentiated from a syndrome characterized by hemolysis, elevated liver enzymes, and low PLT (hemolysis, elevated liver enzymes, and low platelets [HELLP] syndrome). It develops in approximately 4–12% of women with severe preeclampsia or eclampsia at the end of pregnancy [9, 27]. In CCHF infection, the hemolysis is not seen, and leucopenia is common, whereas in the patients with HELLP syndrome, hemolysis is seen and leucopenia is not common [8].

Another pregnant woman was reported to be infected at the second trimester [8]. She was hospitalized and cured. The risk of the fatal delivery was explained to the family. Intraabdominal fluid was found to be increased and hydrocele was detected at the 38th week of gestation, which was compatible with bleeding or perforation. After her vaginal delivery, the baby was operated with the diagnosis of necrotizing enterocolitis (NEC), but died [8]. A possible horizontal transmission of the CCHF infection from a 27-year-old mother to her child was also reported [25].

16.9. DIFFERENTIAL DIAGNOSIS

The list of infectious and noninfectious etiologies was presented in Table 16-5. The infectious agents to be differentiated could differ according to the endemic reports for each geographic location. Particularly, the differential diagnosis at the prehemorrhagic stage is more difficult. As the disease progress, the clinical picture becomes clearer, and diagnosis becomes easier.

Some disease entities can be ruled out easily in differential diagnosis of CCHF. For instance, the etiology of a patient, categorized as "fever of unknown origin" cannot be a CCHF infection. Because, the clinical picture in CCHF infection starts and ends within 2 weeks period, whereas there should be at least 3 weeks time to be defined as fever of unknown origin.

Low hemoglobin level at the initial phase of the infection is not expected. But, the hemoglobin level may become lower after the bleeding. The pancytopenia is very unlikely. However, pancytopenia was noted in brucellosis and vitamin B12 deficiency (Table 16-5).

Since jaundice was not reported except in one case, the CCHF infection can easily be differentiated from many causes of viral hepatitis, such as hepatitis A, B, E, and yellow fever (Table 16-5). The pneumonia and encephalitis in CCHF infection were not reported till now. Renal failure could only be seen as a complication during hospital stay.

Some infections can coexist, such as brucellosis or Q fever, as it was reported from Turkey [16]. There is no known relapse of the infection, and biphasic course of the disease was not observed in recent studies as it was noted in Soviet literature [18].

Table 16-5. Differential diagnosis of CCHF

Disease categories			
Infections	Geographic location	Transmission	Differentials with CCHF (clinical or laboratory findings)
Brucellosis	Worldwide, particularly Mediterranean basin; the Arabian peninsula; the Indian subcontinent; and in parts of Mexico, Central, and South America		Pancytopenia, Wright agglutination
Q fever	Worldwide	Tick	Serology (ELISA or IFAT)
Rickettsiosis	Worldwide	Tick	Weil-Felix test
Ehrlichiosis	America, Europe, Middle East, Southeast Asia	Tick	Serology (ELISA)
Lyme disease	Worldwide, mainly northern hemisphere	Tick	Serology (ELISA), Western blot
Leptospirosis	Worldwide		Hepatorenal failure, agglutination test
Salmonellosis	Worldwide		Widal test
Tick-borne encephalitis	Northern hemisphere	Tick	ELISA
Malaria	Worldwide	Mosquito	Peripheral smear
Viral hepatitis (HAV, HBV, HCV, HEV)	Worldwide		Jaundice, bilirubinemia
Other viral hemorrhagic infections			
Arenaviridae			
South America HFs	Argentine, Bolivia, Brazil, and Venezuela	Interhuman	Neurologic symptoms
Lassa fever	West Africa	Interhuman	
Other Bunyaviridae			
Rift Valley fever	Sub-saharan Africa	Mosquito	Mosquito
HF with renal syndrome (HFRS)	Worldwide		Renal findings, serology, PCR
Hantavirus pulmonary syndrome (HPS)	America	Interhuman	Pulmonary findings, serology, PCR
Filoviridae			
Marburg and Ebola	Africa, Philippines	Interhuman	
Filaviviridae			
Yellow fever	Africa, South America	Mosquito	Jaundice
Dengue	Tropics and subtropics, worldwide	Mosquito	

(*Continued*)

Table 16-5. Differential diagnosis of CCHF—Cont'd

Disease categories			
Infections	Geographic location	Transmission	Differentials with CCHF (clinical or laboratory findings)
Kyasanur forest disease (KFD)	India	Tick	
Omsk HF	Western Siberia	Tick	
Al Khumrah	Middle East, Africa	Tick (?), mosquito (?)	
Non-infectious reasons			
Vitamin B12 deficiency	Worldwide		Pancytopenia, and B12 level in serum
Febrile neutropenia	Worldwide		Underlying disease
Hematological malignancies			
Toxications	Worldwide		History
HELLP syndrome			Underlying conditions, history
Thrombotic thrombocytopenic purpura			Underlying conditions
DIC			Underlying conditions

REFERENCES

1. Alavi-Naini R, Moghtaderi A, Koohpayeh HR, Sharifi-Mood B, Naderi M, Metanat M, Izadi M (2006) Crimean-Congo hemorrhagic fever in Southeast of Iran. J Infect 52:378–382
2. Bakhtiari K, Meijers JC, de Jonge E, Levi M (2004) Prospective validation of the International Society of Thrombosis and Haemostasis scoring system for disseminated intravascular coagulation. Crit Care Med 32:2416–2421
3. Bakir M, Ugurlu M, Dokuzoguz B, Bodur H, Tasyaran MA, Vahaboglu H (2005) Crimean-Congo haemorrhagic fever outbreak in Middle Anatolia: a multicentre study of clinical features and outcome measures. J Med Microbiol 54:385–389
4. Buehler JW (1998) Surveillance. In: Rothman KJ, Greenland S (eds) Modern Epidemiology. Lippincott-Raven, Philadelphia, PA, pp 435–457
5. Burt FJ, Swanepoel R, Shieh WJ, Smith JF, Leman PA, Greer PW, Coffield LM, Rollin PE, Ksiazek TG, Peters CJ, Zaki SR (1997) Immunohistochemical and in situ localization of Crimean-Congo hemorrhagic fever (CCHF) virus in human tissues and implications for CCHF pathogenesis. Arch Pathol Lab Med 121:839–846
6. CDC (1990) Case definitions for public health surveillance. Morb Mortal Wkly Rep 39:1–40
7. Celikbas A, Ergonul O, Dokuzoguz B, Eren S, Baykam N, Polat-Duzgun A (2005) Crimean-Congo hemorrhagic fever infection simulating acute appendicitis. J Infect 50:363–365
8. Celikbas A, Ergonul O, Yildirim U, Zenciroglu A, Erdogan D, Ziraman I, Yilmaz N, Saracoglu F, Demirel N, Cakmak O, Dokuzoguz B (2006) Intrauterine infection of Crimean-Congo haemorrhagic fever: the courses of two episodes. In: 15th European Congress of Clinical Microbiology and Infectious Diseases (ECCMID), 12;4, Nice, France, p 1665

9. Egerman RS, Sibai BM (1999) HELLP syndrome. Clin Obstet Gynecol 42:381–389
10. Ergonul O, Celikbas A, Dokuzoguz B, Eren S, Baykam N, Esener H (2004) Characteristics of patients with Crimean-Congo hemorrhagic fever in a recent outbreak in Turkey and impact of oral ribavirin therapy. Clin Infect Dis 39:284–287
11. Ergonul O (2006) Crimean-Congo haemorrhagic fever. Lancet Infect Dis 6:203–214
12. Ergonul O, Celikbas A, Baykam N, Eren S, Dokuzoguz B (2006) Analysis of risk-factors among patients with Crimean-Congo haemorrhagic fever virus infection: severity criteria revisited. Clin Microbiol Infect 12:551–554
13. Ergonul O, Tuncbilek S, Baykam N, Celikbas A, Dokuzoguz B (2006) Evaluation of serum levels of interleukin (IL)-6, IL-10, and tumor necrosis factor-alpha in patients with Crimean-Congo hemorrhagic fever. J Infect Dis 193:941–944
14. Ergonul O, Zeller H, Menekse S, Celikbas A, Eren S, Baykam N, Dokuzoguz B (2006) The attack and the infection rate of Crimean-Congo haemorrhagic fever virus infection in an endemic region. In: 15th European Congress of Clinical Microbiology and Infectious Diseases (ECCMID), 1–4 April 2006, Nice, France, p 1666
15. Goldfarb LG, Chumakov MP, Myskin AA, Kondratenko VF, Reznikova OY (1980) An epidemiological model of Crimean hemorrhagic fever. Am J Trop Med Hyg 29:260–264
16. Gozalan A, Akin L, Rolain JM, Tapar FS, Oncul O, Yoshikura H, Zeller H, Raoult D, Esen B (2004) Epidemiological evaluation of a possible outbreak in and nearby Tokat province. Mikrobiyol Bul 38:33–44
17. Güngör O, Eroğlu KO, Güven A, Kalayci AG, Duru F (2006) Crimean-Congo Hemorrhagic fever among children [Turkish]. National Pediatrics Conference, Antalya, Turkey, p 281
18. Hoogstraal H (1979) The epidemiology of tick-borne Crimean-Congo hemorrhagic fever in Asia, Europe, and Africa. J Med Entomol 15:307–417
19. Karti SS, Odabasi Z, Korten V, Yilmaz M, Sonmez M, Caylan R, Akdogan E, Eren N, Koksal I, Ovali E, Erickson BR, Vincent MJ, Nichol ST, Comer JA, Rollin PE, Ksiazek TG (2004) Crimean-Congo hemorrhagic fever in Turkey. Emerg Infect Dis 10:1379–1384
20. Martini GA (1969) Marburg agent disease: in man. Trans R Soc Trop Med Hyg 63:295–302
21. McCormick JB, King IJ, Webb PA, Johnson KM, O'Sullivan R, Smith ES, Trippel S, Tong TC (1987) A case-control study of the clinical diagnosis and course of Lassa fever. J Infect Dis 155:445–455
22. Ozkurt Z, Kiki I, Erol S, Erdem F, Yilmaz N, Parlak M, Gundogdu M, Tasyaran MA (2006) Crimean-Congo hemorrhagic fever in Eastern Turkey: clinical features, risk factors and efficacy of ribavirin therapy. J Infect 52:207–215
23. Peters CJ, Zaki SR (2002) Role of the endothelium in viral hemorrhagic fevers. Crit Care Med 30:S268–S273
24. Rothman KJ (2002) Epidemiology: An introduction. Oxford University Press, New York
25. Saijo M, Tang Q, Shimayi B, Han L, Zhang Y, Asiguma M, Tianshu D, Maeda A, Kurane I, Morikawa S (2004) Possible horizontal transmission of Crimean-Congo hemorrhagic fever virus from a mother to her child. Jpn J Infect Dis 57:55–57
26. Schwarz TF, Nsanze H, Ameen AM (1997) Clinical features of Crimean-Congo haemorrhagic fever in the United Arab Emirates. Infection 25:364–367
27. Stone JH (1998) HELLP syndrome: hemolysis, elevated liver enzymes, and low platelets. JAMA 280:559–562
28. Swanepoel R, Shepherd AJ, Leman PA, Shepherd SP, McGillivray GM, Erasmus MJ, Searle LA, Gill DE (1987) Epidemiologic and clinical features of Crimean-Congo hemorrhagic fever in southern Africa. Am J Trop Med Hyg 36:120–132
29. Swanepoel R, Gill DE, Shepherd AJ, Leman PA, Mynhardt JH, Harvey S (1989) The clinical pathology of Crimean-Congo hemorrhagic fever. Rev Infect Dis 11 (Suppl 4):S794–S800
30. Tanir G, Ergonul O, Tuygun N, Golabi P, Korten V (2007) Crimean-Congo haemorrhagic fever infection among children. In: 17th European Congress of Clinical Microbiology and Infectious Diseases, Munich, Germany

31. van Eeden PJ, van Eeden SF, Joubert JR, King JB, van de Wal BW, Michell WL (1985) A nosocomial outbreak of Crimean-Congo haemorrhagic fever at Tygerberg Hospital. Part II. Management of patients. S Afr Med J 68:718–721
32. Wallach J (2000) Interpretation of Diagnostic Tests, 7th edn. Lippincott Williams & Wilkins, Philadelphia, PA
33. Watts DM, Ksiasek TG, Linthicum KJ, Hoogstraal H (1988) Crimean-Congo hemorrhagic fever. In: Monath TP (ed.) The Arboviruses: Epidemiology and Ecology. CRC Press, Boca Raton, FL
34. Watts DM, Ussery MA, Nash D, Peters CJ (1989) Inhibition of Crimean-Congo hemorrhagic fever viral infectivity yields in vitro by ribavirin. Am J Trop Med Hyg 41:581–585

CHAPTER 17

COMPARATIVE PATHOGENESIS OF CRIMEAN-CONGO HEMORRHAGIC FEVER AND EBOLA HEMORRHAGIC FEVER

MIKE BRAY, M.D., M.P.H.

Biodefense Clinical Research Branch, Division of Clinical Research, National Institute of Allergy and Infectious Diseases National Institutes of Health, Room 5128, 6700A Rockledge Dr. Bethesda, MD 20892. Tel.: 301 451 5123; Fax: 301 480 2319; E-mail: mbray@niaid.nih.gov

17.1. INTRODUCTION

Crimean-Congo hemorrhagic fever (CCHF) virus has been called "the Asian Ebola virus" – an epithet that recognizes the close clinical resemblance of CCHF and Ebola hemorrhagic fever (EHF), and also suggests that the two illnesses share similar underlying mechanisms [38]. CCHF and EHF both present difficult challenges to pathophysiology research, because they occur principally in regions lacking a modern medical infrastructure and because the high virulence of their causative agents requires laboratory studies to be performed under Biosafety Level 4 (BSL-4) containment. Efforts to elucidate the pathogenesis of CCHF have been even further handicapped by the failure of the virus to cause disease in laboratory animals other than suckling mice. By contrast, models of EHF in adult mice, guinea pigs, and nonhuman primates have been employed extensively for pathogenesis studies [7, 13, 24]. Detailed examination of clinical and laboratory parameters, pathologic changes, and innate immune responses in cynomolgus macaques over the entire course of fatal EHF has been especially valuable in elucidating how the pathogen overcomes host defenses to cause rapidly overwhelming infection. These findings are leading to novel approaches to postexposure prophylaxis and therapy [10, 14, 20, 35].

Despite the lack of animal models of CCHF, much can still be learned about its pathogenesis through properly designed prospective studies of patients and through in vitro experiments employing virus-infected human cells. This chapter discusses how recent progress in elucidating the underlying mechanisms of EHF could help to guide such research, by providing hypotheses for testing.

O. Ergonul and C. A. Whitehouse (eds.), Crimean-Congo Hemorrhagic Fever, 221–231.

The first section reviews the basic features of viral hemorrhagic fever (VHF), of which EHF and CCHF are prime examples. Following a comparison of the clinical course, laboratory features, immune responses, and pathological findings in the two diseases, the current model of EHF pathogenesis is presented, together with a discussion of evidence that similar mechanisms are responsible for the major features of CCHF. The chapter concludes with recommendations for future clinical and laboratory studies.

17.2. VIRAL HEMORRHAGIC FEVER

Of the enormous range of viruses that humans encounter over the course of life, only a small percentage are capable of causing rapidly lethal disease in an immunocompetent individual. The most striking examples are the hemorrhagic fever viruses (HFV), a group of some 20 different positive- or negative-sense RNA viruses from four different families, that includes CCHFV (a bunyavirus) and EV (a filovirus) [5, 22]. These agents are maintained in a variety of animal species, in which they apparently cause either asymptomatic infection or only mild illness. Human disease results from accidental contact with the blood or tissues of an infected animal, inhalation of aerosolized excretions, or the bite of a blood-feeding arthropod; it is a "dead-end" event that plays no role in the continued existence of the pathogen.

Despite their biological diversity, the HFV produce a similar syndrome in humans, characterized by the sudden onset of fever, headache, and other nonspecific signs and symptoms, followed by the development of coagulation defects and a progressive fall in blood pressure that in fatal cases leads to multiorgan failure intractable shock. As discussed below, these changes appear to be largely or entirely the consequence of the release of cytokines, chemokines, and other proinflammatory mediators from virus-infected monocytes/macrophages [5]. All of the HFV are apparently capable of replicating to high titer in macrophages at their point of entry, resulting in viremia and infection of similar cells in lymphoid organs and other tissues throughout the body. This capacity for rapid dissemination suggests that the HFV are able to block human type I IFN responses, which in turn helps to explain why all of them are zoonotic RNA viruses. Because their replication cycle requires the generation of double-stranded RNA, which is the primary stimulus for the induction of type I interferon, these agents must have acquired the ability to suppress that response while coevolving with their reservoir hosts [25, 31]. Those RNA viruses that are maintained in humans, such as measles virus, presumably block the interferon system sufficiently to facilitate viral replication and person-to-person spread, but without permitting severe illness that might kill the host. The HFV, by contrast, have not adapted to humans, suggesting that their remarkable virulence results from a fortuitous ability of their interferon antagonist proteins to profoundly suppress human interferon responses, leading to overwhelming infection [5]. The ability of these agents to enter and replicate within a variety of cell types may also contribute to their virulence. Fortunately, the lack of coevolution with humans has also prevented the development of effective transmission mechanisms, so the HFV do not spread efficiently from person to person. Disease outbreaks therefore tend to consist of single infections or small, self-limited clusters of cases, except when poor sanitary practices in hospitals catalyze large epidemics.

17.3. COMPARISON OF CCHF AND EHF

CCHF and EHF closely resemble each other in their clinical signs and symptoms, changes in standard laboratory tests, immune responses to infection, and patterns of tissue damage [12, 15, 16, 33, 40, 42]. Both begin abruptly with fever, headache and muscle aches, nausea, vomiting and diarrhea, and a variety of other nonspecific signs and symptoms. Within a few days, the clinical course becomes dominated by a progressive fall in blood pressure, leading to obtundation and shock, and by coagulation defects, which are usually manifested by mild hemorrhagic phenomena such as conjunctival hemorrhages and easy bruising, but in severe cases may result in massive bleeding from the gastrointestinal tract and other sites. The case fatality rate of CCHF has ranged from less than 10% to greater than 60% in a number of large outbreaks, while that of EHF in African epidemics has consistently exceeded 50% [15, 33]. In both diseases, death usually occurs during the second week of illness.

CCHF and EHF also cause similar changes in standard laboratory tests. Beginning with the onset of illness, patients show marked changes in blood cell counts, with neutropenia, a fall in lymphocyte count, and the appearance of atypical lymphocytes, increasing numbers of immature neutrophils and progressive thrombocytopenia [15, 33, 39]. The plasma levels of the liver-associated enzymes alanine and aspartate aminotransferase (ALT and AST) are increasingly elevated over the course of illness, with a rise in the AST/ALT ratio, but jaundice does not occur. Severe cases of CCHF are characterized by prolonged coagulation times, a decrease in plasma fibrinogen and the presence of circulating fibrin degradation products, features diagnostic of disseminated intravascular coagulation (DIC) [39]. The same changes have been demonstrated in Ebola-infected nonhuman primates, but adequate data from human EHF patients are currently lacking.

Limited evaluation of innate and adaptive immune responses has also given similar results in the two diseases. Serum samples from severely ill CCHF and EHF patients and from Ebola-infected macaques have shown elevated levels of the proinflammatory cytokines interleukin-6 (IL-6) and tumor necrosis factor (TNF)-α [17, 27, 41]. The role of the humoral response to infection also appears to be similar in the two diseases, since fatally infected patients typically show no evidence of virus-specific antibodies, while detection of a virus-specific IgM or IgG response is predictive of survival [15, 16, 30, 32, 37].

It is in the area of tissue pathology that the absence of an animal model has been the greatest handicap to research on CCHF, since in contrast to the large amount of data on EHF that has been obtained by studying infected animals, histologic studies of CCHF have been limited to tissues collected postmortem from a small number of patients [11, 30]. The latter have revealed viral replication in macrophages, lymphocyte depletion in the spleen and other lymphoid tissues, multifocal hepatic necrosis, and the presence of viral antigen and RNA in endothelial cells. By comparison, detailed studies of EHF in nonhuman primates have revealed that both macrophages and dendritic cells (DC) are major targets of infection; that Ebola virus infects and causes necrosis of hepatocytes and a range of other parenchymal cells; that lymphocytes remain uninfected, but undergo massive elimination through apoptosis; and that endothelial cells become sites of viral replication only during the terminal phase of illness [21, 23, 24].

17.4. UNDERLYING MECHANISMS OF EHF

Studies in laboratory animals have been instrumental in revealing that the severe illness and high mortality of EHF result from a combination of direct injury to virus-infected tissues and the indirect effects of host responses to viral replication (Fig. 17-1) [33]. The following sections describe the roles of specific cell types in bringing about the EHF syndrome, and present evidence that similar mechanisms underly the clinical features of CCHF.

Monocytes/macrophages are the central players in the pathogenesis of EHF (Fig. 17-1) [8]. Ebola virus replicates to high titer and spreads quickly among these cells, causing the release of proinflammatory cytokines, chemokines, and other potent vasoactive substances such as nitric oxide (NO). The rapidly accelerating release of these mediators produces the abrupt onset of fever and other nonspecific signs and symptoms at the commencement of illness; as the disease progresses, their cumulative effects are responsible for systemic vasodilatation and increased vascular permeability that lead to hypotension, multiorgan failure, and shock. At the same time, production of cell-surface tissue factor (TF) by the same infected macrophages triggers the extrinsic coagulation pathway, leading to local fibrin deposition, consumption of coagulation factors, and DIC [23, 24, 27, 33].

Although the lack of animal models has restricted the study of inflammatory responses in CCHF, a recent report has documented the presence of high levels of IL-6 and TNF-α in the serum of severely ill patients, and the demonstration of widespread viral infection of mononuclear phagocytes in tissues recovered at autopsy suggests that the release of mediators from these cells resembles that in EHF [11, 17, 30]. As discussed below, further characterization of the role of systemic inflammation in the pathogenesis of CCHF could be pursued through additional testing of sequential blood samples from patients and by infecting primary human macrophages in the laboratory. It will be particularly important to assess the production of TF by infected cells, as its presence would indicate

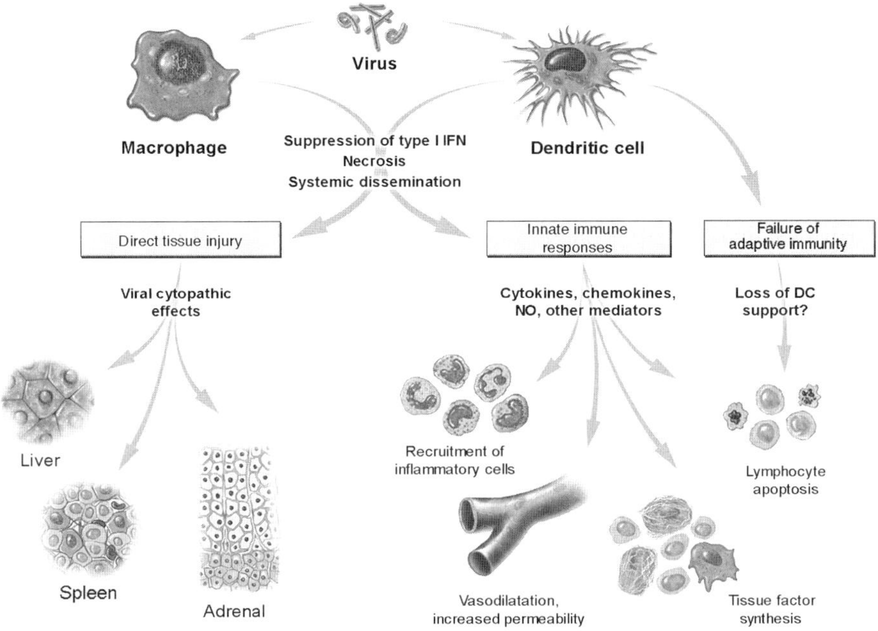

Fig. 17-1. Pathogenetic mechanisms of Ebola hemorrhagic fever, based on data from human cases and intensive studies of lethal infection of rodents and nonhuman primates. The virus initially replicates in macrophages and dendritic cells (DC), causing their necrosis. Suppression of type I interferon responses permits rapid viral dissemination to similar cells in the liver, spleen, lymph nodes, and other tissues throughout the body. Infected cells release proinflammatory cytokines, chemokines, and other mediators and produce cell-surface tissue factor, causing a systemic inflammatory syndrome with a diffuse increase in vascular permeability and disseminated intravascular coagulation. The virus also spreads to parenchymal cells in the liver, adrenal cortex, and other organs, causing multifocal necrosis, while destruction of macrophages and DC causes extensive injury to the spleen and other lymphoid tissues. Lymphocytes remain uninfected, but undergo massive programmed cell death, apparently brought about through the pro-aptoptic effects of inflammatory mediators and the loss of normal support signals from virus-infected dendritic cells. NO: nitric oxide.

the potential for specific interventions that have shown promise when tested in Ebola-infected macaques.

Dendritic cells are also important sites of Ebola virus replication. Immature DC normally respond to the uptake of foreign material by ceasing phagocytosis, migrating to regional lymph nodes, and presenting antigen to lymphocytes to initiate a specific immune response. DC infected with Ebola virus, by contrast, fail to undergo this maturation process and are unable to trigger lymphocyte activation [4, 34]. Viral infection of DC may therefore be a critical factor determining the outcome of EHF, since it could explain the absence of a virus-specific antibody response in those dying of the disease [32, 37]. The fact that fatal cases of CCHF are also characterized by failure to produce antibodies to the virus suggests that acquired impairment of adaptive immunity also occurs in this disease. Infected DC have not been described histologically in the limited amount of postmortem tissue that has been recovered from CCHF patients, but archived material could be reassessed for involvement of these cells.

Endothelial cells play an indirect role in the pathogenesis of EHF, through their activation by cytokines, chemokines, and other mediators. Early theories of VHF pathogenesis attributed increased vascular permeability and coagulopathy to direct viral infection of blood vessel linings or damage by viral proteins, leading to such designations as "capillary toxicosis" for the illness [42]. However, the autopsy findings on which those concepts were based now appear to reflect events in the terminal phase of illness, rather than early steps in disease development. Thus, careful electron microscopic examination of blood vessels of groups of nonhuman primates killed at daily intervals over the entire course of lethal Ebola infection found no evidence of infected endothelial cells during the midcourse of illness, when coagulopathy had already developed, but some infected cells were seen in moribund animals, especially in hepatic vessels [23].

Examination of tissues collected from CCHF patients at autopsy has also shown the presence of viral antigen and RNA within endothelial cells [11]. Although the absence of samples from earlier time points prevents a firm conclusion, analogy with EHF suggests that this is a phenomenon of the late stage of illness, when endothelial cells have undergone prolonged stimulation by inflammatory mediators, altering their permissivity to viral infection. The ability of EV, CCHFV, and other viruses to replicate in primary human endothelial cells may reflect a similar modification of their susceptibility to infection under conditions of in vitro culture [1, 26].

Hepatocytes are among the variety of parenchymal cells that are sites of Ebola virus replication. The agent is apparently first taken up by fixed hepatic macrophages (Kupffer cells), from which it spreads to the surrounding parenchyma, causing multiple foci of necrosis [33]. EHF is therefore characterized by increasing plasma levels of the liver-associated enzymes AST and ALT. Perhaps because of the focal nature of injury, or because viral infection does not spread to bile ducts, jaundice does not develop. Many other cell types in a broad range of tissues, including such basic elements as fibroblasts, are sites of Ebola virus replication; the resulting widespread necrosis is apparently responsible for popular descriptions of the virus "liquifying" its victims. This extensive destruction of infected cells serves as a further stimulus driving the intense systemic inflammatory response [28]. CCHF is also known to feature extensive infection of hepatocytes, with an increase in circulating liver enzymes, but limited autopsy studies have not defined the extent of involvement of other parenchymal cells.

Lymphocytes are among the few cell types that remain uninfected during the course of EHF, but they nevertheless undergo extensive destruction through apoptosis [3, 19, 21]. The susceptibility of this cell type to programmed cell death may reflect the fact that the immune system employs the same mechanism to eliminate self-reactive lymphocytes and reduce the number of effector T and B cells at the conclusion of an acute immune response. Lymphocytes may thus be especially sensitive to the variety of pro-apoptotic stimuli that occur during the course of EHF, including high plasma levels of TNF-α, Fas ligand, NO, and other substances [27]. In Ebola-infected nonhuman primates, lymphocyte apoptosis, principally involving T cells and NK cells, is reflected in progressive lymphopenia during the early course of infection and in the dramatic loss of these cells from the spleen, thymus, and lymph nodes seen at necropsy [18, 19, 36]. Serum samples from fatally infected individuals in Africa have also shown biochemical changes indicative of intravascular apoptosis [2, 3].

CCHF patients also show a progressive lymphopenia, and postmortem tissues display marked lymphoid depletion, suggesting that extensive loss of lymphocytes through programmed cell death also occurs in this disease. This hypothesis is supported by observations that some degree of lymphocyte apoptosis is a common feature of many, and perhaps all severe infections, since it is regularly observed in patients with sepsis, caused a variety of bacterial pathogens, and has been reproduced in animal models of septic shock [29, 30].

17.5. SHARED MECHANISMS WITH OTHER SEVERE INFECTIONS

Even though the fulminant course and high mortality of CCHF and EHF rank them among the most severe diseases of humans, they appear to share basic pathogenetic mechanisms with many other infectious processes, both common and obscure. The most important overlap is with septic shock caused by a variety of bacterial species, in which the major features of illness are caused not by microbial invasion of tissues, but by the host response to infection [9]. Highly lethal gram-negative bacterial infections such as meningococcemia and plague represent particularly striking variants of this process. Overwhelming infections by other viral species, such as the hemorrhagic forms of smallpox and varicella, also appear to represent cases in which rapid dissemination of a pathogen triggers a massive inflammatory response that is the actual cause of death [6].

Severe systemic inflammatory syndromes such as EHF provide examples of a host response that is beneficial when confined to a localized setting, but harmful when it occurs throughout the organism. In the tissues surrounding an infected wound, for example, proinflammatory mediators such as IL-6 and TNF-α help to contain and eliminate an invading pathogen by attracting leukocytes from the bloodstream, aiding the passage of antibodies, complement and other plasma proteins through capillary walls, and triggering coagulation to block the further spread of pathogens – effects that combine to produce the swelling, warmth, and other features of a local inflammatory process. During Ebola virus infection, release into plasma of the same mediators from macrophages in tissues throughout the body results in diffuse vasodilatation and increased vascular permeability, mobilization of leukocytes from the bone marrow, activation of the endothelium with increased adhesion of cells and platelets, and stimulation of coagulation pathways. It is the combination of these systemic effects that produces the progressive hypotension and shock, coagulopathy, and hemorrhage that characterize EHF [9].

17.6. GOALS FOR PATHOGENESIS RESEARCH

Pathogenesis studies of CCHF will continue to be hampered by the absence of animal models, but recent advances in understanding the underlying mechanisms of EHF suggest a number of hypotheses that could be tested using blood samples from patients, tissues recovered from fatal cases at autopsy or infection of primary human cells in a BSL-4 containment laboratory. These are listed below, together with suggested experimental approaches.

- *CCHFV-infected macrophages release proinflammatory cytokines, chemokines, NO and other mediators.* As described for Ebola virus infection, this can be tested in vitro through infection of primary human macrophages obtained from peripheral blood mononuclear cells (PBMC) harvested from donated blood. Assays should include detection of specific mRNA, immunohistochemical staining of fixed cells, enzyme-linked immunosorbent assay (ELISA) tests for individual protein products in cell supernatants and testing for nitrite in cell medium [27].
- *The DIC of CCHF is induced through TF production by infected macrophages.* TF synthesis by virus-infected primary human macrophages can be assessed by the methods just listed. Archived tissues from fatal cases of CCHF could also be reexamined for the presence of fibrin deposits on the surface of infected macrophages. Blood samples from patients could also be screened for the presence of circulating membrane microparticles expressing TF [24].
- *Infection of dendritic cells by CCHFV impairs their immune function.* Production of primary human myeloid DC from PBMC and characterization of their function in the absence or presence of viral infection has been described for Ebola and Lassa virus infection [4, 34]. Methods employed in those studies should be repeated with CCHFV.
- *The lymphoid depletion seen in severe CCHF cases results from apoptosis of uninfected cells.* As noted above, reexamination of lymphoid tissues recovered at autopsy, using specific histological methods, could determine whether the "lymphoid necrosis" that has been reported in fatal CCHF cases is the result of programmed cell death. Blood samples from patients could also be tested for the presence of markers of apoptosis.

17.7. CONCLUSION

There is no reason to believe that CCHF and EHF are identical diseases, but the striking similarities in their clinical, laboratory, immunological, and pathological features suggest that they share fundamental underlying mechanisms, and that they may potentially be managed through intervention strategies that target the same host responses. Continuing progress in understanding the pathogenesis of EHF should be used to guide future studies of CCHF, to elucidate the basis of virulence of the "Asian Ebola virus."

REFERENCES

1. Andersson I, Lundkvist A, Haller O, Mirazimi A (2006) Type I interferon inhibits Crimean-Congo hemorrhagic fever virus in human target cells. J Med Virol 78:216–222
2. Baize S, Leroy EM, Georges-Courbot MC, Capron M, Lansoud-Soukate J, Debre P, Fisher-Hoch SP, McCormick JB, Georges AJ (1999) Defective humoral responses and extensive intravascular apoptosis are associated with fatal outcome in Ebola virus-infected patients. Nat Med 5:423–426
3. Baize S, Leroy EM, Mavoungou E, Fisher-Hoch SP (2000) Apoptosis in fatal Ebola infection. Does the virus toll the bell for immune system? Apoptosis 5:5–7
4. Bosio CM, Aman MJ, Grogan C, Hogan R, Ruthel G, Negley D, Mohamadzadeh M, Bavari S, Schmaljohn A (2003) Ebola and Marburg viruses replicate in monocyte-derived dendritic cells without inducing the production of cytokines and full maturation. J Infect Dis 188:1630–1638
5. Bray M (2005) Pathogenesis of viral hemorrhagic fever. Curr Opin Immunol 17:399–403
6. Bray M, Buller M (2004) Looking back at smallpox. Clin Infect Dis 38:882–889
7. Bray M, Davis K, Geisbert T, Schmaljohn C, Huggins J (1998) A mouse model for evaluation of prophylaxis and therapy of Ebola hemorrhagic fever. J Infect Dis 178:651–661
8. Bray M, Geisbert TW (2005) Ebola virus: the role of macrophages and dendritic cells in the pathogenesis of Ebola hemorrhagic fever. Int J Biochem Cell Biol 37:1560–1566
9. Bray M, Mahanty S (2003) Ebola hemorrhagic fever and septic shock. J Infect Dis 188:1613–1617
10. Bray M, Paragas J (2002) Experimental therapy of filovirus infections. Antiviral Res 54:1–17
11. Burt FJ, Swanepoel R, Shieh WJ, Smith JF, Leman PA, Greer PW, Coffield LM, Rollin PE, Ksiazek TG, Peters CJ, Zaki SR (1997). Immunohistochemical and in situ localization of Crimean-Congo hemorrhagic fever (CCHF) virus in human tissues and implications for CCHF pathogenesis. Arch Pathol Lab Med 121:839–846
12. Bwaka MA, Bonnet MJ, Calain P, Colebunders R, De Roo A, Guimard Y, Katwidi DR, Kibadi K, Kipasa MA, Kuvula KJ, Mapanda BB, Massamba M, Mupapa KD, Muyembe-Tamfum JJ, Ndaberey E, Peters CJ, Rollin PE, Van den Enden E (1999) Ebola hemorrhagic fever in Kikwit, Democratic Republic of the Congo: clinical observations in 103 patients. J Infect Dis 179 (Suppl 1):S1–7
13. Connolly BM, Steele KE, Davis KJ, Geisbert TW, Kell WM, Jaax NK, Jahrling PB (1999) Pathogenesis of experimental Ebola virus infection in guinea pigs. J Infect Dis 179 (Suppl 1):S203–217
14. Daddario-DiCaprio KM, Geisbert TW, Stroher U, Geisbert JB, Grolla A, Fritz EA, Fernando L, Kagan E, Jahrling PB, Hensley LE, Jones SM, Feldmann H (2006) Postexposure protection against Marburg haemorrhagic fever with recombinant vesicular stomatitis virus vectors in non-human primates: an efficacy assessment. Lancet 367:1399–1404
15. Ergonul O (2006) Crimean-Congo haemorrhagic fever. Lancet Infect Dis 203–214
16. Ergonul O, Celikbas A, Dokuzoguz B, Eren S, Baykam N, Esener H (2004) Characteristics of patients with Crimean-Congo hemorrhagic fever in a recent outbreak in Turkey and impact of oral ribavirin therapy. Clin Infect Dis 39:284–287
17. Ergonul O, Tuncbilek S, Baykam N, Celikbas A, Dokuzoguz B (2006) Evaluation of serum levels of interleukin (IL)-6, IL-10, and tumor necrosis factor-alpha in patients with Crimean-Congo hemorrhagic fever. J Infect Dis 193:941–944
18. Geisbert TW, Hensley LE (2004) Ebola virus: new insights into disease aetiopathology and possible therapeutic interventions. Expert Rev Mol Med 6:1–24
19. Geisbert TW, Hensley LE, Gibb TR, Steele KE, Jaax NK, Jahrling PB (2000) Apoptosis induced in vitro and in vivo during infection by Ebola and Marburg viruses. Lab Invest 80:171–186
20. Geisbert TW, Hensley LE, Jahrling PB, Larsen T, Geisbert JB, Paragas J, Young HA, Fredeking TM, Rote WE, Vlasuk GP (2003) Treatment of Ebola virus infection with a recombinant inhibitor of factor VIIa/tissue factor: a study in rhesus monkeys. Lancet 362:1953–1958

21. Geisbert TW, Hensley LE, Larsen T, Young HA, Reed DS, Geisbert JB, Scott DP, Kagan E, Jahrling PB, Davis KJ (2003) Pathogenesis of Ebola hemorrhagic fever in cynomolgus macaques: evidence that dendritic cells are early and sustained targets of infection. Am J Pathol 163:2347–2370
22. Geisbert TW, Jahrling PB (2004) Exotic emerging viral diseases: progress and challenges. Nat Med 10:S110–121
23. Geisbert TW, Young HA, Jahrling PB, Davis KJ, Kagan E, Hensley LE (2003) Mechanisms underlying coagulation abnormalities in Ebola hemorrhagic fever: overexpression of tissue factor in primate monocytes/macrophages is a key event. J Infect Dis 188:1618–1629
24. Geisbert TW, Young HA, Jahrling PB, Davis KJ, Larsen T, Kagan E, Hensley LE (2003) Pathogenesis of Ebola hemorrhagic fever in primate models: evidence that hemorrhage is not a direct effect of virus-induced cytolysis of endothelial cells. Am J Pathol 163:2371–2382
25. Haller O, Kochs G, Weber F (2006) The interferon response circuit: induction and suppression by pathogenic viruses. Virology 344:119–130
26. Harcourt BH, Sanchez A, Offermann MK (1999) Ebola virus selectively inhibits responses to interferons, but not to interleukin-1beta, in endothelial cells. J Virol 73:3491–3496
27. Hensley LE, Young HA, Jahrling PB, Geisbert TW (2002) Proinflammatory response during Ebola virus infection of primate models: possible involvement of the tumor necrosis factor receptor superfamily. Immunol Lett 80:169–179
28. Hotchkiss RS, Karl IE (2003) The pathophysiology and treatment of sepsis. N Engl J Med 348:138–150
29. Hotchkiss RS, Osmon SB, Chang KC, Wagner TH, Coopersmith CM, Karl IE (2005) Accelerated lymphocyte death in sepsis occurs by both the death receptor and mitochondrial pathways. J Immunol 174:5110–5118
30. Hotchkiss RS, Tinsley KW, Karl IE (2003) Role of apoptotic cell death in sepsis. Scand J Infect Dis 35:585–592
31. Joubert JR, King JB, Rossouw DJ, Cooper R (1985) A nosocomial outbreak of Crimean-Congo haemorrhagic fever at Tygerberg Hospital. Part III. Clinical pathology and pathogenesis. S Afr Med J 68:722–728
32. Korth MJ, Kash JC, Furlong JC, Katze MG (2005) Virus infection and the interferon response: a global view through functional genomics. Methods Mol Med 116:37–55
33. Ksiazek TG, Rollin PE, Williams AJ, Bressler DS, Martin ML, Swanepoel R, Burt FJ, Leman PA, Khan AS, Rowe AK, Mukunu R, Sanchez A, Peters CJ (1999) Clinical virology of Ebola hemorrhagic fever (EHF): virus, virus antigen, and IgG and IgM antibody findings among EHF patients in Kikwit, Democratic Republic of the Congo, 1995. J Infect Dis 179 (Suppl 1):S177–187
34. Mahanty S, Bray M (2004) Pathogenesis of filoviral haemorrhagic fevers. Lancet Infect Dis 4:487–498
35. Mahanty S, Hutchinson K, Agarwal S, McRae M, Rollin PE, Pulendran B (2003) Cutting edge: impairment of dendritic cells and adaptive immunity by Ebola and Lassa viruses. J Immunol 170:2797–2801
36. Paragas J, Geisbert TW (2006) Development of treatment strategies to combat Ebola and Marburg viruses. Expert Rev Anti Infect Ther 4:67–76
37. Reed DS, Hensley LE, Geisbert JB, Jahrling PB, Geisbert TW (2004) Depletion of peripheral blood T lymphocytes and NK cells during the course of ebola hemorrhagic Fever in cynomolgus macaques. Viral Immunol 17:390–400
38. Sanchez A, Lukwiya M, Bausch D, Mahanty S, Sanchez AJ, Wagoner KD, Rollin PE (2004) Analysis of human peripheral blood samples from fatal and nonfatal cases of Ebola (Sudan) hemorrhagic fever: cellular responses, virus load, and nitric oxide levels. J Virol 78:10370–10377
39. Smego RA, Jr., Sarwari AR, Siddiqui AR (2004) Crimean-Congo hemorrhagic fever: prevention and control limitations in a resource-poor country. Clin Infect Dis 38:1731–1735

40. van Eeden PJ, Joubert JR, van de Wal BW, King JB, de Kock A, Groenewald JH (1985) A nosocomial outbreak of Crimean-Congo haemorrhagic fever at Tygerberg Hospital. Part I. Clinical features. S Afr Med J 68:711–717
41. Villinger F, Rollin PE, Brar SS, Chikkala NF, Winter J, Sundstrom JB, Zaki SR, Swanepoel R, Ansari AA, Peters CJ (1999) Markedly elevated levels of interferon (IFN)-gamma, IFN-alpha, interleukin (IL)-2, IL-10, and tumor necrosis factor-alpha associated with fatal Ebola virus infection. J Infect Dis 179 (Suppl 1):S188–191
42. Whitehouse CA (2004) Crimean-Congo hemorrhagic fever. Antiviral Res 64:145–160

CHAPTER 18

LABORATORY DIAGNOSIS OF CRIMEAN-CONGO HEMORRHAGIC FEVER

HERVÉ ZELLER, PH.D

Unité de Biologie des Infections Virales Emergentes, Institut Pasteur, 21 avenue Tony Garnier, 69365 Lyon, Cedex 07, France. Tel.: 33 (0)4 37 28 24 57; Fax: 33 (0)4 37 28 24 51; E-mail: zeller@cervi-lyon.inserm.fr

18.1. INTRODUCTION

Crimean-Congo hemorrhagic fever virus (CCHFV) can induce in humans a severe multisystem syndrome associated with fever, shock, and hemorrhages. In absence of specific clinical symptoms, physicians need a rapid and reliable diagnosis to reinforce the measures of safety (barrier nursing), and possibly to initiate quickly a suitable antiviral treatment. Equally, a differential diagnosis with other agents responsible of hemorrhagic fevers according to epidemiological features has to be undertaken [18, 51].

The usual approach for CCHF diagnosis combines the detection of the viral RNA genome and/or the antigen and the detection of specific IgM antibodies in human serum or blood. Therefore diagnosis is hampered by the problems of handling suspected specimens which require the highest levels of biological containment. Unfortunately in many endemic areas, laboratory capacities are limited and there is no possibility to detect quickly a CCHFV infection.

18.2. BIOSAFETY AND PROCESSING

18.2.1. Biosafety

CCHFV is classified among the biosafety level 4 (BSL-4) agents such as other viruses (e.g. Ebola, Marburg, Lassa) responsible for viral hemorrhagic fevers (VHF) due its propensity for human-to-human transmission and high risks for laboratory-acquired infections, and the lack of a specific safe vaccine. The virus also belongs to a list of agents with the potential for dual use (i.e. bioterrorism) which includes additional specific regulations in several countries limiting the exchange of biological materials. Nevertheless, CCHFV is not as much of a

O. Ergonul and C. A. Whitehouse (eds.), Crimean-Congo Hemorrhagic Fever, 233–243.

concern in terms of its potential use as a bioweapon due to barriers to large-scale production, which precludes its development as mass casualty weapon [4].

Special attention is needed for transportation and handling of biological specimens for CCHF diagnosis. Due to the variety of host range (e.g. humans, birds, domestic animals, rodents) and tick vectors, samples types are diverse. For humans, the laboratory diagnosis commonly is performed on whole blood, plasma, serum, or other body fluids and/or biopsy material [49, 52]. Specimens must be transported to the laboratory in clearly labeled ("Infectious Risk") appropriate containers: triple package with United Nations (UN) specification marks printed directly on packaging according to International Air Transport Association (IATA) regulations [26]. Specimens of human origin with suspicion of acute CCHF infection have to be considered as potentially infectious assigned UN 2814 [26]. An individual sheet with clinical and epidemiological information for each patient must be included. The receiving laboratory should be notified in advance before specimens are dispatched.

18.2.2. Laboratory processing

Laboratory processing requires a high security BSL-4 laboratory for cultivation of the virus. In laboratories without such facility, validated techniques for inactivation of the biological specimens are needed prior diagnosis. Current contingency plans for VHF samples recommend the use of heat, gamma irradiation, or Triton X-100 to inactivate samples before handling [33]. As with other bunyaviruses, CCHFV is sensitive to disinfectants such as sodium hypochlorite, formaldehyde, or beta-propiolactone (0.1% at 4°C for 3 days) or lipid solvents [28, 30]. It is labile in frozen material subjected to repeated freezing and thawing. Serum specimens may be inactivated by heat treatment at 60°C for 60 min [2]. Tissues may be fixed in 10% buffered formalin or other tissue fixatives that are suitable for use prior to viral detection [9]. Treatment in acetone (85–100%), glutaraldehyde (1% or greater), or 10% buffered formalin for 15 min is satisfactory. Reagents for diagnosis issued from BSL-4 facilities are irradiated prior use to avoid any risks of contamination.

18.3. DIAGNOSIS OF CCHF

Direct and indirect approaches are combined for the diagnosis of CCHF infection; specifically, viral isolation, detection of viral genome or antigen, and detection of specific antibodies. Likewise leukopenia, particularly neutropenia, thrombocytopenia, high levels of liver enzymes alanine aminotransferase (ASL) and aspartate aminotransferase (AST), and lactate dehydrogenase (LDH) are regularly reported in patients with CCHF [17, 52].

18.3.1. Direct diagnosis

18.3.1.1. Viral isolation

CCHFV has been isolated most frequently by intracranial (i.c.) inoculation of newborn suckling mice; the standard protocol used for years for most of the arboviruses [28]. The virus causes illness and death in suckling mice by i.c. or intraperitonneal (i.p.) inoculations in 4–6 days and in weaning mice in 5–7 days. In vitro, a large variety of cells have been used for viral cultivation: primary chicken embryo, human embryo, primary green monkey kidney cells, or continuous cell lines as CF- 1 (*Cercopithecus aethiops*), Vero or VeroE6 (African green monkey kidney, *C. aethiops*), SW13 (human small cell carcinoma of adrenal cortex), or LLC-MK2 (rhesus monkey, *Macaca mulatta*), CER (derived from hamster) [49, 59]. Cell lines from ticks have been proposed but they are not readily accessible, and their replication is tedious requiring specific expertise for cultivation [58]. In most CCHFV strains, there is no discernible evidence of a cytopathogenic effect and viral replication is detected in 2–6 days by indirect immunofluorescence assay (IFA) using a specific CCHF mouse hyperimmune ascitic fluid (MHIAF) or monoclonal antibodies against the nucleocapsid protien [3, 21, 41, 43]. However, a few cross-reactions with the closely related nairovirus Hazara, as well as Dugbe and Nairobi sheep disease viruses has been reported [12]. Usually, monoclonal antibodies against the NP cross-react with different CCHF strains; in addition, some cross-reactions with another nairovirus, Qualyub virus, has been observed [3]. Detection by reverse transcriptase polymerase chain reaction (RT-PCR) is also used but the assay can also detect residual CCHF genome. Other, "classical" approaches such as immunodiffusion, complement fixation, and inhibition hemagglutination test were original used for CCHF diagnosis [13, 14, 54].

The SW-13 cell line has been extensively used for isolations of strains from Asia, Russia, and Africa, producing plaques in 4 days [59]. Nevertheless, strains of CCHFV differ in their ability to replicate and/or produce plaques in several cell lines. Although cell cultures were less sensitive than the i.c. inoculation into newborn suckling mice for viral isolation from clinical specimens, they produced diagnostic results much more rapidly [49]. For specimens from 26 CCHF patients in South Africa, virus was isolated from 20 patients by mouse inoculation and from only 11 by cell culture. Isolation of some CCHF viral strains from field-collected ticks, however, can only be obtained by suckling mice inoculation [49, 62].

Viral titrations are can be performed by mice i.c. inoculation (0.02 mL with the 50% i.c. lethal dose (LD_{50}), fluorescence focus assay, or plaque assay in CER or SW13 cells. However, the in vitro methods have been shown to give viral titres 10- to 100-fold lower than those obtained using suckling mice inoculation [49].

Viremia is detectable at the onset of clinical symptoms and can persist for a long period (up to day 12 [5, 10, 52]), even when specific IgM antibodies appear, and the patient is afebrile. Severely ill patients develop a more intense viremia, with a maximum $10^{6.2}$ mouse LD_{50} recorded [10]. In experimentally infected ostriches, viremia was detected during the first 4 days postinfection only, and antibodies were initially detected from day 5–13 [53]. In hare blood, the virus persisted for 15 days with the highest titer (3.6 log LD_{50}/0.02 mL) on day 4 [25].

18.3.1.2. Antigen detection

The detection of CCHFV antigen is a useful rapid technique for the diagnosis of acute infections [16, 47]. The viral antigen can be detected by immunocapture enzyme-linked immunosorbent assay (ELISA) or reverse passive hemagglutination assay (RPHA) [47]. For immunocapture of CCHFV antigen, plates are coated with a CCHF MHIAF or monoclonal antibodies directed against the nucleoprotein (NP) of African or Chinese strains [3, 21, 42, 47]. ELISA is more sensitive, as shown in a comparative study, with the detection of CCHFV antigen in 29 of 49 sera from patients versus 20 of 49 sera by RPHA or the SW-13 cell culture plaque assay to detect and quantify the presence of CCHF viral antigen in ticks [32, 47]. In parallel with viremia, CCHFV antigenemia was detected more frequently in fatal human cases (9/11) than in nonfatal cases (9/17) [47]. In patients who survived the disease, antigenemia can be detected reliably only if specimens are taken within 5 days of the onset of disease probably due to immune complex formation.

Immunochemistry and in situ hybridization have been used for the detection of CCHFV in formalin-fixed paraffin-embedded tissues. The virus distribution was examined from 12 fatal CCHF cases in South Africa; 10 liver tissues were positive by immunohistochemistry, and five were positive by in situ hybridization [9].

18.3.1.3. Molecular detection: RT-PCR

PCR is a very sensitive method for the identification of an agent and possible quantification of viral load in a sample within a few hours. Viral RNA is extracted from potentially infectious material using a detergent-based protocol, which renders it noninfectious, at which point it can be handled in a BSL-2 laboratory. Techniques usually combine the RT step with specific amplification, minimizing the risks of contaminations. For diagnostic purposes, assays are typically based on consensus nucleotide sequences primarily on the S segment, which codes for the NP. Some of the first assays were designed based on the S segment of the reference IbAr 10200 strain (virus isolated in 1970 from ticks *Hyalomma excavatum* in Nigeria) with R2/F3 primers (amplicons 536 base pairs) and nested F3/R2 primers (amplicons 259 base pairs) [Smith JF and Lofts R, unpublished data; 45]. Restriction fragment length polymorphism had been used for preliminary analysis prior to sequence analysis [62]. The comparative analysis of sequences on the S segment allows determining the origin and possible source of infections [24, 39, 45]. The first round of RT-PCR assay can fail to detect nucleic acid in a small proportion of serum samples; such result emphasizes the need of a nested PCR assay for direct diagnosis despite the risks of contamination and longer duration [5].

One-step real time RT-PCR assays using primers to the same NP gene have been developed but the development of these assays has been hampered by the high diversity of genome sequence [17, 20, 61]. Initially, only SYBR Green dye intercalation had been implemented in the early assays [17]. These assays are more rapid than the conventional RT-PCR; they provide sensitivity comparable to that of the nested PCR with a low risk of contamination, and the possible quantification of viral load is an added benefit. Sensitivity, using a plasmid standard, has been reported to range from 10^2–10^7 copies/mL [19].

CCHFV RNA can be detected in a patient's serum several days following the onset of disease. At least in the case of human immunodeficiency virus (HIV) RNA, collection of blood in EDTA tubes has been shown to ensure high PCR efficiency [15]. In a retrospective study of 80 specimens from 38 patients with suspected VHF from southern Africa, 19 cases of CCHF were confirmed. From 45 specimens samples of patients with CCHF infection collected on day 8–16 post onset of the disease, 28 were positive for genome detection (62.2%), and 5 (11.1%) for viral isolation due to progressive clearance of the infective virus while nucleic acid remains demonstrable in most cases [5]. The virus had been isolated on Vero76 cells or suckling mouse brain in two specimens collected on day 11 and 13 after onset of illness and the viral genome had been detected in 12 samples from day 13 to 16 [5].

18.3.2. Indirect diagnosis

18.3.2.1. Indirect serological diagnosis

The serological diagnosis of CCHF infection is based on the detection of specific IgM and IgG antibodies induced by the immune response principally to the NP which is recognized as the predominant antigen [36]. One serotype of CCHFV has been described and the comparison of CCHF viral strains from diverse geographic origins has shown a large antigenic homogeneity by fluorescent focus neutralization [56]. Seroconversion with detection of CCHF IgM antibodies or a ≥ fourfold increase in antibody titer between two successive blood samples is evidence of a recent infection [51, 52]. The serological diagnosis is valid after several days post onset of the disease; nevertheless the antibody response rarely is observed in fatal cases [46]. Various techniques have been developed for antibody detection (listed in order of decreasing sensitivity): ELISA, inhibition of PRHA; IFA, reduction of fluorescence focus, complement fixation, and immunodiffusion [6, 13, 16, 48, 58].

18.3.2.2. Indirect immunofluorescence assay

The IFA is convenient for making a rapid serodiagnosis of the disease [6, 45]. Slides of CCHFV-infected cells are prepared in a BSL-4 facility and fixed prior use. A screening dilution of serum specimens can range from 1:25 to 1:50. The presence of immunoglobulins is revealed with a fluorescein isothiocyanate

(FITC) conjugated anti-species IgG antibody [52]. For IgM detection, treatment of human sera for removal of rheumatoid factor with commercial reagent has been reported. A biotin–streptavidin IFA had been shown to detect IgM antibody earlier in approximately 25% of the patients tested [6].

Immunofluorescence, indicating the presence of IgM, can be detected as early as day 4 of illness in patients who survived the disease [6]. The IFA with native antigens has been applied for a rapid differential serological diagnosis of the African VHF Ebola Zaire, Ebola Sudan, Marburg, Rift valley fever, Lassa, and CCHF [27, 31]. Limitations of the use of such an assay in Africa appear to be due to the lack of specificity and expertise for interpretation [57]. For example, sera from patients with malaria or from malaria-affected areas often exhibited nonspecific fluorescence at low titers in both IgG and IgM assays [6]. The use of cell cultures in BSL-4 facilities is a major handicap for the production of suitable reagents. Recently, however, this obstacle has been overcome by the production of recombinant proteins derived from the N protein [34]. HeLa cells expressing the CCHF NP of the Chinese strain 8402 has been cultivated as a source of antigen for IFA [40] (see Chapter 10). Likewise, the Semliki Forest virus replicon has been used as another expression vector to produce recombinant antigen derived from the NP of the Nigerian strain IbAr 10200 [22].

18.3.2.3. Reverse passive hemagglutination inhibition

RPHA inhibition assays have been performed using CCHFV antibody-sensitized, glutaraldehyde-fixed sheep erythrocytes and sucrose–acetone antigen of infected suckling mice brain. Such assays have been used for CCHFV antibody detection in wild animals and in humans [46, 51].

18.3.2.4. Enzyme-linked immunosorbent assay

The ELISA is the most common technique for CCHFV antibody detection and is reported to be more sensitive than IFA [8, 35]. Techniques have been developed using a native antigen produced in suckling mouse brain or with cell culture. A common source of CCHFV antigen had been the sucrose–acetone treated suckling mouse brain suspension or a crude suckling mouse brain suspension inactivated by β-propiolactone or heating [14, 23, 61]. Likewise absence of residual infectivity had been reported in a suckling mouse brain suspension inactivated by heating (1 h at 60°C) [43]. All native antigens have to be produced in a BSL-4 laboratory and irradiated prior to use. A CCHFV VeroE6 cell slurry antigen is generally used for CCHFV IgM antibody detection by immunocapture on plates coated with specific anti-μ serum [29]. Then the serum sample, CCHFV antigen, specific CCHFV antibody (MHIAF), anti-species conjugate, and a chromogenic substrate are successively added. For CCHFV IgG detection, most of the techniques use a sandwich ELISA with capture of the CCHFV antigen in plates coated with a CCHFV MHIAF.

These assays have been extensively used in animal serosurveys [60, 62]. A competitive ELISA using a rabbit anti-CCHFV peroxidase conjugate has also been useful for the testing of a large variety of domestic and wild animal species [9, 46, 53].

In three out of 72 confirmed CCHF patients from southern Africa, cross-reactivity by ELISA with the closely related Hazara, Nairobi sheep disease, and Dugbe viruses has been reported. Otherwise, serum from one out of 162 other patients reacted specifically with Dugbe antigen. The latter patient suffered from febrile illness with prolonged thrombocytopenia [7]. In West Africa, the CCHFV suckling mouse brain crude antigen used in serology did not cross-react with the nairoviruses circulating in the region (Dugbe, Bandia and Bakel viruses); no cross-reaction or few weak positive reactions were reported in animal sera from Senegal [43, 60].

Recombinant proteins derived from the N protein originated from Greece AP92, China 8402, and Nigeria IbAr 10200 strains have been produced as previously reported [22, 34, 41]. The derived assays by ELISA G and M (sandwich) or IFA are highly sensitive and specific [22, 40]. When tested with laboratory animal sera representing all seven serogroups of nairoviruses, the only reactive sera were those raised to CCHFV and more weakly, Hazara virus [34].

Usually IgM and IgG CCHFV antibodies are detected 4–5 days post onset of symptoms [6]. In those who survived the disease in southern Africa, antibodies were detectable in 10%, 65%, 83%, 94%, and 100% of the patients at day 4, 6, 7, 8, and 9, respectively [6, 48]. Very few patients who died during day 4–9 post onset showed an antibody response [52]. The IgM titer is maximal 2–3 weeks after onset of the disease, and the IgM antibodies generally disappear within 4 months. The IgG antibodies remain detectable for several years [6, 43, 48]. Thus, the persistence of IgM antibody activity in ruminants would be of shorter duration than in humans [23, 62].

18.3.2.5. Neutralization assay

Nairoviruses generally induce weaker neutralizing antibody responses than do members of the other genera of the *Bunyaviridae* family, and thus, the neutralization assay is generally not used for CCHF diagnosis [38]. Few serological studies on neutralizing CCHFV antibodies in patients have been published. Neutralizing antibodies detected by fluorescent focus technique were present in five patients on day 6–10 after disease onset, with maximum titers ranging from 1:16 to 1:256. By the 5th month, neutralizing antibody titers were low, with maximum titres of 1:4 to 1:32. These results contrasted with those obtained from ELISA in which an increase between the 6th week and the 5th month was seen [48]. On the other hand, the administration of immune plasma did not appear to eliminate demonstrable viremia or improve the chances of survival. In fact, neutralization of CCHFV likely depends not only on the properties of the antibody, but on host cell factors as well [1].

18.4. DIFFERENTIAL DIAGNOSIS

The combined approach of viral genome detection and IgM detection is recommended for diagnosis of CCHF in humans. CCHFV has an extended geographical distribution from southern Europe to Russia and western China, in the Middle East and Africa. Other viral etiologies have to be investigated according to the origin of the patient and the risks of potential exposure. These would include Alkhurma and Rift Valley fever in the Middle East; Omsk hemorrhagic fever in Russia; Kyasanur Forest disease in India; hantaviruses in Europe and Asia; Lassa, Ebola, Marburg, Rift Valley fever, yellow fever in Africa; and dengue in various locations [18, 51]. In tropical and subtropical countries, malaria is the most important alternative diagnosis to be excluded in cases of suspected VHF. Differential diagnosis also would include hepatitis viruses, *Neisseria meningitidis*, leptospirosis, borreliosis, typhoid, rickettsiosis, and Q fever (*Coxiella burnetii*). Prior the confirmation of CCHFV infection in humans, usually multiple samples have been drawn and sent to various laboratories for clinical chemistry, hematology, and bacteriology in accordance with standard procedures, but with no particular safety precautions [55]. In such conditions, risk of exposure to CCHFV is low, but constant awareness is required.

Some broad-range approaches for the rapid detection of VHF agents have been designed by the use of unique multiplex assays, such as MassTag-PCR [37]. Additionally, direct visualization by electron microscopy can be used when the identity of an agent is totally unknown.

REFERENCES

1. Ahmed AA, McFalls JM, Hoffmann C, Filone CM, Stewart SM, Paragas J, Khodjaev S, Shermukhamedova D, Schmaljohn CS, Doms RW, Bertolotti-Ciarlet A (2005) Presence of broadly reactive and group-specific neutralizing epitopes on newly described isolates of Crimean-Congo hemorrhagic fever virus. J Gen Virol 86:3327–3336
2. Bhagat CI, Lewer M, Prins A, Beilby JP (2000) Effects of heating plasma at 56°C for 30 minutes and 60°C for 60 minutes on routine biochemistry analytes. Ann Clin Biochem 37:802–804
3. Blackburn NK, Besselaar TG, Sheperd AJ, Swanepoel R (1987) Preparation and use of monoclonal antibodies for identifying Crimean-Congo hemorrhagic fever virus. Am J Trop Med Hyg 37:392–397
4. Borio L, Inglesby T, Peters CJ, Schmaljohn AL, Hughes JM, Jahrling PB, Ksiazek T, Johnson KM, Meyerhoff A, O'Toole T, Ascher MS, Bartlett J, Breman JG, Eitzen EM Jr, Hamburg M, Hauer J, Henderson DA, Johnson RT, Kwik G, Layton M, Lillibridge S, Nabel GJ, Osterholm MT, Perl TM, Russell P, Tonat K; Working Group on Civilian Biodefense (2002) Hemorrhagic fever viruses as biological weapons: medical and public health management. JAMA 287:2391–2405
5. Burt FJ, Leman PA, Smith JF, Swanepoel R (1998) The use of a reverse transcription-polymerase chain reaction for the detection of viral nucleic acid in the diagnosis of Crimean-Congo haemorrhagic fever. J Virol Methods 70:129–137
6. Burt FJ, Leman PA, Abbott JC, Swanepoel R (1994) Serodiagnosis of Crimean-Congo haemorrhagic fever. Epidemiol Infect 113:551–562
7. Burt FJ, Spencer DC, Leman PA, Patterson B, Swanepoel R (1996) Investigation of tick-borne viruses as pathogens of humans in South Africa and evidence of Dugbe virus infection in a patient with prolonged thrombocytopenia. Epidemiol Infect 116:353–361

8. Burt FJ, Swanepoel R, Braack LE (1993) Enzyme-linked immunosorbent assays for the detection of antibody to Crimean-Congo haemorrhagic fever virus in the sera of livestock and wild vertebrates. Epidemiol Infect 111:547–557
9. Burt FJ, Swanepoel R, Shieh WJ, Smith JF, Leman PA, Greer PW, Coffield LM, Rollin PE, Ksiazek TG, Peters CJ, Zaki SR (1997) Immunohistochemical and in situ localization of Crimean-Congo hemorrhagic fever (CCHF) virus in human tissues and implications for CCHF pathogenesis. Arch Pathol Lab Med 121:839–846
10. Butenko AM, Chumakov MP (1990) Isolation of Crimean-Congo hemorrhagic fever virus from patients and from autopsy specimens. Arch Virol (Suppl 1):295–301
11. Casals J, Tignor GH (1980) The *Nairovirus* genus. Serological relationships. Intervirol 14:144–147
12. Casey HL (1965) Standard diagnostic complement fixation method and adaptation to microtest. Public Health Monogr 74:1–34
13. Clarke D, Casals J (1958) Techniques for hemagglutination and hemagglutination-inhibition with arthropod-borne viruses. Am J Trop Med Hyg 7:561–573
14. Dickover RE, Herman SA, Saddiq K, Wafer D, Dillon M, Bryson YJ (1998) Optimization of specimen-handling procedures for accurate quantification of levels of human immunodeficiency virus RNA in plasma by reverse transcriptase PCR. J Clin Microbiol 36:1070–1073
15. Donets MA, Rezapkin GV, Ivanov AP, Tkachenko EA (1982) Immunosorbent assays for diagnosis of Crimean-Congo hemorrhagic fever (CCHF). Am J Trop Med Hyg 31:156–162
16. Drosten C, Gottig S, Schilling S, Asper M, Panning M, Schmitz H, Günther S (2002) Rapid detection and quantification of RNA of Ebola and Marburg viruses, Lassa virus, Crimean-Congo hemorrhagic fever virus, Rift Valley fever virus, dengue virus, and yellow fever virus by real-time reverse transcription-PCR. J Clin Microbiol 40:2323–2330
17. Drosten C, Kummerer BM, Schmitz H, Gunther S (2003) Molecular diagnostics of viral hemorrhagic fevers. Antiviral Res 57:61–87
18. Drosten C, Minnak D, Emmerich P, Schmitz H, Reinicke T (2002) Crimean-Congo hemorrhagic fever in Kosovo. J Clin Microbiol 40:1122–1123
19. Duh D, Saksida A, Petrovec M, Dedushaj I, Avsic-Zupanc T (2006). Novel one-step real-time RT-PCR assay for rapid and specific diagnosis of Crimean-Congo hemorrhagic fever encountered in the Balkans. J Virol Methods 133:175–179
20. Gaidamovich SIa, Mel'nikova EE, Shutkova TM, Novokhatskii AS, Turchinskaia AL (1989) Characteristics of the monoclonal antibodies induced by the Crimean hemorrhagic fever virus. Vopr Virusol 34:201–204
21. Garcia S, Chinikar S, Coudrier D, Billecocq A, Hooshmand B, Crance JM, Garin D, Bouloy M (2006) Evaluation of a Crimean-Congo hemorrhagic fever virus recombinant antigen expressed by Semliki forest suicide virus for IgM and IgG antibody detection in human and animal sera collected in Iran. J Clin Microbiol 35:154–159
22. Gonzalez JP, Camicas JL, Cornet JP, Wilson ML (1998) Biological and clinical responses of West African sheep to Crimean-Congo haemorrhagic fever virus experimental transmission. Res Virol 149:445–455
23. Hewson R, Chamberlain J, Mioulet V, Lloyd G, Jamil B, Hasan R, Gmyl A, Gmyl L, Smirnova SE, Lukashev A, Karganova G, Clegg C (2004) Crimean-Congo haemorrhagic fever virus: sequence analysis of the small RNA segments from a collection of viruses world wide. Virus Res 102:185–189
24. Hoogstraal H (1979) The epidemiology of tick-borne Crimean-Congo hemorrhagic fever in Asia, Europe and Africa. J Med Entomol 15:307–417
25. International Air Transport Association (2005) http://www.iata.org/NR/ContentConnector/CS2000/SiteInterface/sites/whatwedo/dangerousgoods/file/Section3.6.2Mar05.pdf
26. Johnson KM, Elliott LH, Heymann DL (1981) Preparation of polyvalent viral immunofluorescent intracellular antigens and use in human serosurveys. J Clin Microbiol 14:527–29
27. Karabatsos N (1985) International catalog of arboviruses including certain other viruses of vertebrates. American Society of Tropical Medicine and Hygiene, San Antonio, TX

28. Karti SS, Odabasi Z, Korten V, Yilmaz M, Sonmez M, Caylan R, Akdogan E, Eren N, Koksal I, Ovali E, Erickson BR, Vincent MJ, Nichol ST, Comer JA, Rollin PE, Ksiazek TG (2004) Crimean-Congo hemorrhagic fever in Turkey. Emerg Inf Dis 10:1379–1384
29. Kolman J (1970) Some physical and chemical properties of Uukuniemi virus, strain Potepli-63. Acta Virol 14:159–62
30. Kurata T, Hondo R, Sato S, Oda A, Aoyama Y, McCormick JB (1983) Detection of viral antigens in formalin-fixed specimens by enzyme treatment. Ann N Y Acad Sci 420:192–207
31. Logan TM, Linthicum KJ, Moulton JR, Ksiazek TG (1993) Antigen-capture enzyme-linked immunosorbent assay for detection and quantification of Crimean-Congo hemorrhagic fever virus in the tick, *Hyalomma truncatum*. J Virol Methods 42:33–44
32. Loutfy MR, Assmar M, Hay Burgess DC, Kain KC (1998) Effects of viral hemorrhagic fever inactivation methods on the performance of rapid diagnostic tests for *Plasmodium falciparum*. J Infect Dis 178:1852–1855
33. Marriott AC, Polyzoni T, Antoniadis A, Nuttall PA (1994) Detection of human antibodies to Crimean-Congo haemorrhagic fever virus using expressed viral nucleocapsid protein. J Gen Virol 75:2157–2161
34. Meegan JM, Yedloutschnig RJ, Peleg BA, Shy J, Peters CJ, Walker JS, Shope RE (1987) Enzyme-linked immunosorbent assay for detection of antibodies to Rift valley fever virus in ovine and bovine sera. Am J Vet Res 48:1138–1141
35. Nichol ST (2001) Bunyaviruses. In: Knipe DM, Howley PM (eds) Fields Virology, Lippincott Raven, Philadelphia, pp 1603–1633
36. Palacios G, Briese T, Kapoor V, Jabado O, Liu Z, Venter M, Zhai J, Renwick N, Grolla A, Geisbert TW, Drosten C, Towner J, Ju J, Paweska J, Nichol ST, Swanepoel R, Feldmann H, Jahrling PB, Lipkin WI (2006) MassTag polymerase chain reaction for differential diagnosis of viral hemorrhagic fever. Emerg Infect Dis 12:692–695
37. Peters CJ, LeDuc JW (1991) Bunyaviridae: bunyaviruses, phleboviruses and related viruses. In: Belshe RB (ed.) Textbook of human virology, Mosby Year Book Inc, St Louis, pp 571–614
38. Rodriguez LL, Maupin GO, Ksiazek TG, Rollin PE, Khan AS, Schwarz TF, Lofts RS, Smith JF, Noor AM, Peters CJ, Nichol ST (1997) Molecular investigation of a multisource outbreak of Crimean-Congo hemorrhagic fever in the United Arab Emirates. Am J Trop Med Hyg 57:512–518
39. Saijo M, Quing T, Niikura M, Maeda A, Ikegami T, Sakai K, Prehaud C, Kurane I, Morikawa S (2002) Immunofluorescence technique using HeLa cells expressing recombinant nucleoprotein for detection of immunoglobulin G antibodies to Crimean-Congo hemorrhagic fever virus. J Clin Microbiol 40:372–375
40. Saijo M, Tang Q, Shimayi B, Han L, Zhang Y, Asiguma M, Tianshu D, Maeda A, Kurane I, Morikawa S (2005) Recombinant nucleoprotein-based serological diagnosis of Crimean-Congo hemorrhagic fever virus infections. J Med Virol 75:295–299
41. Saijo M, Tang Q, Shimayi B, Han L, Zhang Y, Asiguma M, Tianshu D, Maeda A, Kurane I, Morikawa S (2005) Antigen-capture enzyme-linked immunosorbent assay for the diagnosis of Crimean-Congo hemorrhagic fever using a novel monoclonal antibody. J Med Virol 77:83–88
42. Saluzzo JF, Le Guenno B (1987) Rapid diagnosis of human Crimean-Congo hemorrhagic fever and detection of the virus in naturally infected ticks. J Clin Microbiol 25:922–924
43. Schwarz TF, Nsanze H, Longson M, Nitschko H, Gilch S, Shurie H, Ameen A, Zahir AR, Acharya UG, Jager G (1996) Polymerase chain reaction for diagnosis and identification of distinct variants of Crimean-Congo haemorrahgic fever virus in the United Arab Emirates. Am J Trop Med Hyg 55:190–196
44. Shepherd AJ, Leman PA, Swanepoel R (1989) Viremia and antibody response of small African and laboratory animals to Crimean-Congo hemorrhagic fever virus infection. Am J Trop Med Hyg 40:541–547
45. Shepherd AJ, Swanepoel R, Gill DE (1988) Evaluation of enzyme-linked immunosorbent assay and reversed passive hemagglutination for detection of Crimean-Congo hemorrhagic fever virus antigen. J Clin Microbiol 26:347–353

46. Sheperd AJ, Swanepoel R, Leman PA (1989) Antibody response in Crimean-Congo hemorrhagic fever. Rev Infect Dis 11 (Suppl 4):S801–S806
47. Sheperd AJ, Swanepoel R, Leman PA, Sheperd SP (1986) Comparison of methods for isolation and titration of Crimean-Congo hemorrhagic fever virus. J Clin Microbiol 24: 654–656
48. Swanepoel R (1994) Crimean-Congo haemorrhagic fever. In: Coetzer JAW, Thomson GR, Tustin RC (eds) Infectious diseases of livestock, vol. I. Oxford University Press, Capetown, pp 723–729
49. Swanepoel R, Gill DE, Sheperd AJ, Leman PA (1989) The clinical pathology of Crimean-Congo hemorrhagic fever. Rev Infect Dis 11 (Suppl 4):S794–S800
50. Swanepoel R, Leman PA, Burt FJ, Jardine J, Verwoerd DJ, Capua I, Bruckner GK, Burger WP (1998) Experimental infection of ostriches with Crimean-Congo haemorrhagic fever virus. Epidemiol Infect 121:427–432
51. Swanepoel R, Struthers JK, McGillivray GM (1983) Reversed passive hemagglutination and inhibition with Rift Valley fever and Crimean-Congo hemorrhagic fever viruses. Am J Trop Med Hyg 32:610–617
52. Tarantola A, Nabeth P, Tattevin P, Michelet C, Zeller H, Incident Management Group (2006) Lookback exercise with imported Crimean-Congo hemorrhagic fever, Senegal and France. Emerg Infect Dis 12:1424–1426
53. Tignor GH, Smith AL, Casals J, Ezeokoli CD, Okoli J (1980) Close relationship of Crimean hemorrhagic fever-Congo (CHF-C) virus strains by neutralizing antibody assays. Am J Trop Med Hyg 29:675–685
54. Van der Waals FW, Pomeroy KL, Goudsmit J, Asher DM, Gajdusek DC (1986) Hemorrhagic fever virus infections in an isolated rainforest area of Central Liberia. Limitations of the indirect immunofluorescence slide test for antibody screening in Africa. Trop Geogr Med 38:209–214
55. Varma MGR, Pudney M, Leake CJ (1975) The establishment of three cell lines from the tick *Rhipicephalus appendiculatus* (Acari: Ixodidae) and their infection with some arboviruses. J Med Entomol 11:698–706
56. Watts DM, Ksiazek TG, Linthicum KJ, Hoogstraal H (1988) Crimean-Congo Hemorrhagic Fever. In: Monath TP (ed.) The arboviruses: epidemiology and ecology, vol. II. CRC Press, Boca Raton, FL, pp 177–222
57. Wilson ML, LeGuenno B, Guillaud M, Desoutter D, Gonzalez JP, Camicas JL. (1990) Distribution of Crimean-Congo hemorrhagic fever viral antibody in Senegal: environmental and vectorial correlates. Am J Trop Med Hyg 43:557–566
58. Yapar M, Aydogan H, Pahsa A, Besirbellioglu BA, Bodur H, Basustaoglu AC, Guney C, Avci IY, Sener K, Setteh MH, Kubar A (2005) Rapid and quantitative detection of Crimean-Congo hemorrhagic fever virus by one-step real-time reverse transcriptase-PCR. Jpn J Infect Dis 58:358–362
59. Zeller HG, Cornet JP, Camicas JL (1997) Crimean Congo hemorrhagic fever in ticks (Acari: Ixodidae) and ruminants: field observations of an epizootic in Bandia, Senegal (1989–1992). J Med Entomol 34:511–516

CHAPTER 19

TREATMENT OF CRIMEAN-CONGO HEMORRHAGIC FEVER

ONDER ERGONUL, M.D., M.P.H.[1], ALI MIRAZIMI, PH.D.[2], AND DIMITER S. DIMITROV, PH.D., SC.D.[3]

[1]*Marmara University, School of Medicine, Infectious Diseases and Clinical Microbiology Department, Altunizade, Istanbul, Turkey. E-mail: onderergonul@yahoo.com*
[2]*Center for Microbiological Preparedness, Swedish Institute for infectious disease control, 171 82 Solna/Sweden. Tel.: +46 8 457 2573; Fax: +46 8 307957; E-mail: Ali.Mirazimi@smi.ki.se*
[3]*Protein Interactions Group, CCR Nanobiology Program, Center for Cancer Research, NCI-Frederick, NIH, Building 469, Room 105, P.O. Box B, Miller Drive, Frederick, M.D. 21702-1201, USA. Tel.: +301-846-1352; Fax: +301-846-5598; E-mail: dimitrov@ncifcrf.gov (or dimitrov@helix.nih.gov)*

19.1. CURRENT THERAPY: RIBAVIRIN USE AND HEMATOLOGICAL SUPPORT

Onder Ergonul

19.1.1. Ribavirin

19.1.1.1. Introduction

Ribavirin is a synthetic purine nucleoside analog with a modified base and D-ribose sugar, also known as virazol, first synthesized by Sidwell and colleagues in 1972 [43, 49] (Fig. 19-1). It is of particular interest, because it was the first synthetic nucleoside to exhibit broad spectrum antiviral activity, and it is one of few antiviral drugs in clinical use effective against agents other than HIV and herpesviruses [43]. It inhibits the replication of a wide range of RNA and DNA viruses in vitro, including orthomyxo, paramyxo, arena, bunya, flavi, herpes, adeno, pox, and retroviruses [49]. In current clinical practice, ribavirin is commonly used for certain viral infections (Table 19-1). Most notably, it is used in combination with interferon-α for treatment of HCV infection [66]. Ribavirin aerosol is used for treatment of pediatric infection by respiratory syncytial virus [19]. It is the only antiviral drug that could be also used in viral hemorrhagic fever syndromes. Besides Crimean-Congo hemorrhagic fever (CCHF), it is used in Lassa fever [70]. Viruses in the

O. Ergonul and C. A. Whitehouse (eds.), Crimean-Congo Hemorrhagic Fever, 245–260.

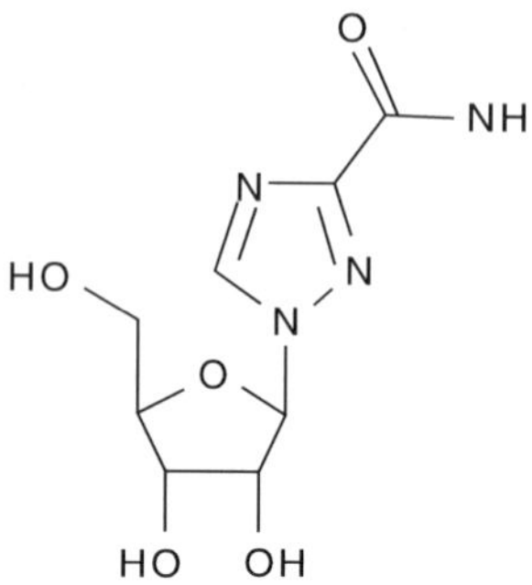

Fig. 19-1. The chemical structure of ribavirin.

Table 19-1. The activity spectrum of ribavirin

In vitro activity
Orthomyxo, paramyxo, arena, bunya, flavi, herpes, adeno, pox, and retroviruses
In vivo benefits
HCV
RSV
Influenza A and B
VHF: Lassa, hemorrhagic fever with renal syndrome, Rift Valley, CCHF
SARS

Bunyaviridae family are generally sensitive to ribavirin [92]. A prospective, randomized, double-blind, placebo-controlled trial of 242 patients with serologically confirmed Hantaan virus in the People's Republic of China found a sevenfold decrease in mortality among ribavirin-treated patients [54], other studies did not confirm these benefits. Ribavirin was found to be effective against CCHF virus (CCHFV) in vitro [99, 104].

19.1.1.2. Mechanisms of action

More than 30 years since its discovery, the mechanism of action of ribavirin still remains controversial. Ribavirin is clinically administered as the nucleoside. Adenosine kinase is the cellular enzyme responsible for conversion to ribavirin monophospate (RMP) (Box 19-1).

The antiviral mechanism of ribavirin is not fully defined but relates to alteration of cellular nucleotide pools and inhibition of viral messenger RNA synthesis [97]. Intracellular phosphorylation to the mono, di, and triphosphate derivatives is mediated by host cell enzymes. In both uninfected and respiratory syncytial virus (RSV)-infected cells the predominant derivative is the triphosphate, which has an intracellular elimination half life of less than 2 h [76]. Ribavirin triphosphate is generally the predominant metabolite [76].

The broad spectrum antiviral activity of ribavirin can potentially be attributed to its multiple mechanisms of action. As a purine analog, it can function in multiple cellular and viral processes. An important aspect of the antiviral activity of ribavirin may stem from the ability to act via multiple mechanisms simultaneously [43]. Five distinct mechanisms were suggested. Some of these actions potentiate others [49]. Ribavirin monophosphate competitively inhibits cellular inosine-5′- phosphate dehydrogenase and interferes with the synthesis of guanosine triphosphate (GTP) and thus nucleic acid synthesis in general. Ribavirin triphosphate also competitively inhibits the GTP dependent 5′-capping of viral messenger RNA. Ribavirin is an inhibitor of inosine monophosphate dehydrogenase, an inhibitor of RNA capping, a polymerase inhibitor, a lethal mutagen, and an immunomodulatory agent [43].

19.1.1.2.1. Ribavirin as an immunomodulatory agent

Ribavirin has also been postulated to act via another indirect antiviral mechanism, by enhancing the host T-cell response. This conclusion stems from observations in hepatitis C virus (HCV) infected patients that ribavirin can reduce serum alanine aminotransferase (ALT) levels without significantly reducing levels of circulating HCV RNA as determined via PCR [25]. The ribavirin is thought to induce a switch in T-helper cell phenotype from type 2 to type 1 [57]. The T-helper type 1 response is associated with cellular immunity and with expression of IL-2, γ-interferon, and tumor necrosis factor-α [71].

In vitro inhibitory concentrations of ribavirin may reversibly inhibit macromolecular synthesis and proliferation of uninfected cells, suppress lymphocyte responses [50], and alter cytokine profiles in vitro. However, this effect was not studied in vivo. The cytokines IL-6 and TNF-α were found to be higher among fatal CCHF patients, whereas there was no significant difference in the levels of IL-10 between the favorable and fatal cases [29]. However, the immunomodulatory effect of ribavirin was not studied on CCHFV yet.

19.1.1.3. In vitro activity of ribavirin against CCHFV

In an in vitro study, ribavirin was shown to inhibit the viral activity, and some CCHF viral strains appeared more sensitive than others [104]. In contrast, a dose of ribavirin at least nine times greater was required to induce a comparable inhibitory effect on the yields of Rift Valley fever virus, for which the drug has been shown to inhibit replication in monkeys and rodents [104]. After intraperitoneal infection of infant mice with CCHF virus, virus titers in liver remained significantly higher than in other organs except serum. Within the liver, virus antigen was first found by immunofluorescence assays (IFA) in Kupffer cells followed by more extensive hepatic spread. Later, virus was found in other organs including brain and heart. Ribavirin treatment significantly reduced infant mouse mortality and extended the geometric mean time to death. Ribavirin treatment reduced CCHF virus growth in liver and significantly decreased, but did not prevent, viremia [99].

19.1.1.4. Clinical observational studies on ribavirin

The clinical data related to effectiveness of ribavirin stems from the studies on chronic infection with the HCV infection. The effect of ribavirin use in CCHFV infection was not evidenced by randomized clinical trials. The effectiveness of its use was described by observational studies [26, 33, 67, 74]. In the first clinical report in 1995, the observation was limited to three health-care workers (HCW), who had been infected with CCHFV infection [33]. All three patients were severely ill with low platelet and white blood cell counts, raised aspartate transaminase, and evidence of impaired haemostasis. According to severity criteria defined by Swanepoel [95], all had an estimated probability of death of 90% or more. The patients became afebrile, and their haematological and biochemical abnormalities returned to normal within 48 h of ribavirin treatment, and all made a complete recovery. This finding was found to be encouraging for the use of ribavirin in CCHFV infections, but did not constitute evidence of efficacy. After this study, ribavirin was introduced to be used in CCHFV infections.

Mardani et al, compared the fatality rate among patients suspected of having CCHF who received treatment with oral ribavirin and those who did not [67]. As many as 97 (69.8%) of 139 treated patients suspected of having CCHF survived, and 61 (88.9%) of 69 treated patients with confirmed CCHF survived. The efficacy of oral ribavirin was 80% among patients with confirmed CCHF and 34% among patients suspected of having CCHF. This study used historical control, and did not stratify the patients according to the severity of the patients.

Ergonul et al, described the role of ribavirin therapy for 35 confirmed CCHF patients. About 86% of the patients were considered to have severe cases of CCHF. Eight patients were given ribavirin, and all eight of them survived. The study results suggested the use of ribavirin especially for the severe cases [26] (Fig. 19-2).

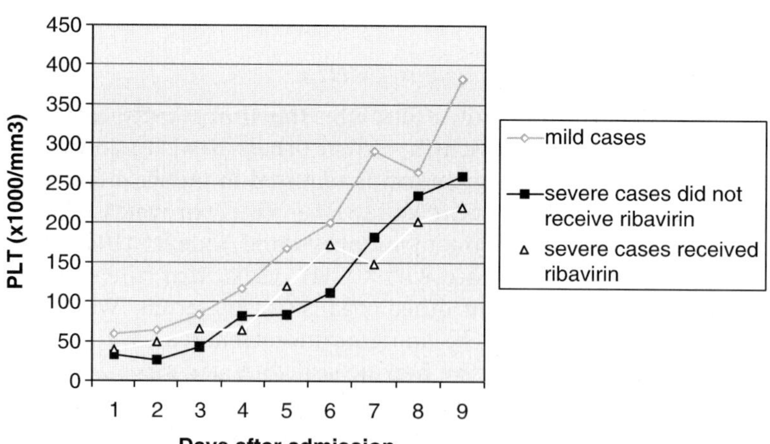

Fig. 19-2. The course of platelet count after admission.

The cases in this study were grouped according to the severity criteria defined by Swanepoel [95].

Ozkurt et al. described the efficacy of ribavirin therapy for CCHF among 60 confirmed CCHF patients in Eastern Turkey. Mean recovery time was shorter in the cases treated with ribavirin than those of control. But, the need for blood and blood product, mean hospitalization duration, fatality rates, and hospital expenditure values were not significantly different between two group of patients, who received ribavirin or not [74].

Almost all the authors claimed that, they could not perform a randomized clinical trial (RCT), because of ethical constraints. The lack of randomization is the main criticism for these observational studies. However, sometimes the observational studies could give qualified information, if they could be well designed [101]. But, the researchers should be aware of the potential confounders. While performing the clinical outcome studies on ribavirin use in CCHF, there are significant confounders;

1. Severity of the infection might differ. There are mild and severe forms of the disease, because of several reasons. (see also Chapter 16).
2. Number of days from onset might differ. Some patients could get the ribavirin at an earlier, prehemorrhagic phase of the infection, whereas some patients could get the ribavirin at a later phase, hemorrhagic phase of the infection.
3. The severity of the gastrointestinal symptoms might differ. Some patients cannot get drug via oral route, because of severe hematemesis. This parameter is important in pharmacological effectivity.

If the ribavirin was saved for the severe cases, an observational comparative study between the ribavirin given and not given group would have a misclassification bias in favor of not using ribavirin. The patients who would not receive ribavirin would be usually the mild cases. A well designed observational study should minimize the effect of these confounders to avoid the misclassification bias.

According to the clinical observation of the cases, the effective treatment for CCHF could be considered hypothetically in two phases (Fig. 19-3). The first phase, starts from the exposure to the virus, and characterized by the viremia, which usually lasts 5–15 days. In clinical terms, this period starts from the onset of prehemorrhagic symptoms (fever, myalgia, nausea, and vomiting), and ends up with the bleeding from various sites. The antiviral effect of the ribavirin most likely occurs in the first phase. The second phase is characterized by the decline of viremia, but bleeding from the various sites starts. In this phase, cytokines are released extensively, and the coagulation cascade is disrupted in some patients, and disseminated intravascular coagulation was noted [29]. The antiviral action of the ribavirin is less likely in this phase. The immunomodulatory effect is not known yet. Alternative drugs targeted to disseminated intravascular coagulation (DIC) or sepsis could be considered in the second phase of the disease course.

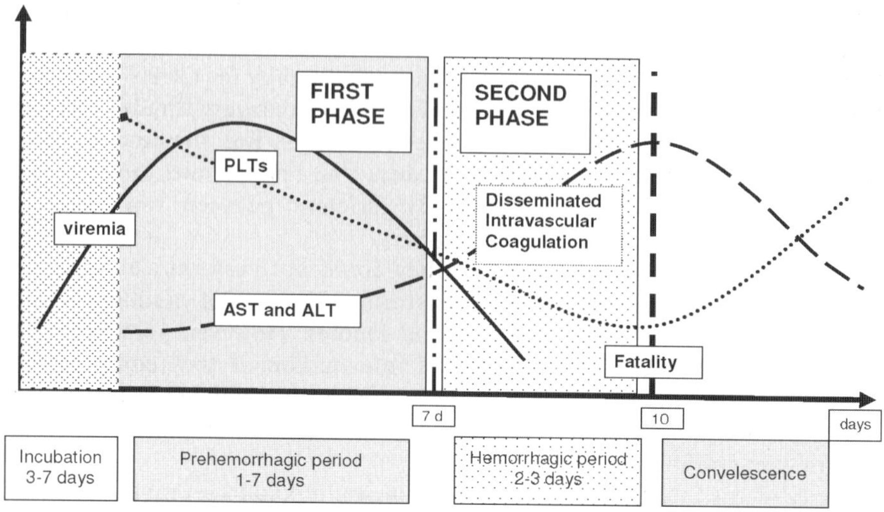

Fig. 19-3. The probable role of the ribavirin and alternatives in Crimean-Congo haemorrhagic fever (CCHF) infection course. The effective treatment for CCHF should be considered in two phases. The first phase, starts from the exposure to the virus, and characterized by the viremia, which usually lasts 5–15 days. In clinical terms, this period starts from the onset of prehemorrhagic symptoms (fever, myalgia, nausea, and vomiting), and ends up with the bleeding from various sites. Ribavirin is most likely to be effective in the first phase. The second phase is characterized by the decline of viremia, and clinically the onset of bleeding from the various sites. The immunologic mechanisms, such as cytokine storm, disrupt the coagulation cascade in some patients, and disseminated intravascular coagulation occurs. For these patients other treatment alternatives should be considered. Most likely, the ribavirin is less effective in this phase. Ribavirin could be effective in second phase, mainly because of its immunomodulatory effect. However, immunomodulation was not suggested as the main action for ribavirin.

19.1.1.5. Absorption, distribution, and elimination

Ribavirin is actively taken up by gastrointestinal nucleoside transporters located in the proximal small bowel, and oral bioavailability averages approximately 50% [49]. Extensive accumulation occurs in plasma, and steady state is reached by about 4 weeks. Food increases plasma levels substantially, so ingestion with food may be prudent [49]. Following single or multiple oral doses of 600 mg and 1,200 mg, peak plasma concentrations average 0.8 μg/mL and 3.7 μg/mL, respectively. After intravenous doses of 1,000 mg and 500 mg, plasma concentrations average approximately 24 μg/mL and 17 μg/mL, respectively.

The apparent volume of distribution is large (~10 L/kg) due to ribavirin's uptake into cells. Plasma protein binding is negligible. The elimination of ribavirin is complex. The plasma half life averages 30–40 h after a single dose but increases to approximately 200–300 h at steady state. Ribavirin triphosphate concentrates in erythrocytes, and red blood cell levels gradually decrease with a $t_{1/2}$ of about 40 days. Hepatic metabolism and renal excretion of ribavirin and its metabolites are the principal routes of elimination. Hepatic metabolism involves deribosylation and hydrolysis to yield a triazole carboxamide. Ribavirin clearance decreases threefold in those with advanced renal insufficiency (CL_{cr} 10–30 mL/min); the drug should be used cautiously in patients with creatinin clearance of less than 50 mL/min [49].

Oral and intravenous forms are available in many countries, and the dosage was given in Box 19-2. The total duration of treatment was defined as 10 days [33]. If the laboratory results get better, then the ribavirin could be stopped. The effectivness of IV and oral forms were not compared yet.

Emergence of viral resistance to ribavirin has not been documented in most viruses but has been reported in Sindbis and HCV [107], although it has been possible to select cells that do not phosphorylate it to active forms.

19.1.1.6. Adverse events

Systemic ribavirin causes dose-related reversible anemia due to extravascular hemolysis and suppression of bone marrow [53]. Associated increases occur in reticulocyte counts and in serum bilirubin, iron, and uric acid concentrations. High ribavirin triphosphate levels may cause oxidative damage to membranes, leading to erythrophagocytosis by the reticuloendothelial system [49]. The half life of ribavirin metabolites is relatively short in cultured fibroblasts and lymphoblasts, although the nucleotides are much more stable in erythrocytes [76]. This accumulation of ribavirin in erythrocytes is responsible for the reversible hemolytic anemia that is a side effect of clinical ribavirin therapy [1].

Bolus intravenous infusion may cause rigors. About 20% of chronic hepatitis C patients receiving combination interferon–ribavirin therapy discontinue treatment

Box 19-1. The action mechanism of antiviral agents against RNA viruses
The replication strategy of the RNA viruses relies either on enzymes in the virion (the whole infective viral particle) to synthesize its mRNA or on the viral RNA serving as its own mRNA. The mRNA is translated into various viral proteins, including RNA polymerase, which directs the synthesis of more viral mRNA [97].

Box 19-2. The dosage of ribavirin in CCHF infections
About 30 mg/kg as an initial loading dose, then 15 mg/kg every 6 h (4 × 1 g) for 4 days, and 7.5 mg/kg every 8 h (4 × 0.5 g) for 6 days was recommended [27].

early because of side effects. In addition to interferon toxicity, oral ribavirin increases the risk of fatigue, cough, rash, pruritus, nausea, insomnia, dyspnea, depression, and particularly anemia. About 8% of patients require ribavirin dose reduction because of anemia. Hemolytic anemia, hypocalcemia, and hypomagnesemia were reported in patients, who received ribavirin because of severe acute respiratory syndrome (SARS) [17, 63].

However, no adverse event related to ribavirin therapy was noted among CCHF patients. Mainly because of two reasons, (1) acute course of the disease, that might not allow time to observe the side effects, and (2) overshadowing of the disease findings, which are the same with the potential adverse events, such as anemia.

Preclinical studies indicate that ribavirin is teratogenic, embryotoxic, oncogenic, and possibly gonadotoxic. In vivo genotoxicity of ribavirin among three patients with CCHF were studied [98]. The micronucleus and the sister chromatid exchange test were found to be higher among all three patients, during and right after the ribavirin therapy. A month later, the test results became normal. This finding revealed that ribavirin has a reversible in vivo genotoxic effect in humans [98]. To prevent possible teratogenic side effects, up to 6 months is required for washout following cessation of long-term treatment [49]. The use of ribavirin is contraindicated among pregnant women. Ribavirin is in Food and Drug Administration (FDA) pregnancy category X.

19.1.2. Supportive therapy

Potential bleeding foci of the patients should be considered and conservative measures should be taken, such as use of histamine receptor blockers for peptic ulcer patients, avoidance of intramuscular injections, and not using aspirin or other drugs with actions on the coagulation system. Non-steroidal anti-inflammatory drugs should be avoided. Fluid and electrolyte balance should also be monitored meticulously.

Supportive therapy is the essential part of the case management. It includes the administration of thrombocytes, fresh frozen plasma (FFP), and sometimes erythrocyte preparations. The replacement therapy with these blood products should be performed by checking the complete blood count, which should be done one or two times a day. In clinical practice, checking the thrombocyte level once a day would be sufficient [27].

Thrombocyte solution or FFP are the products to be replaced according to the deficit of the individual patient. Figure 19-3 depicts the tremendous amount of thrombocyte solution and FFP that were given to the patients in Turkey [28]. The fatal cases received significantly ($p < 0.001$) higher amount of thrombocyte suspensions and FFP (Fig. 19-4).

Sometimes, it is hard to decide where to start or stop the blood product replacement. Therefore, we reviewed the general rules of platelet and FFP use by adapting to CCHFV infection.

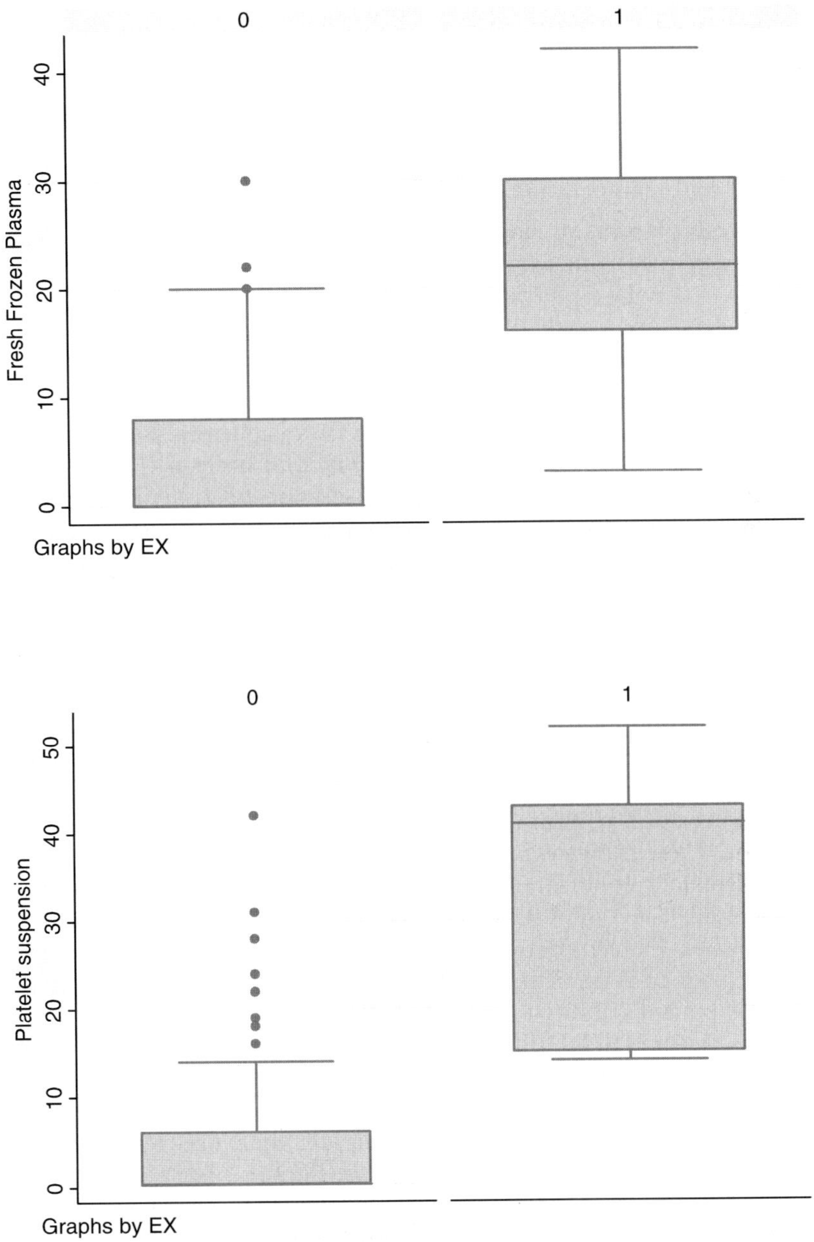

Fig. 19-4. The number of units of (A) fresh frozen plasma or (B) platelet suspensions given to CCHF patients. The box plots on the right represents total units given to patients who scummed to the disease and, on the left, those who survived.

19.1.2.1. Platelets

Random-donor platelets (RDPs): Platelet concentrates are prepared by separating platelets from a single unit of whole blood so that a minimum of 5.5×10^{10} platelets are suspended in approximately 50 mL of plasma (1 U). Platelets should be separated from whole blood within 8 h of collection. Depending on the container in which RDPs are stored, the shelf life varies from 3 to 5 days when stored at room temperature.

Pheresis–apheresis platelets or single-donor platelets (SDPs): Platelets collected by apheresis techniques from a single donor, using blood cell separators, contain a minimum of 3.0×10^{11} platelets suspended in 200 to 400 mL of plasma. Apheresis platelets have a shelf life ranging from 24 h to 5 days and are stored at room temperature. These products account for most platelet transfusions in the developed countries.

HLA-matched platelet concentrates: If the donor of an apheresis platelet concentrate is selected because the donor's human leukocyte antigen (HLA) type is matched to the recipient's HLA type (due to the development of alloimmunization to HLA), the apheresis platelet product is considered to be an HLA-matched platelet component.

Use of platelet transfusions is indicated to control active bleeding or to prevent hemorrhage associated with a deficiency in platelet number or function. Platelets are used prophylactically to prevent bleeding when the platelet count is less than 10,000–20,000/μL. There is a growing trend to reduce the prophylactic platelet transfusion "trigger" to counts as low as 5,000–10,000/μL in stable patients without significant hemorrhage [69].

Dose and infusion rate:

RDPs The average adult platelet concentrate dose is one unit of RDP per 12 kg of body weight. There is a growing trend in the United States to define a standard "dose" of platelet concentrates. This standard varies among different institutions from 4 U, 6 U, or 8 U of platelet concentrates to be infused per transfusion episode. Platelet units usually are pooled into one bag by the transfusion service prior to issue. They should be infused within 4 h at a rate that depends on the patient's ability to tolerate the volume. RDPs may be "volume reduced; that is, the pooled RDPs undergo centrifugation that allows separation of platelets from the platelet poor plasma. Excess platelet poor plasma is removed, and the therapeutic dose then is concentrated into approximately 100 mL. Volume-reduced platelets are infused more rapidly than nonvolume-reduced platelets and must be infused within 4 h of pooling [69].

Pheresis platelet-SDPs. One SDP concentrate is considered a therapeutic dose, equivalent to six units of RDP. SDPs should be infused as rapidly as possible, depending on the patient's ability to tolerate the infused volume.

HLA-matched platelets. The dose and infusion rate are similar to those of SDPs.

Expected outcome: Platelet transfusions should result in prevention or resolution of bleeding caused by thrombocytopenia or platelet dysfunction. As a general rule, platelet counts should be obtained 18–24 h post-infusion. The average-sized adult who receives 1 U of RDP per 12 kg of body weight or one SDP dose should have a posttransfusion platelet increment of 30,000–50,000 μL. Patients who have smaller or no increment at 18–24 h post-infusion should have platelet counts performed 10–60 min after the next platelet transfusion. If the 10 to 60 min posttransfusion increment is minimal or not increased at all, the possibility of refractoriness caused by alloimmunization should be entertained. Antibody-related platelet destruction is often related to the development of HLA-specific antibodies in response to foreign donor HLA antigens [79]. The patient then should be considered for HLA-matched platelet transfusions or crossmatched platelet concentrates. Poor posttransfusion platelet survival, in addition to being caused by alloimmunization, is often seen in conjunction with fever, sepsis, disseminated intravascular coagulation, and others. Patients in these settings are considered to be refractory on a nonimmunologic basis and are not expected to benefit from HLA-matched platelet transfusions. Patients who have life-threatening hemorrhage may require larger platelet doses to be therapeutic. Currently, controversy is growing as to whether patients requiring prophylactic platelet transfusions would benefit more from repetitive small-dose platelet transfusions or be better off using extra high-dose platelet transfusions. Randomized, well-controlled studies to answer this issue have not yet been done [69].

In CCHF, since disseminated intravascular coagulation occurs in the disease course, platelet destruction is expected, and the rapid increment of PLT level after transfusion may not be observed.

19.1.2.2. Fresh frozen plasma

FFP is plasma that is separated from whole blood and frozen within 8 h of collection and has normal levels of all clotting factors and anticoagulants. Three different types of FFP are available for transfusion: (i) Standard FFP is as just described; (ii) Donor-retested FFP is plasma that is donated, frozen, and stored for 112 days. The donor then returns, and if all infectious disease testing is still negative, the original unit is released into the inventory. Donor-retested plasma has the value of closing the window period for the infectious diseases that are currently tested for HIV, HVC; (iii) pooled solvent detergent-treated plasma is frozen, and plasma has been sent to a commercial company to be treated by a solvent detergent process that will prevent the transmission of envelope viruses (i.e., HIV, HCV). The value of this product is that it will prevent transmission of any envelope virus by blood transfusion. It is a pooled product, and recipients are exposed to 1,500–2,500 donors with each transfusion of a unit of frozen plasma [69].

FFP is indicated to replenish clotting factors in patients with demonstrated deficiencies, such as prothrombin time or partial thromboplastin time greater than 1.5 times normal, international normalized ratio (INR) > 1.6. FFP is most commonly used in the setting of acquired coagulopathy, such as in patients with liver disease, DIC, or excess warfarin effect [69].

The average adult dose is determined by the clinical situation and the underlying disease process. It is reasonable to administer plasma at a dose of 10–15 mL/kg of body weight (2–4 U of FFP) followed by laboratory evaluation to determine responsiveness and to decide the interval between doses. The infusion rate is determined on the basis of the patient's clinical need and hemodynamic status. Plasma does not contain red cells, and therefore, cross-matching is not required. The ABO type of the donor should be compatible with the recipient. FFP is thawed at 37°C and must be transfused with 24 h of thawing if used for coagulation factor replacement. Plasma can be thawed and maintained in the refrigerated state for up to 5 days, but there is some loss of coagulation factor V and a greater decrease in factor VIII [69].

Expected outcomes: Improvement of coagulation factor deficiency is expected as assessed by the prothrombin time (INR), partial thromboplastin time, or specific factor assays [69].

The other alternatives could be considered the possible pathogenetic mechanisms of the infection. The therapeutic agents, that were considered in hemophagocystosis and DIC could also be studied in CCHF infection. The agents considered in DIC are shown in Table 19-2. Corticosteroids accompanied with ribavirin were reported to be useful at the early stage of CCHF [60]. However, this experience was limited to six patients, only.

An algorithm for the management of the cases was suggested in Table 19-3.

Table 19-2. Treatment modalities for disseminated intravascular coagulation [35] (Box 19-3)

1. Replacement therapy	Fresh-frozen plasma
2. Anticoagulants	Unfractionated and low molecular weight heparin
	Danaparoid sodium
	Recombinant hirudin
	Recombinant tissue factor pathway inhibitor
	Recombinant nematode anticoagulant protein c2
3. Restoration of anticoagulant pathways	Antithrombin
	Recombinant human activated protein C
4. Other agents	Recombinant activated factor VII
	Antifibrinolytic agents
	Antiselectin antibodies
	Recombinant interleukin-10
	Monoclonal antibodies against TNF and CD14

Table 19-3. An algorithm for the case management

Evaluation of the suspected case
Clinical symptoms (fever, myalgia, bleeding from various sites)
Patient history
i. Referral from endemic area
ii. Outdoor activities (picnic, tracking, etc.) in endemic area
iii. History of tick exposure
iv. Exposure to potentially viremic domestic animal blood
Laboratory tests (low platelet and high white blood cell count, elevated AST, ALT, LDH, CPK)
Preventive measures
a. Isolate the patient
b. Inform and educate colleagues and staff
c. Use the barrier precautions
Investigations for confirmation
Serum for PCR (early in disease) and ELISA (late in disease or convalescence)
a. IgM positivity or PCR positive confirms diagnosis, IgG positivity cannot
b. Sera for differential diagnosis
Decision making for therapy
1. Ribavirin
2. Do not neglect other causes of clinical picture. Starting doxycycline or equivalent should be considered
3. Hematological support
a. Fresh frozen plasma to improve hemostasis
b. Thrombocyte solutions
4. Respiratory support
Follow-up
1. No relapse occurs after the disease. Therefore there is no need for the follow up of the cases
2. HCWs exposed to the virus should be followed up with complete blood counts and biochemical tests for 14 days

Box 19-3. Disseminated intravascular coagulation and CCHF

Disseminated intravascular coagulation (DIC) results from activation of coagulation in the vascular tree. Accelerated platelet consumption is almost always seen. DIC can be distinguished from immune thrombocytopenia by finding prolongation of the prothrombin and partial th romboplastin times, decreased plasma fibrinogen, and elevated plasma fibrin–fibrinogen split products. DIC can be seen with infections (e.g., viral, Rickettsial, bacterial, malarial infections); obstetric catastrophes (abruptio placentae and the retained dead fetus syndrome); malignancies; trauma; and vascular abnormalities such as giant hemangiomas and aortic aneurysms. Generally, the treatment should be directed toward correcting the underlying cause. Support with plasma and platelet transfusions may be required for bleeding complications until the cause has been corrected. The thrombocytopenia gradually resolves as the infection is controlled [35, 108].

19.2. OLD AND NEW TREATMENT STRATEGIES

Ali Mirazimi

19.2.1. Introduction

Historically, treatment for Crimean-Congo hemorrhagic fever (CCHF) is limited and very poorly studied. However hemorrhagic fever virus infections can be approached by the following different therapeutic strategies [6]: (i) administration of high-titered neutralizing antibodies and/or (ii) treatment with antiviral drugs.

In more recent times, an immunotherapy approach has been described that uses passive transfer of CCHF convalescence sera from recovered patients [102]. The specific use of immunotherapy as a therapeutic approach will be discussed in Section 19.1.

Depending on the site of interactions and their molecular target, anti-viral drugs may be classified as follows: orotidine monophosphate decarboxylase inhibitors (i.e. pryrazofurin) [5, 96], inosine monophosphate dehydogenease inhibitors (i.e. ribavirin and its derivates) [11, 26, 89, 92], cytidine triphosphate Synthase inhibitors (i.e. cyclopentenylcytosine) [5, 68], S-adenosylhomocysteine hydrolase inhibitors (i.e. neplanocin A) [5], polyanionic substances (i.e. sulfated polymers) [8–10] and interferons and immunomodulaterors [90, 91].

Currently, there is no specific antiviral drug against CCHF, however, reports exist which describe a potential antiviral effect of ribavirin [26, 99]. It should be mentioned that no randomized, controlled studies exist, which confirms the potency of ribavirin against CCHF.

Another interesting area of treatment is the antiviral activity of interferons and immunomodulators against viral infections. Almost all hemorrhagic fever viruses have been shown to be sensitive to interferons [4, 7, 20, 23, 48]. However, in order to develop an agent, whose function exploits the antiviral activities of interferon, more knowledge of the antiviral mechanism of interferon against CCHFV is required. The role of interferon and the interferon-induced antiviral proteins will be discussed in Section 19.2.2.

During past years, several groups have become interested in studying CCHFV as it has a potential for being abused as a bioterrorism weapon; such research has increased our knowledge of the basic biology of the CCHFV [3, 4, 13, 34, 40, 46, 62, 85, 103]. This may lead to an improvement in therapy; an example could be the finding of inhibitors of the viral protein processing. During the last decade, the potential of RNA interference as a therapeutic agent against virus infection has been discussed. Recently many studies have displayed the efficacy of chemically synthesized short interference RNA as a tool to combat a diverse group of viruses. However it should be mentioned that before RNA interference can be used as therapeutic agents several questions need to be addressed, such as: (1) the transient effect of short interference RNA, (2) delivery methods, and (3) cell uptake [15].

19.2.2. Interferons and interferon-stimulated antiviral proteins

Host responses to infection comprise two general categories: (1) Innate immunity, a rapid and phylogenetically nonspecific response to infection [61], and (2) adaptive immunity, long lived, and highly-specific immune response to infection [39, 75].

Innate immunity has been described as an essential immune response against emerging viruses. Central to the innate immune response is a large heterogeneous group of peptide mediators (interferons, chemokines, interleukins, and growth factors), these molecules are part of a complex network that regulates the immune and inflammatory responses [14, 61, 82, 100].

Interferons are produced by leukocytes and fibroblasts in response to viral infection [44, 45] and results in the induction of antiviral pathways within a period of hours postinfection. The interferons have a long history of clinical application against some viral diseases [18]. As early as 1986, interferon-α was demonstrated to have a potent antiviral effect in patients with chronic hepatitis C [86].

In recent times, several groups have studied the antiviral activities of interferons against hemorrhagic fever viruses. These studies clearly demonstrate that interferons have significant antiviral activities against most of all hemorrhagic fever viruses in vitro and animal models. However, to date, no clinical data has addressed the effect of interferons against viral hemorrhagic fever and in particular CCHFV. Moreover, the use of interferons as a therapeutic agent against hemorrhagic fever diseases poses significant clinical challenges. A better understanding of interferons action is essential if current therapy is to be optimized and also if new strategies therapy are to be approached.

Isaacs and Lindenmann have discovered the antiviral activity of interferon class of molecules almost 50 years ago [58, 59]. Today interferons are divided into two groups, interferon types I and II. The type I interferons are major players in antiviral defense against almost all known viruses [48]. The type I interferons includes interferon-α, -β, and -λ [12, 87]. There are at least 14 subtypes of interferon-α genes, but only one β and λ gene. All type I interferons have antiviral activity as well as antiproliferative and immunomodulatory activities. Virus-infected or stimulated cells upregulate and secretes interferon type I. The secreted interferons induce other cells to express potent antiviral proteins and to activate additional antiviral mechanisms that will limit the viral spread. The antiviral activity of interferon-α and -β is not directly coupled to the interferon molecule per se. Rather, these molecules act indirectly, by inducing interferon-stimulated genes (ISGs), which in turn establish an antiviral phase in the target cells.

Type I interferons trigger the expression of a large number of ISGs. These ISGs have antiviral, antiproliferative, and immunomodulatory functions [21, 22]. Interferon-induced proteins include transcription factors, enzymes, cytokines, chemokines, and glycoproteins, although a large number of molecules need further characterization. Until now, a few antiviral proteins have been studied in detail. The best-characterized interferon-induced antiviral proteins are the Mx GTPase [64], the 2′-5′ oligoadenylate synthetases (2–5 OAS)/RNaseL [93], and the protein kinase R (PKR) [105].

MxA has been found to be one of the major antiviral proteins against a large group of viruses, including viruses of the Ortomyxo, Paramyxo, Togoviridae, Rhabdoviridae, Picrnaviridae, and Hepdnaviridae families [36–38, 47, 52, 65]. MxA belongs to the dynamin super family of large GTPases, whose members function in a variety of intracellular transport process. Recent studies have clearly demonstrated that MxA has a significant antiviral activity against CCHFV [3, 4]. Andersson and coworkers demonstrated that MxA colocalizes and interacts with CCHFV nucleocapsid protein in the perinuclear region of infected cells; it was therefore suggested that this interaction inhibits the virus replication process. Similar results have been observed for other members of Bunyaviridae [37, 64, 81]. These results together imply that all animal bunyaviridae may be restricted in their intracellular growth by MxA, and probably by the same mechanism.

Another recent report by Andersson and co-workers showed that ISGs, induced by interferon-α in human endothelial and hepatoma cells was sufficient to inhibit CCHFV growth significantly [4]. Yet, despite this effect, why does CCHFV still run its devastating course? It has been estimated that CCHFV causes mortality at rates belying tween 30–50% in infected patients. The most likely explanation is that interferon is induced at insufficient levels or too late in the course of infection to put MxA and other antiviral proteins in place to combat the CCHFV when it is urgently needed.

Additional proteins with potentially important antiviral activities are ISG20 [30–32], P56 [55, 56], RNA-specific adenosine deaminase 1 (ADAR 1) [83], promyelocytic leukemia protein (PML) [80] and guanylate-binding protein 1 (GBP-1) [2]. ISG20, an interferon-induced exonuclease, specifically degrades ssRNA [31]. Expression of ISG20 inhibits viral replication of vesicular stomatitis virus and human immunodeficiency virus [30, 32] in cell culture. Recent studies by Weber and Mirazimi have clearly shown that ISG20 has an antiviral activity against CCHFV by so far unknown mechanism (unpublished).

There is a critical need to identify new effective treatments for viral hemorrhagic fever and in particular CCHFV. One of the interesting issues is the interferons and other immunomodulators. Our present knowledge of the antiviral activity of interferon system is still limited. However, a better understanding of the complicated interaction between viruses and innate immune response will help to design new antiviral treatments and therapy.

19.3. ANTIBODIES TO CCHFV FOR PROPHYLAXIS AND TREATMENT

Dimiter S. Dimitrov

Infections with CCHFV elicit antibodies to the virus that are not present several (about 5–9) days after onset of illness [78, 88]. Patients who have died of CCHF do not usually develop a measurable antibody response [78]. It is likely that even if such patients did have low levels of CCHFV-specific IgG they still could be reinfected. For example, it was shown for West African sheep, which play a central role in the maintenance cycle of CCHF virus in disease-endemic areas because they serve as host for both the virus and the tick vector, that even sheep that were infected previously and had anti-CCHF virus IgG can be reinfected and transmit the virus [42, 106]. CCHFV-specific IgM remains elevated for 40 days after infection [42] but there are no reports whether such elevated levels could protect from infection or reinfection. However, such antibodies could play a protective role in secondary transmissions. The virulence of the virus is likely diminished in such transmissions possibly due to existence of subpopulations of virus adapted to a host that are selected after passage through another vertebrate host; such subpopulations seem to be less virulent and might have an altered capacity of transmission [41]. This is consistent with the observation that infectivity from secondary cases is unusual [94] unless heavy exposure to virus contaminated tissue has occurred [1, 73, 77]. However, the efficacy of treatment in such secondary transmissions is difficult to evaluate because of lack of controls and because the risk for fatal outcome of the secondary case is not so high. There is currently no specific antiviral therapy for CCHF approved for use in humans by the FDA.

A formalin-inactivated suckling mouse brain-based vaccine has been used in Bulgaria and other parts of Eastern Europe and the former Soviet Union for protection from CCHFV infections. In the Rostov region of the former Soviet Union, 1,500 persons received the vaccine and showed a high frequency of detectable antibodies. Likewise, vaccine was given to several hundred human volunteers in Bulgaria, with resulting high antibody induction. With the relatively small target population of persons at risk for contracting CCHFV, the large-scale development and production of a CCHF vaccine by modern standards seems unlikely.

There was an early recognition of the possible benefits of treatments using serum prepared from the blood of recovered CCHF patients or gammaglobulin obtained from immunization of horses [51]. In more recent times, Bulgarian investigators suggested that immunotherapy treatment of seven patients with severe CCHF via passive simultaneous transfer of two different specific immunoglobulin preparations, CCHF-bulin (for intramuscular use) and CCHF-venin (for intravenous use), prepared from the plasma of CCHF survivor donors boosted with one dose of CCHF vaccine, resulted in quick recovery of all patients [102]. No side effects were observed and the patients were discharged

O. Ergonul and C. A. Whitehouse (eds.), Crimean-Congo Hemorrhagic Fever, 261–269.

in good health. Although there was no control group of cases of the same severity (an average of 25 cases per year are reported in Bulgaria), and firm evidence of its value is lacking, previous experience suggested that in such cases the prognosis is unfavorable even when the patients are treated with the intramuscular formulation (CCHF-bulin) – for 4 years (1985–1988) 15 such patients have died. The investigators attributed the success of this approach to the quick distribution of the antibodies through the circulation that is important for acute infections combined with a sustained and even increasing antibody concentration from the muscle depot. They suggested that the intravenous preparation be used for treatment of all cases of CCHF.

Advances in development of monoclonal antibodies and antibody engineering have raised hopes for new candidate drugs for prevention and treatment of CCHFV infections. Immunotherapy of CCHF has been based on polyclonlal antibodies obtained from serum of immunized animals or humans. Although such preparations could be effective, the use of monoclonal antibodies could allow better control of the composition of the therapeutic preparations and in some cases could be more effective. For example, the humanized monoclonal antibody Synagis, which is the first and only monoclonal antibody yet licensed for an infectious disease, is the preferred of the two available licensed products in most situations (the second one is the RSV-IGIV, RespiGam, which is a solution of IgG enriched in neutralizing antibodies to RSV). However, the mAbs recognize a single epitope that may limit their usefulness against pathogens that exhibit genetic variations [16]. This problem could be solved by generating mAbs against conserved epitopes and/or use of mAb cocktails [16, 24].

Recently, mouse monoclonal antibodies (mmAbs) specific for the two CCHFV envelope glycoproteins (Envs), G_N and G_C, were developed [13]. In neutralization assays on SW-13 cells, mmAbs to G_C, but not to G_N, prevented CCHFV infection. However, only a subset of G_C mmAbs protected mice in passive-immunization experiments, while some non-neutralizing G_N mmAbs efficiently protected animals from a lethal CCHFV challenge. Thus, neutralization of CCHFV likely depends not only on the properties of the antibody, but on host cell factors as well. In addition, non-neutralizing antibody-dependent mechanisms, such as antibody-dependent cell-mediated cytotoxicity, may be involved in the in vivo protection seen with some of the mmAbs.

Mouse antibodies could induce anti-mouse antigen immune responses and should by humanized if used in humans. Humanization decreases but does not eliminate the possibility for immunogenic effects and in some cases is difficult to do. We have initiated an experimental program aimed at the development of fully human mAbs (hmAbs) that recognize G_N and G_C and neutralize the virus by using our phage display libraries. G_N and G_C mediate the virus entry into susceptible cells, and have been recently expressed and characterized as membrane-associated proteins [13, 84, 103]. Currently there are no available human monoclonal antibodies (hmAbs) directed to epitopes on G_N and G_C.

We plan to use two different procedures for Env expression from non-infectious material and presentation in native conformations: (i) mammalian cell surface associated functional Envs, and (ii) recombinant expressed soluble Envs. Production of Envs in close to native conformation is of critical importance, for in many instances the ability to elicit potent neutralizing activity to oligomeric Envs depends on the retention of their native three-dimensional structure. As our second method for presentation of Env antigens, we use recombinant soluble Envs we have been developing. The ability to produce soluble forms of such glycoproteins using recombinant mammalian expression systems often allows for production of the protein in a form that is unperturbed in tertiary and quaternary structure. For example, viral Envs are invariably oligomeric, composed of homo or heterotrimeric or tetrameric forms, and the systems using high expression vectors and tissue culture often allows for rapid production of such native oligomeric versions of these types of proteins. Such soluble constructs can be more easily purified and they are quite useful in analyzing the structure and oligomeric nature of the proteins. Thus significant amounts of purified soluble Envs from all these viruses will be available for screening of PDLs.

The panning of PDLs with the soluble Envs and screening of the highest affinity binders will be performed by using standard protocols. We and others have successfully used these protocols for identification of hmAbs directed against epitopes on the HIV Env [72, 110–112]. Recently, we developed a novel methodology, sequential antigen panning (SAP), based on alternating various Envs during panning of PDLs, which was used to identify four new potent broadly HIV-neutralizing human monoclonal antibody Fabs (m12,14,16,18) [109]. We plan to use the SAP methodology for the Envs. If there are conserved neutralization epitopes on these viruses and the SAP approach is successful, we could obtain a hmAb(s) that is effective against several or all isolates in addition to nhmAbs specific to individual isolates. In preliminary experiments we identified three hmAbs that bind specifically to the Envs expressed by recombinant vaccinia virus and are being characterized for neutralizing activity by our collaborator R. Flick and epitope mapping (Zhu et al., in preparation).

Acknowledgment

We had the opportunity to observe the effectivity of ribavirin in clinical practice and discuss with my colleagues, Başak Dokuzoğuz, M.D., Aysel Çelikbaş, MD, Nurcan Baykam, MD, and Şebnem Eren, M.D. from Ankara Numune Education and Research Hospital, Ankara, Turkey. I am grateful to my colleagues for their comments and elaborations. We also thank Mahmut Bayik, MD, Professor of Hematology, and Volkan Korten, MD, Professor of Infectious Diseases from Marmara University, School of Medicine, for his support and comments on hematological approach to the CCHF patients.

REFERENCES

1. Altaf A, Luby S, Ahmed AJ, Zaidi N, Khan AJ, Mirza S, McCormick J, Fisher-Hoch S (1998) Outbreak of Crimean-Congo haemorrhagic fever in Quetta, Pakistan: contact tracing and risk assessment. Trop Med Int Health 3:878–882
2. Anderson SL, Carton JM, Lou J, Xing L, Rubin BY (1999) Interferon-induced guanylate binding protein-1 (GBP-1) mediates an antiviral effect against vesicular stomatitis virus and encephalomyocarditis virus. Virology 256:8–14
3. Andersson I, Bladh L, Mousavi-Jazi M, Magnusson KE, Lundkvist A, Haller O, Mirazimi A (2004) Human MxA protein inhibits the replication of Crimean-Congo hemorrhagic fever virus. J Virol 78:4323–4329
4. Andersson I, Lundkvist A, Haller O, Mirazimi A (2006) Type I interferon inhibits Crimean-Congo hemorrhagic fever virus in human target cells. J Med Virol 78:216–222
5. Andrei G, De Clercq E (1990) Inhibitory effect of selected antiviral compounds on arenavirus replication in vitro. Antiviral Res 14:287–299
6. Andrei G, De Clercq E (1993) Molecular approaches for the treatment of hemorrhagic fever virus infections. Antiviral Res 22:45–75
7. Asper M, Sternsdorf T, Hass M, Drosten C, Rhode A, Schmitz H, Gunther S (2004) Inhibition of different Lassa virus strains by alpha and gamma interferons and comparison with a less pathogenic arenavirus. J Virol 78:3162–3169
8. Baba M, Nakajima M, Schols D, Pauwels R, Balzarini J, De Clercq E (1988) Pentosan polysulfate, a sulfated oligosaccharide, is a potent and selective anti-HIV agent in vitro. Antiviral Res 9:335–343
9. Baba M, Pauwels R, Balzarini J, Arnout J, Desmyter J, De Clercq E (1988) Mechanism of inhibitory effect of dextran sulfate and heparin on replication of human immunodeficiency virus in vitro. Proc Natl Acad Sci USA 85:6132–6136
10. Baba M, Snoeck R, Pauwels R, de Clercq E (1988) Sulfated polysaccharides are potent and selective inhibitors of various enveloped viruses, including herpes simplex virus, cytomegalovirus, vesicular stomatitis virus, and human immunodeficiency virus. Antimicrob Agents Chemother 32:1742–1745
11. Bakir M, Ugurlu M, Dokuzoguz B, Bodur H, Tasyaran MA, Vahaboglu H (2005) Crimean-Congo haemorrhagic fever outbreak in Middle Anatolia: a multicentre study of clinical features and outcome measures. J Med Microbiol 54:385–389
12. Bekisz J, Schmeisser H, Hernandez J, Goldman ND, Zoon KC (2004) Human interferons alpha, beta and omega. Growth Factors 22:243–251
13. Bertolotti-Ciarlet A, Smith J, Strecker K, Paragas J, Altamura LA, McFalls JM, Frias-Staheli N, Garcia-Sastre A, Schmaljohn CS, Doms RW (2005) Cellular localization and antigenic characterization of Crimean-Congo hemorrhagic fever virus glycoproteins. J Virol 79:6152–6161
14. Borish LC, Steinke JW (2003) 2. Cytokines and chemokines. J Allergy Clin Immunol 111:S460–475
15. Bradbury J (2005) RNA interference: new drugs on the horizon. Drug Discov Today 10:1266–1267
16. Casadevall A (2002) Passive antibody administration (immediate immunity) as a specific defense against biological weapons. Emerg Infect Dis 8:833–841
17. Chiou HE, Liu CL, Buttrey MJ, Kuo HP, Liu HW, Kuo HT, Lu YT (2005) Adverse effects of ribavirin and outcome in severe acute respiratory syndrome: experience in two medical centers. Chest 128:263–272
18. Clemens MJ (2003) Interferons and apoptosis. J Interferon Cytokine Res 23:277–292
19. Cooper AC, Banasiak NC, Allen PJ (2003) Management and prevention strategies for respiratory syncytial virus (RSV) bronchiolitis in infants and young children: a review of evidence-based practice interventions. Pediatr Nurs 29:452–456
20. De Clercq E (2005) Interferon: ten stories in one. A short review of some of the highlights in the history of an almost quinquagenarian. Acta Microbiol Immunol Hung 52:273–289

21. de Veer MJ, Holko M, Frevel M, Walker E, Der S, Paranjape JM, Silverman RH, Williams BR (2001) Functional classification of interferon-stimulated genes identified using microarrays. J Leukoc Biol 69:912–920
22. Der SD, Zhou A, Williams BR, Silverman RH (1998) Identification of genes differentially regulated by interferon alpha, beta, or gamma using oligonucleotide arrays. Proc Natl Acad Sci USA 95:15623–15628
23. Diamond MS, Harris E (2001) Interferon inhibits dengue virus infection by preventing translation of viral RNA through a PKR-independent mechanism. Virology 289:297–311
24. Dimitrov DS (2004) Virus entry: molecular mechanisms and biomedical applications. Nat Rev Microbiol 2:109–122
25. Dusheiko G, Main J, Thomas H, Reichard O, Lee C, Dhillon A, Rassam S, Fryden A, Reesink H, Bassendine M, Norkrans G, Cuypers T, Lelie N, Telfer P, Watson J, Weegink C, Sillikens P, Weiland O (1996) Ribavirin treatment for patients with chronic hepatitis C: results of a placebo-controlled study. J Hepatol 25:591–598
26. Ergonul O, Celikbas A, Dokuzoguz B, Eren S, Baykam N, Esener H (2004) Characteristics of patients with Crimean-Congo hemorrhagic fever in a recent outbreak in Turkey and impact of oral ribavirin therapy. Clin Infect Dis 39:284–287
27. Ergonul O (2006) Crimean-Congo haemorrhagic fever. Lancet Infect Dis 6:203–214
28. Ergonul O, Celikbas A, Baykam N, Eren S, Dokuzoguz B (2006) Analysis of risk-factors among patients with Crimean-Congo haemorrhagic fever virus infection: severity criteria revisited. Clin Microbiol Infect 12:551–554
29. Ergonul O, Tuncbilek S, Baykam N, Celikbas A, Dokuzoguz B (2006) Evaluation of serum levels of interleukin (IL)-6, IL-10, and tumor necrosis factor-alpha in patients with Crimean-Congo hemorrhagic fever. J Infect Dis 193:941–944
30. Espert L, Degols G, Gongora C, Blondel D, Williams BR, Silverman RH, Mechti N (2003) ISG20, a new interferon-induced RNase specific for single-stranded RNA, defines an alternative antiviral pathway against RNA genomic viruses. J Biol Chem 278:16151–16158
31. Espert L, Rey C, Gonzalez L, Degols G, Chelbi-Alix MK, Mechti N, Gongora C (2004) The exonuclease ISG20 is directly induced by synthetic dsRNA via NF-kappaB and IRF1 activation. Oncogene 23:4636–4640
32. Espert L, Degols G, Lin YL, Vincent T, Benkirane M, Mechti N (2005) Interferon-induced exonuclease ISG20 exhibits an antiviral activity against human immunodeficiency virus type 1. J Gen Virol 86:2221–2229
33. Fisher-Hoch SP, Khan JA, Rehman S, Mirza S, Khurshid M, McCormick JB (1995) Crimean Congo-haemorrhagic fever treated with oral ribavirin. Lancet 346: 472–475
34. Flick R, Flick K, Feldmann H, Elgh F (2003) Reverse genetics for Crimean-Congo hemorrhagic fever virus. J Virol 77:5997–6006
35. Franchini M, Lippi G, Manzato F (2006) Recent acquisitions in the pathophysiology, diagnosis and treatment of disseminated intravascular coagulation. Thromb J 4:4
36. Frese M, Kochs G, Meier-Dieter U, Siebler J, Haller O (1995) Human MxA protein inhibits tick-borne Thogoto virus but not Dhori virus. J Virol 69:3904–3909
37. Frese M, Kochs G, Feldmann H, Hertkorn C, Haller O (1996) Inhibition of bunyaviruses, phleboviruses, and hantaviruses by human MxA protein. J Virol 70:915–923
38. Frese M, Pietschmann T, Moradpour D, Haller O, Bartenschlager R (2001) Interferon-alpha inhibits hepatitis C virus subgenomic RNA replication by an MxA-independent pathway. J Gen Virol 82:723–733
39. Garcia KC, Teyton L, Wilson IA (1999) Structural basis of T cell recognition. Annu Rev Immunol 17:369–397
40. Garcia S, Chinikar S, Coudrier D, Billecocq A, Hooshmand B, Crance JM, Garin D, Bouloy M (2006) Evaluation of a Crimean-Congo hemorrhagic fever virus recombinant antigen expressed by Semliki Forest suicide virus for IgM and IgG antibody detection in human and animal sera collected in Iran. J Clin Virol 32:154–159

41. Gonzalez JP, Wilson ML, Cornet JP, Camicas JL (1995) Host-passage-induced phenotypic changes in Crimean-Congo haemorrhagic fever virus. Res Virol 146:131–140
42. Gonzalez JP, Camicas JL, Cornet JP, Wilson ML (1998) Biological and clinical responses of West African sheep to Crimean-Congo haemorrhagic fever virus experimental infection. Res Virol 149:445–455
43. Graci JD, Cameron CE (2006) Mechanisms of action of ribavirin against distinct viruses. Rev Med Virol 16:37–48
44. Gresser I (1997) Wherefore interferon? J Leukoc Biol 61:567–574
45. Griffin DE (2003) Immune responses to RNA-virus infections of the CNS. Nat Rev Immunol 3:493–502
46. Haferkamp S, Fernando L, Schwarz TF, Feldmann H, Flick R (2005) Intracellular localization of Crimean-Congo Hemorrhagic Fever (CCHF) virus glycoproteins. Virol J 2:42
47. Haller O, Frese M, Kochs G (1998) Mx proteins: mediators of innate resistance to RNA viruses. Rev Sci Tech 17:220–230
48. Haller O, Kochs G, Weber F (2006) The interferon response circuit: Induction and suppression by pathogenic viruses. Virology 344:119–130
49. Hayden FG (2006) Antiviral Agents. In: Brunton LL, Lazo JS, Parker KL (eds) Goodman and Gilman's: The Pharmacological Basis of Therapeutics. Mc Graw Hill, New York, pp 1265–1267
50. Heagy W, Crumpacker C, Lopez PA, Finberg RW (1991) Inhibition of immune functions by antiviral drugs. J Clin Invest 87:1916–1924
51. Hoogstraal H (1979) The epidemiology of tick-borne Crimean-Congo hemorrhagic fever in Asia, Europe, and Africa. J Med Entomol 15:307–417
52. Horisberger MA (1995) Interferons, Mx genes, and resistance to influenza virus. Am J Respir Crit Care Med 152:S67–71
53. Huggins JW (1989) Prospects for treatment of viral hemorrhagic fevers with ribavirin, a broad-spectrum antiviral drug. Rev Infect Dis 11 (Suppl 4):S750–761
54. Huggins JW, Hsiang CM, Cosgriff TM, Guang MY, Smith JI, Wu ZO, LeDuc JW, Zheng ZM, Meegan JM, Wang QN, et al. (1991) Prospective, double-blind, concurrent, placebo-controlled clinical trial of intravenous ribavirin therapy of hemorrhagic fever with renal syndrome. J Infect Dis 164:1119–1127
55. Hui DJ, Bhasker CR, Merrick WC, Sen GC (2003) Viral stress-inducible protein p56 inhibits translation by blocking the interaction of eIF3 with the ternary complex eIF2.GTP.Met-tRNAi. J Biol Chem 278:39477–39482
56. Hui DJ, Terenzi F, Merrick WC, Sen GC (2005) Mouse p56 blocks a distinct function of eukaryotic initiation factor 3 in translation initiation. J Biol Chem 280:3433–3440
57. Hultgren C, Milich DR, Weiland O, Sallberg M (1998) The antiviral compound ribavirin modulates the T helper (Th) 1/Th2 subset balance in hepatitis B and C virus-specific immune responses. J Gen Virol 79 (Pt 10):2381–2391
58. Isaacs A, Lindenmann J (1957) Virus interference. I. The interferon. Proc R Soc Lond B Biol Sci 147:258–267
59. Isaacs A, Lindenmann J, Valentine RC (1957) Virus interference. II. Some properties of interferon. Proc R Soc Lond B Biol Sci 147:268–273
60. Jabbari A, Besharat S, Abbasi A, Moradi A, Kalavi K (2006) Crimean-Congo hemorrhagic fever: case series from a medical center in Golestan province, Northeast of Iran (2004). Indian J Med Sci 60:327–329
61. Janeway CA, Jr., Medzhitov R (2002) Innate immune recognition. Annu Rev Immunol 20:197–216
62. Kinsella E, Martin SG, Grolla A, Czub M, Feldmann H, Flick R (2004) Sequence determination of the Crimean-Congo hemorrhagic fever virus L segment. Virology 321:23–28
63. Knowles SR, Phillips EJ, Dresser L, Matukas L (2003) Common adverse events associated with the use of ribavirin for severe acute respiratory syndrome in Canada. Clin Infect Dis 37:1139–1142

64. Kochs G, Janzen C, Hohenberg H, Haller O (2002) Antivirally active MxA protein sequesters La Crosse virus nucleocapsid protein into perinuclear complexes. Proc Natl Acad Sci USA 99:3153–3158
65. Leroy M, Pire G, Baise E, Desmecht D (2005) Expression of the interferon-alpha/beta-inducible bovine Mx1 dynamin interferes with replication of rabies virus. Neurobiol Dis 21:515–521
66. Mangia A, Santoro R, Minerva N, Ricci GL, Carretta V, Persico M, Vinelli F, Scotto G, Bacca D, Annese M, Romano M, Zechini F, Sogari F, Spirito F, Andriulli A (2005) Peginterferon alfa-2b and ribavirin for 12 vs. 24 weeks in HCV genotype 2 or 3. N Engl J Med 352:2609–2617
67. Mardani M, Jahromi MK, Naieni KH, Zeinali M (2003) The efficacy of oral ribavirin in the treatment of Crimean-Congo hemorrhagic fever in Iran. Clin Infect Dis 36:1613–1618
68. Marquez VE, Lim MI, Treanor SP, Plowman J, Priest MA, Markovac A, Khan MS, Kaskar B, Driscoll JS (1988) Cyclopentenylcytosine. A carbocyclic nucleoside with antitumor and antiviral properties. J Med Chem 31:1687–1694
69. Mazza JJ (2002) Manual of Clinical Hematology. Lippincott Williams & Wilkins, Philadelphia
70. McCormick JB, King IJ, Webb PA, Scribner CL, Craven RB, Johnson KM, Elliott LH, Belmont-Williams R (1986) Lassa fever. Effective therapy with ribavirin. N Engl J Med 314:20–26
71. Mosmann TR, Sad S (1996) The expanding universe of T-cell subsets: Th1, Th2 and more. Immunol Today 17:138–146
72. Moulard M, Phogat SK, Shu Y, Labrijn AF, Xiao X, Binley JM, Zhang MY, Sidorov IA, Broder CC, Robinson J, Parren PW, Burton DR, Dimitrov DS (2002) Broadly cross-reactive HIV-1-neutralizing human monoclonal Fab selected for binding to gp120-CD4-CCR5 complexes. Proc Natl Acad Sci USA 99:6913–6918
73. Nabeth P, Cheikh DO, Lo B, Faye O, Vall IO, Niang M, Wague B, Diop D, Diallo M, Diallo B, Diop OM, Simon F (2004) Crimean-Congo hemorrhagic fever, Mauritania. Emerg Infect Dis 10:2143–2149
74. Ozkurt Z, Kiki I, Erol S, Erdem F, Yilmaz N, Parlak M, Gundogdu M, Tasyaran MA (2006) Crimean-Congo hemorrhagic fever in Eastern Turkey: clinical features, risk factors and efficacy of ribavirin therapy. J Infect 52:207–215
75. Padlan EA (1994) Anatomy of the antibody molecule. Mol Immunol 31:169–217
76. Page T, Connor JD (1990) The metabolism of ribavirin in erythrocytes and nucleated cells. Int J Biochem 22:379–383
77. Papa A, Bino S, Llagami A, Brahimaj B, Papadimitriou E, Pavlidou V, Velo E, Cahani G, Hajdini M, Pilaca A, Harxhi A, Antoniadis A (2002) Crimean-Congo hemorrhagic fever in Albania, 2001. Eur J Clin Microbiol Infect Dis 21:603–606
78. Papa A, Bozovi B, Pavlidou V, Papadimitriou E, Pelemis M, Antoniadis A (2002) Genetic detection and isolation of Crimean-Congo hemorrhagic fever virus, Kosovo, Yugoslavia. Emerg Infect Dis 8:852–854
79. Perrota PL, Pisciotto PT, Snyder EL (2003) Platelets and Related Products. In: Hillyer CD, Silberstein LE, Ness PM, Anderson KC, Roush KS (eds) Blood Banking and Transfusion Medicine. Churchill Livingstone, Philadelphia, pp 181–205
80. Regad T, Chelbi-Alix MK (2001) Role and fate of PML nuclear bodies in response to interferon and viral infections. Oncogene 20:7274–7286
81. Reichelt M, Stertz S, Krijnse-Locker J, Haller O, Kochs G (2004) Missorting of LaCrosse virus nucleocapsid protein by the interferon-induced MxA GTPase involves smooth ER membranes. Traffic 5:772–784
82. Rubinstein M, Dinarello CA, Oppenheim JJ, Hertzog P (1998) Recent advances in cytokines, cytokine receptors and signal transduction. Cytokine Growth Factor Rev 9:175–181
83. Samuel CE (2001) Antiviral actions of interferons. Clin Microbiol Rev 14:778–809
84. Sanchez AJ, Vincent MJ, Nichol ST (2002) Characterization of the glycoproteins of Crimean-Congo hemorrhagic fever virus. J Virol 76:7263–7275

85. Sanchez AJ, Vincent MJ, Erickson BR, Nichol ST (2006) Crimean-Congo hemorrhagic fever virus glycoprotein precursor is cleaved by Furin-like and SKI-1 proteases to generate a novel 38-kilodalton glycoprotein. J Virol 80:514–525
86. Sasaki N, Matsui A, Momoi M, Tsuda F, Okamoto H (1997) Loss of circulating hepatitis C virus in children who developed a persistent carrier state after mother-to-baby transmission. Pediatr Res 42:263–267
87. Sen GC (2001) Viruses and interferons. Annu Rev Microbiol 55:255–281
88. Shepherd AJ, Swanepoel R, Leman PA (1989) Antibody response in Crimean-Congo hemorrhagic fever. Rev Infect Dis 11 (Suppl 4):S801–806
89. Sidwell RW, Huffman JH, Khare GP, Allen LB, Witkowski JT, Robins RK (1972) Broad-spectrum antiviral activity of Virazole: 1-beta-D-ribofuranosyl-1,2,4-triazole-3-carboxamide. Science 177:705–706
90. Sidwell RW, Huffman JH, Smee DF, Gilbert J, Gessaman A, Pease A, Warren RP, Huggins J, Kende M (1992) Potential role of immunomodulators for treatment of phlebovirus infections of animals. Ann N Y Acad Sci 653:344–355
91. Sidwell RW, Huffman JH, Barnard DL, Smee DF, Warren RP, Chirigos MA, Kende M, Huggins J (1994) Antiviral and immunomodulating inhibitors of experimentally-induced Punta Toro virus infections. Antiviral Res 25:105–122
92. Sidwell RW, Smee DF (2003) Viruses of the Bunya- and Togaviridae families: potential as bioterrorism agents and means of control. Antiviral Res 57:101–111
93. Silverman RH (1994) Fascination with 2–5A-dependent RNase: a unique enzyme that functions in interferon action. J Interferon Res 14:101–104
94. Suleiman MN, Muscat-Baron JM, Harries JR, Satti AG, Platt GS, Bowen ET, Simpson DI (1980) Congo/Crimean haemorrhagic fever in Dubai. An outbreak at the Rashid Hospital. Lancet 2:939–941
95. Swanepoel R, Gill DE, Shepherd AJ, Leman PA, Mynhardt JH, Harvey S (1989) The clinical pathology of Crimean-Congo hemorrhagic fever. Rev Infect Dis 11(Suppl 4):S794–800
96. Sweeney MJ, Gerzon K, Harris PN, Holmes RE, Poore GA, Williams RH (1972) Experimental antitumor activity and preclinical toxicology of mycophenolic acid. Cancer Res 32:1795–1802
97. Tam RC, Lau JY, Hong Z (2001) Mechanisms of action of ribavirin in antiviral therapies. Antivir Chem Chemother 12:261–272
98. Tatar A, Ozkurt Z, Kiki I (2005) Genotoxic effect of ribavirin in patients with Crimean-Congo hemorrhagic fever. Jpn J Infect Dis 58:313–315
99. Tignor GH, Hanham CA (1993) Ribavirin efficacy in an in vivo model of Crimean-Congo hemorrhagic fever virus (CCHF) infection. Antiviral Res 22:309–325
100. Tosi MF (2005) Innate immune responses to infection. J Allergy Clin Immunol 116:241–249; quiz 250
101. Vandenbroucke JP (2004) When are observational studies as credible as randomised trials? Lancet 363:1728–1731
102. Vassilenko SM, Vassilev TL, Bozadjiev LG, Bineva IL, Kazarov GZ (1990) Specific intravenous immunoglobulin for Crimean-Congo haemorrhagic fever. Lancet 335:791–792
103. Vincent MJ, Sanchez AJ, Erickson BR, Basak A, Chretien M, Seidah NG, Nichol ST (2003) Crimean-Congo hemorrhagic fever virus glycoprotein proteolytic processing by subtilase SKI-1. J Virol 77:8640–8649
104. Watts DM, Ussery MA, Nash D, Peters CJ (1989) Inhibition of Crimean-Congo hemorrhagic fever viral infectivity yields in vitro by ribavirin. Am J Trop Med Hyg 41:581–585
105. Williams BR (1999) PKR; a sentinel kinase for cellular stress. Oncogene 18:6112–6120
106. Wilson ML, Gonzalez JP, Cornet JP, Camicas JL (1991) Transmission of Crimean-Congo haemorrhagic fever virus from experimentally infected sheep to Hyalomma truncatum ticks. Res Virol 142:395–404
107. Young KC, Lindsay KL, Lee KJ, Liu WC, He JW, Milstein SL, Lai MM (2003) Identification of a ribavirin-resistant NS5B mutation of hepatitis C virus during ribavirin monotherapy. Hepatol (Baltimore, MD) 38:869–878

108. Zeerleder S, Hack CE, Wuillemin WA (2005) Disseminated intravascular coagulation in sepsis. Chest 128:2864–2875
109. Zhang MY, Shu Y, Phogat S, Xiao X, Cham F, Bouma P, Choudhary A, Feng YR, Sanz I, Rybak S, Broder CC, Quinnan GV, Evans T, Dimitrov DS (2003) Broadly cross-reactive HIV neutralizing human monoclonal antibody Fab selected by sequential antigen panning of a phage display library. J Immunol Methods 283:17–25
110. Zhang MY, Shu Y, Rudolph D, Prabakaran P, Labrijn AF, Zwick MB, Lal RB, Dimitrov DS (2004) Improved breadth and potency of an HIV-1-neutralizing human single-chain antibody by random mutagenesis and sequential antigen panning. J Mol Biol 335:209–219
111. Zhang MY, Shu Y, Sidorov I, Dimitrov DS (2004) Identification of a novel CD4i human monoclonal antibody Fab that neutralizes HIV-1 primary isolates from different clades. Antiviral Res 61:161–164
112. Zhang MY, Xiao X, Sidorov IA, Choudhry V, Cham F, Zhang PF, Bouma P, Zwick M, Choudhary A, Montefiori DC, Broder CC, Burton DR, Quinnan GV, Jr., Dimitrov DS (2004) Identification and characterization of a new cross-reactive human immunodeficiency virus type 1-neutralizing human monoclonal antibody. J Virol 78:9233–9242

SECTION V

PREVENTION AND CONTROL

CHAPTER 20

RISK GROUPS AND CONTROL MEASURES FOR CRIMEAN-CONGO HEMORRHAGIC FEVER

CHRIS A. WHITEHOUSE, PH.D.*

Diagnostic Systems Division, United States Army Institute of Infectious Diseases, Fort Detrick, Frederick, MD 21702-5011, USA. Tel.: +1-301-619-2098; Fax: +1-301-619-2492; E-mail: chris.whitehouse@amedd.army.mil

20.1. RISK FACTORS

As with many vector-borne, zoonotic diseases, the ecology of the pathogen–reservoir–vector cycle predisposes certain individuals to infection with Crimean-Congo hemorrhagic fever virus (CCHFV). Hence, there are several groups of individuals who are considered to be at risk of contracting Crimean-Congo hemorrhagic fever (CCHF), specifically, people from endemic areas who are susceptible to tick bite, particularly from *Hyalomma* spp. ticks. These include individuals who work outdoors, particularly those who work with large domestic animals. This fact was exemplified during a recent CCHF outbreak in Turkey where 90% of the infected patients were farmers [1]. Although CCHFV has been isolated from numerous species of ticks (see Chapter 12), those of the *Hyalomma* genus are considered the primary vector in CCHF enzootic and endemic areas. The distribution of CCHFV coincides precisely with the distribution of *Hyalomma* ticks [6]; therefore, there appears to be little or no risk in areas outside the known distribution of these ticks. Exposures such as crushing-infected ticks and butchering-infected animals have also been a frequent source of CCHFV infection. Other groups who are at risk include those caring for CCHF patients. In fact, the risk of nosocomial infection in health-care workers is well documented and can be extremely high. This can be exemplified by a nosocomial outbreak that occurred at Tygerberg Hospital in South Africa where 33% of the medical staff who had contact with CCHF patients through accidental needle sticks developed

*The views, opinions, and findings contained herein are those of the authors and should not be construed as an official Department of the Army position, policy, or decision unless so designated by other documentation.

O. Ergonul and C. A. Whitehouse (eds.), Crimean-Congo Hemorrhagic Fever, 273–280.

CCHF and another 8.7% contracted disease by other contacts with the patients' blood [14]. Nosocomial infections with CCHFV will be discussed in greater detail in Chapter 21. In addition to health-care workers who care for CCHF patients, laboratory workers, and researchers working directly with the virus in the laboratory, or with infected vectors and/or animals, are at a higher risk of infection.

20.2. CONTROL MEASURES

Prevention and control of CCHF are achieved by avoiding or minimizing the exposure to the virus. As mentioned above, there are several groups of individuals who are considered at risk for contracting CCHF, including those persons in endemic areas who are susceptible to tick bites (particularly from *Hyalomma* ticks) and persons exposed to infected animals or animal tissues. Thus, it is not surprising that control measures have mainly focused on the control of ticks in endemic areas and on personal protective measures for persons handling virus-infected materials. Universal precautions to protect against nosocomial infections will be discussed in detail in Chapter 21 by Tarantola et al.

20.2.1. Control of ticks

The risk of CCHFV infection of humans may be reduced by implementing efficient tick control. A variety of techniques have been developed to control ticks of medical or veterinary importance. Some of these include the use of chemical acaricides, biological control methods (e.g. tick parasites or entomopathogenic fungi), habitat modification, repellent, and more recently, molecular methods (e.g. vaccines or pheromone-baited devices to attract and kill ticks [so-called tick decoys]). However, despite the obvious need, efficient tick control remains a formidable, and often costly, endeavor.

20.2.1.1. Chemical acaricides

Strategies to kill ticks rely mostly on area wide or host-targeted chemical acaricide applications. However, many of these have environmental, operational, and cost-effectiveness hurdles to overcome. The first insecticide widely used for tick control was arsenic, which was used in the beginning of the 20th century in the southern USA to control vectors of Texas cattle fever, and in Africa to control the tick vectors of east coast fever. The use of arsenic was supplanted by dichloro diphenyl trichloro ethane (DDT) (1,1,1-trichloro-2,2-bis [*p*-chlorophenyl]ethane), which was introduced on a massive scale as a broad-range insecticide after World War II, but was ultimately banned in the 1970s. Other groups of pesticides that have been used to kill ticks include the organophosphorus compounds and the carbamates, both of which act by antagonizing acetylcholinesterase activity (i.e. they disrupt nerve transmission, thereby causing rapid paralysis and death). Natural pyrethrins (from the chrysanthemum plants), and later synthetic pyrethroids are toxic to arthropods and are relatively nontoxic to mammals. Additionally, these compounds are effective repellents.

Host-targeted acaricide control has long been the principal method for controlling ticks on domestic livestock. Early efforts to control ticks on domestic animals led to the development of the "cattle dip" for immersing animals in a trough containing a chemical acaricide. Usually the cattle dip consists of a trench constructed of concrete with high walls to contain several adult cows (Fig. 20-1). Cattle dipping has proven highly effective for controlling several tick-borne diseases and was the primary method of tick control on livestock during the first half of the 20th century. Spraying is another method of applying of chemical acaricides (Fig. 20-2). This has become one of the popular methods of tick control for livestock or pets. In fact, spraying is often preferred over dipping since it is usually less expensive and the procedure is less stressful for the animals. However, one

Fig. 20-1. Cattle going through a tick-treatment bath at an Animal and Plant Health Inspection Service (APHIS) facility in McAllen, Texas, USA. (Courtesy of Scott Bauer.)

Fig. 20-2. Cattle being sprayed with a chemical acaricide solution. (Courtesy of Dr. Telford Work.)

disadvantage is that spraying cannot assure complete coverage of all parts of the animal's body, thus dipping is considered a more effective method. Various other methods of application include "pour-ons" in which acaricide dissolved in a specially formulated liquid is streaked along the back of a cow, sheep, goat, or other domestic animals, and dusts, which are formulated with a fine powder and an acaricide and applied directly to the animal. Another interesting means of controlling ticks and other blood-sucking parasites is the use of systemics. These are compounds that are delivered to the blood stream of the livestock or pet (usually by mixing with food, by injection, or as a time-release implant inserted under the skin). These compounds circulate in the blood stream of the vertebrate host and are ingested by the tick along with the blood meal.

Very little information is available on tick control efforts specifically against *Hyalomma* ticks for preventing CCHF. In general, it is believed that control of CCHF through the application of acaricides to livestock is impractical [3]. Prevention and control of CCHF has been attempted by acaricide treatment of domestic animals in the Rostov region of the former Soviet Union with only limited success. For example, acaricide treatment of cattle with Sevin (carbaryl [VI]) during the period of adult attachment was found to be the most efficient control measure for *Hyalomma marginatum marginatum* in the Astrakhan Oblast [15]. Attempts to control ticks on cattle and in pastures by aerial spraying of acaricides were made during epidemics of CCHF in the Rostov Oblast; however, it was concluded that these measures were not effective for preventing disease [8]. Likewise, habitat modification (e.g. removal of vegetation, irrigation, and plowing) in an attempt to reduce populations of *Hyalomma anatolicum* has been tested with little success.

20.2.1.2. Biological control

Biological control measures typically involve the deliberate introduction of predators, parasites, or pathogens to control populations of a pest or vector species. Some examples of biocontrol agents have included entomopathogenic fungi, parasitic wasps, lycosid spiders, and fire ants. Unfortunately, some of the arthropods also attack other animals, including humans, and mass release of these agents would certainly not be desirable. Certain birds, most notably oxpeckers, are highly efficient predators of ticks, consuming large numbers from the hides of various ungulates [12]. However, whether these could serve as efficient biocontrol agents for disease vectors has not been explored. In addition to entomopathogenic fungi, bacteria, especially *Bacillus thuringiensis*, certain strains of which are pathogenic to various species of insects, have been shown to be pathogenic to ticks in the laboratory [5]. Although many of these biocontrol methods holds promise, further research is needed to determine whether they can be effective under field conditions.

20.2.1.3. Pheromone-mediated control methods

One of the most innovative methods to attempt to control ticks has been the use of tick pheromones in conjunction with acaricides. Tick pheromones are messenger chemicals used by various tick species to control and modulate tick movement and mating behavior. Readers are referred to Sonenshine [12] for a detailed review on tick pheromones and their uses in tick control. In many cases, these strategies incorporate a pheromone–acaricide-impregnated device, which is used to lure and kill host-seeking ticks. In this way, ticks are brought to the acaricide as compared with conventional ticks control techniques where massive amounts of acaricides are dispensed in the hope of killing the attached ticks.

20.2.2. Personal protective measures and laboratory safety

Health-care workers are an important risk group as evidenced by the large number of nosocomial infections often associated with CCHF outbreaks [2, 4, 9, 14]. Infected patients should be isolated and subjected to barrier nursing techniques. Health-care workers should wear protective clothing such as disposable gowns, gloves, masks, goggles, and overshoes. All items should be discarded on leaving the patient's room and be safely disposed of or disinfected. Clinical laboratory tests should be kept to a minimum and performed by experienced staff wearing protective clothing, and automated analyzers must be decontaminated after use with dilute bleach solution [3]. Others who work outdoors in direct contact with large domestic animals are at risk and should take appropriate precautions to check for tick exposure. Interestingly, Russian milkmaids

Fig. 20-3. Russian milkmaids dressed in white garments to better check for the presence of ticks. (Courtesy of Dr. Telford Work.)

have long recognized the possibility of infection with tick-borne diseases, and thus, dress from head –to toe in white garments to better check for the presence of ticks (Fig. 20-3). Laboratory workers handling viral material are also at high risk of contracting the disease as evidenced by several cases of laboratory-acquired CCHF in Africa [7, 11], and several cases in Russia in which aerosol and/or droplet-respiratory route of infection were suspected [6]. For these reasons, in the USA, the Centers for Disease Control and Prevention (CDC) have classified CCHFV as a biosafety level 4 (BSL-4) pathogen [10]. BSL-4 safety practices, safety equipment, and facility design and construction are applicable for work with dangerous and exotic agents that pose a high risk of life-threatening disease, which many be transmitted via the aerosol route and for which there is no available vaccine or therapy. In a BSL-4 laboratory, the worker is completely isolated from potential aerosolized infectious material by working in a full-body, air-supplied positive pressure personnel suit (Fig. 20-4).

20.2.3. Vaccines against CCHF

There is currently no effective vaccine available for CCHF. A suckling mouse brain, formalin-inactivated vaccine has been used in Bulgaria and other parts of Eastern Europe and the former Soviet Union. In the Rostov region of the former Soviet Union, 1,500 persons received the vaccine and showed a high frequency of detectable antibody by the N test [13]. Likewise, vaccine was given

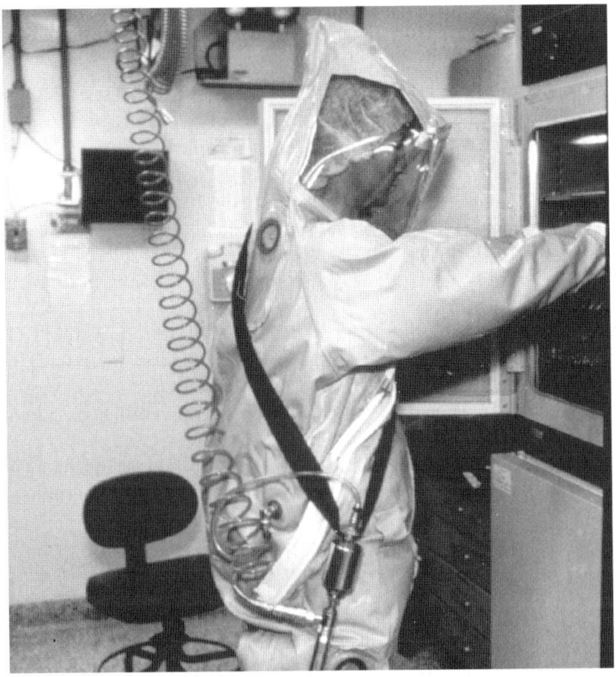

Fig. 20-4. A laboratory worker wearing a full-body protective suit while working in a maximum containment BSL-4 laboratory at the Centers for Disease Control and Prevention. (Courtesy of Troy Hall, CDC, Public Health Image Library.)

to several hundred human volunteers in Bulgaria, with resulting high antibody induction [16]. There are currently no efforts to develop a CCHF vaccine and with the relatively small target population of persons at risk for contracting the disease, the large-scale development and production of a CCHF vaccine by modern standards seems unlikely.

REFERENCES

1. Bakir M, Ugurlu M, Dokuzoguz B, Bodur H, Tasyaran MA, Vahaboglu H (2005) Crimean-Congo haemorrhagic fever outbreak in Middle Anatolia: a multicentre study of clinical features and outcome measures. J Med Microbiol 54:385–389
2. Burney MI, Ghafoor A, Saleen M, Webb PA, Casals J (1980) Nosocomial outbreak of viral hemorrhagic fever caused by Crimean hemorrhagic fever-Congo virus in Pakistan, January 1976. Am J Trop Med Hyg 29:941–947
3. Burt FJ, Swanepoel R (2005) Crimean-Congo hemorrhagic fever. In: Goodman JL, Dennis DT, Sonenshine DE (eds) Tick-Borne Diseases of Humans. ASM Press, Washington, DC, pp 164–175
4. Fisher-Hoch SP, Khan JA, Rehman S, Mirza S, Khurshid M, McCormick JB (1995) Crimean Congo-haemorrhagic fever treated with oral ribavirin. Lancet 346:472–475

5. Hassanain MA, El Garhy MF, Abdel-Ghaffar FA, El Sharaby A, Abdel-Mageed-Kadria N (1997) Biological control studies of soft and hard ticks in Egypt: I. The effect of *Bacillus thuringiensis* varieties on soft and hard ticks (Ixodidae) Parasitol Res 83:209–213
6. Hoogstraal H (1979) The epidemiology of tick-borne Crimean-Congo hemorrhagic fever in Asia, Europe, and Africa. J Med Entomol 15:307–417
7. Kalunda M, Lule M, Sekyalo E, Mukuye A, Mujomba E (1983) Virus isolation and identification. Rep East Afr Virus Res Inst 27:17–20
8. Kondratenkov VF, Shevchenko SF, Perelatov VD, Badalov ME, Ionov SS, Semenov MY, Romanova VA, Lobanov VV, Tekut'ev IV (1970) Two year experiment on application of chemical campaign method against ixodid ticks in Crimean hemorrhagic fever focus of Rosov Oblast. Mater 3 Oblast Nauchn Prakt Konf (Rostov-on-Don, May 1970) p 157 (in Russian; in English, NAMRU3-T550)
9. Papa A, Bino S, Llagami A, Brahimaj B, Papadimitriou E, Pavlidou V, Velo E, Cahani G, Hajdini M, Pilaca A, Harxhi A, Antoniadis A (2002) Crimean-Congo hemorrhagic fever in Albania, 2001. Eur J Clin Microbiol Infect Dis 21:603–606
10. Richmond JY, McKinney RW (1999) Biosafety in Microbiological and Biomedical Laboratories, 4th edn. US DHHS, Washington, DC
11. Simpson DIH, Knight HEM, Courtois G, Williams MC, Weinbren MP, Kibukamusoke JW (1967) Congo virus: a hitherto undescribed virus occurring in Africa. I. Human isolations – clinical notes. East Afr Med J 44:87–92
12. Sonenshine DE (1993) Biology of Ticks, vol. 2. Oxford University Press, New York
13. Tkachenko EA, Butenko AM, Badalov ME, Zavodov TI, Chumakov MP (1971) Investigation of the immunogenic activity of killed brain vaccine against Crimean hemorrhagic fever. Tr Inst Polio Virusn Entsegalitov Akad Med Nauk SSSR 19:119–129 (in Russian; in English, NAMRU3-T931)
14. van Eeden PJ, van Eeden SF, Joubert JR, King JB, van de Wal BW, Michell WL (1985) A nosocomial outbreak of Crimean-Congo haemorrhagic fever at Tygerberg Hospital. Part II. Management of patients. S Afr Med J 68:718–721
15. Vashkov V, Poleshchuk VD (1971) Measures for control of vectors of CHF – *Hyalomma plumbeum plumbeum* Panz. Ticks. Tr Inst Polio Virusn Entsefalitov Akad Med Nauk SSSR 19:239–244 (in Russian; in English, NAMRU3-T983)
16. Vasilenko SM (1973) Results of the investigation on etiology, epidemiologic features and the specific prophylactic of Crimean hemorrhagic fever (CHF) in Bulgaria. Abstract Inv Paper 9, International Congress on Tropical Medicine and Malaria, Athens, October 1973, vol 1, pp 32–33

CHAPTER 21

ESTIMATES AND PREVENTION OF CRIMEAN-CONGO HEMORRHAGIC FEVER RISKS FOR HEALTH-CARE WORKERS

ARNAUD TARANTOLA, MD., M.S.[1], ONDER ERGONUL, M.D., M.P.H.[2], AND PIERRE TATTEVIN, M.D.[3]

[1]*International and Tropical Department, Institut National de Veille Sanitaire, 12 rue du Val d'Osne, 94415 Saint-Maurice, Cedex, France. Tel.: + 33 (0) 1 41 79 67 14; Fax: + 33 (0) 1 41 79 69 65; E-mail: a.tarantola@invs.sante.fr*

[2]*Marmara University, School of Medicine, Infectious Diseases and Clinical Microbiology, Istanbul, Turkey*

[3]*Infectious Diseases and Reanimation Clinic, Pontchaillou University Hospital, rue Le Guilloux, 35033 Rennes, Cedex, France*

21.1. INTRODUCTION

Crimean-Congo hemorrhagic fever virus (CCHFV) is one of the most widespread pathogens causing viral hemorrhagic fever (VHF). A disease with the clinical and epidemiological features of Crimean-Congo hemorrhagic fever (CCHF) has been long known in Central Asia, its first known formal description appearing in a text written circa 1110 in what is now Tadjikistan [43]. It has caused outbreaks in dry steppe, savannah, semidesert and foothill areas of Eastern and Central Europe, most of European and Asian parts of the former USSR, parts of the oriental region, and in Africa from Egypt to South Africa and from Senegal to Kenya. Within 25 years of its first detailed description of clinical and epidemiological features in the Ukraine, the long-known hemorrhagic fever had been described extensively, the causative agent had been isolated and shown to replicate the disease in nonimmune humans, an animal model, and a diagnostic test was developed which showed its area of prevalence. Although CCHFV infection is rarely documented in humans, outbreaks continue to occur and are on the rise in Turkey. But what are the risks of secondary transmission to health-care workers (HCW)?

O. Ergonul and C. A. Whitehouse (eds.), Crimean-Congo Hemorrhagic Fever, 281–294.

21.2. AVAILABLE DATA ON THE TRANSMISSION OF CCHF AND OTHER VHF TO HCW

21.2.1. Transmission of CCHF to HCW

In order to better assess the suitability of management procedures, we have conducted an extensive review of the English-, French- and Spanish-language literature for published VHF cases and secondary transmission. The reference list of retrieved articles and textbooks were systematically checked to identify other relevant articles.

Simpson et al. described five cases of infection in laboratory workers with a pathogen subsequently identified as CCHF virus [63] and the first case of occupational transmission of CCHF and resultant death among HCW delivering clinical care was reported in 1976 [16]. All published occurrences of hospital-managed CCHF cases but the one described above have occurred in endemic countries. To our knowledge there have been 27 published sources on the management of confirmed CCHF cases, which led to 44 documented secondary cases in HCW (Table 21-1). If only the published events with available HCW denominators are considered, 63 initial cases caused 13 secondary cases (2.6%; 95% confidence interval: 1.4–4.4%) in 494 HCW contacts, among which at least five died. Apart from sporadic case reports, several sero-epidemiologic studies have been conducted. One of the earlier studies showed that the antibody prevalence was less than 1% (1 of 128) among the hospital staff who had cared for patients with CCHF [35]. During the CCHF outbreak at Tygerberg Hospital, seven (1.5%) of 459 listed CCHF contacts developed the disease: six were contacts of the index case and only one HCW was a contact of a secondary case [75]. A contact tracing and risk assessment study was conducted around a case documented in Pakistan in 2005 [11]. Two interns acquired CCHF, and one of them died. The authors performed a risk assessment study to explore the possible modes of CCHF infection. They divided the contacts into five groups, (A) percutaneous contact with blood (needlesticks, blood contact to broken skin/mucosa, $n = 4$ contacts), (B) blood contact to unbroken skin, ($n = 35$ contacts), (C) cutaneous contact with body fluids other than blood, ($n = 22$), (D) physical contact with patients without body fluid contact ($n = 50$), and (E) close proximity to the patient without touching ($n = 79$). Only the contacts in group A had the infection with the attack rate of 50% (two out of four HCWs), a rate similar to that found in the Tygeberg outbreak where four of 46 blood contacts (8,7%) and three of nine needle contacts (33%) developed the disease [75]. In Turkey, four HCW were diagnosed with CCHF infection. Among these, only one HCW who had sustained a needlestick injury became the only HCW to have died of CCHF in Turkey (unpublished data).

In a report from Ankara, five HCWs who participated in an abdominal explorative surgery [19] and five HCWs who participated to the delivery of two

Table 21-1. Published cases of CCHF and outcome in documented health-care worker (HCW) contacts

Year	Country	Primary cases	HCW contacts	Secondary/ tertiary HCW cases	Infection	Progression	Exposure	Reference
1976	Pakistan	1	ND*	10	Symptomatic	3 deaths	Hospital care	[17]
1979	Dubai	1	ND	6	Symptomatic	1 death	Hospital care	[66]
1979	Iraq	1	ND	2	Symptomatic	2 deaths	Hospital care	[4]
1984	South Africa	2	35	8	Symptomatic	2 deaths	Hospital care	[21, 75]
1994	Pakistan	1	12	3	Symptomatic	Recovered	Surgery	[5]
1994	Pakistan	3	40	0	NA	NA	NA	[5]
1994	Pakistan	1	ND	3	Symptomatic	Recovered	Surgery	[34]
1995	Oman	2	ND	0	NA	NA	NA	[62]
1999	Iran	3	ND	0	NA	NA	NA	[51]
2000	Kenya	1	ND	0	NA	NA	NA	[28]
2000	Pakistan	1	ND	2	Symptomatic	1 death	Hospital care	[64]
2001	Yugoslavia	1	ND	1	Symptomatic	Recovered	Intubation	[53]
2001	Albania	1	ND	1	Symptomatic	Recovered	Electrocardiogram	[39]
2002	Pakistan	3	154	2	Symptomatic	2 deaths	Muco-cutaneous contact	[10, 11]
2003	Turkey	1	5	0	NA	NA	NA	[19]
2002–2003	Turkey	50	62	0	NA	NA	NA	[32]
2003	Mauritania	1	ND	6	Symptomatic	5 deaths	Hospital care	[52]
2004	Senegal-France	1	181	0	NA	NA	Hospital care	[72]
2005	Turkey	2	5	0	NA	NA	NA	[20]
Total		77	494	44	–	16	–	–

*ND: not documented; NA: not applicable

CCHF-infected women [20] were secondarily followed-up for CCHF antibodies: none of them developed antibodies. In another study reported from Turkey [32], the HCWs exposed to CCHF patients underwent serosurveillance. Since the cause of infection was identified only after they had begun caring for their patient, the protective measures used were limited to simple general precautions, such as using gloves during phlebotomy, and self-protection from all by the use of masks and gowns. None of the 62 HCWs under the risk group acquired CCHFV infection.

All reviewed cases but one, however, were managed in the developing country setting where standard precautions (SP) observance is often lacking [71, 73] (Box 21-1). Even in intermediate, more technically advanced countries such as Turkey where cases CCHF are increasingly diagnosed [9, 33], HCW training and prevention device availability may make standard precautions observance difficult [12, 29].

21.2.2. VHF and airborne transmission

Epidemiological studies of VHF indicate that infection is not readily transmitted from person to person by airborne routes [6, 13, 55]. CCHF and other VHF viruses are transmissible to HCW by contact with infected patients' blood and body fluids (BBF). Airborne transmission of VHF was suspected in one Lassa fever outbreak in 1969 [18], but there has never been any documented case of airborne transmission of that virus to humans. Apart from experimental data on weaponized VHF viruses and suspected airborne transmission of Ebola Reston in monkeys [15, 44, 46, 56], no airborne interhuman transmission of VHF has ever been documented, and all recommendations point to the risk being extremely low if it exists at all. The vast majority of VHF cases are managed in the developing country setting, where unfortunately secondary transmission occurs regularly [36, 58]. To our knowledge, there have been 24 published cases of imported Lassa fever – the VHF with a reputation for airborne transmission – to industrialized western countries [48] and nine instances where contacts were traced and follow-up results were published (Table 21-2). There were 1521 documented close HCW contacts of these 9 cases, only 1 of which was found to seroconvert (estimated transmission risk = 0.06%; 95% confidence interval: 0.0–0.3%). This was a German HCW who presented an asymptomatic seroconversion after providing invasive care with no barrier precautions. Therefore the estimated global risk for transmission of Lassa fever – a "proxy" for VHF supposedly transmissible through respiratory contact – can be estimated at less than 0.1% in a developed setting where standard precautions are routinely observed. In a high-endemicity, high-risk setting, a prospective serological study from Sierra Leone suggested that the hospital staff who cared for Lassa fever patients using simple barrier nursing methods have no higher risk of infection than the local population [35, 41].

Table 21-2. Confirmed imported Lassa fever to western countries, contact tracing in HCW and secondary transmission

Year	Country	Primary cases	HCW contacts	Secondary/ tertiary HCW cases	Progression	Reference
1976	USA	1	552	0	NA*	[79]
1982	United Kingdom	1	159	0	NA	[26]
1989	USA	1	102	0	NA	[42]
2000	The Netherlands	1	123	0	NA	[67]
2000	United Kingdom	1	74	0	NA	[8]
2000	Germany	1	19	0	NA	[47]
2003	Germany	1	65	1	Asymptomatic infection	[38]
2004	USA	1	188	0	NA	[25]
2006	Germany	1	220	0	NA	Personal communication: Dr. Tim Eckmann at the Robert Koch Institute in Germany
Total		9	1502	1		

* NA: not applicable

21.3. EXISTING GUIDELINES FOR THE CONTROL OF SECONDARY VHF TRANSMISSION

Available data has therefore progressively been viewed as increasingly reassuring and the risk of secondary transmission has been said to be lower than initially thought. Successive US recommendations [7, 22, 23] and French documents [60] for the management of patients with suspected VHF, however, have become only more stringent. They recommend isolation precautions seemingly influenced by management procedures for other pathogens with well-described airborne transmission such as multiresistant tuberculosis. Yet existing recommendations [3, 40] suggest isolation of suspected/confirmed VHF patients in negative-pressure rooms or isolation units. These stringent recommendations targeted Lassa, for which airborne transmission was suspected in one instance [18].

Although warranting increased HCW safety at first glance, such stringent recommendations are actually extremely difficult to observe while awaiting test results for every patient in whom VHF may be one of the many possible diagnoses in the current daily clinical setting, even in industrialized countries [78] let alone in the developing world [11]. Indeed, the early stages of VHF mimic those of other much more frequent diseases (such as malaria or dengue) which circulate in areas where VHF occur. These precautions for suspected VHF patients can therefore be seen as excessive, costly, and simply inapplicable in the daily clinical setting. Excessive precautions may delay and hinder diagnostic or therapeutic procedures, causing a loss of chance for patients in whom more common diagnoses will eventually be confirmed. The increasing amount of evidence suggests that reinforced and well-observed SP and a strategy for reducing exposure to BBF will suffice to prevent secondary spread of VHF in the health-care setting, even during outbreaks in endemic countries [35], with limited additional costs and no loss of chance in patients.

21.4. AN INTEGRATED STRATEGY FOR THE CONTROL OF ACCIDENTAL EXPOSURE TO BLOOD AND BODY FLUIDS

Various levels of response are thus needed to control BBF exposure: organizational, institutional, and governmental. Improved control of BBF exposure in HCW must be based on a well-planned, integrated strategy which will include the aspects outlined below.

21.4.1. General considerations

Standard precautions must be the basis for the standard management of all patients in all countries. They are both necessary and sufficient to prevent VHF transmission in HCW. If resources are scarce, then these resources can be prioritized according to the risks associated with individual procedures (such as blood-drawing or catheterization) rather than patient profiles. Adopting a basic management procedure for all patients rather than allocating scarce resources only to one supposedly more "risky" hospital department is more cautious. However, resources must be warranted for managing patients presenting with fever and signs of bleeding in countries endemic for VHF.

21.4.2. Gloves and hand hygiene

Gloves prevent exposure of non-intact skin to BBF, and may reduce the infective inoculum and therefore transmission risks in case of injury [14, 50]. In the surgical setting, the use of double layers of gloves and blunt-tipped needles will dramatically improve HCW safety against bloodborne pathogens [24, 68]. Proper use of gloves and hand hygiene also participates effectively to the decrease transmission of other nosocomial infections during health care [57].

21.4.3. Personal protective equipment and prevention equipment

To our knowledge, there is no documented instance of airborne transmission of VHF. The sole identified risk to date is that of transmission is mucous membrane contact with large droplets projected by infected patients' cough or during intubation procedures. Masks placed on a patient will limit droplet diffusion during cough or hematemesis, and masks worn by HCW will prevent projection to the mouth. HCW must therefore have access at all times to an uninterrupted supply of gloves and surgical masks. At best, this will be completed by access to protective eyewear in the form of goggles or face shields. In case of procedure which may generate an aerosol, HCW should consider wearing an N95 or FFP2 respirator (European Norm (EN) 61010-1).

21.4.4. Criteria for choice and proper use of sharps containers

Sharps containers are the first and foremost safety equipment which should be available at all times to all units. Their desired qualities are now well standardized. They are best chosen by HCW themselves. They must be resistant to sharps, be waterproof, and must remain sealed and resistant even if they fall. Their size and structure should be adapted to the type and quantity of sharps eliminated. Users should have a sharps container within reach during care and be able to single-handedly eliminate devices entirely immediately after use, or use features to disconnect used needles. There should be a visual for level of fullness as well as a handle or grip for transportation. Once full, sharps containers must be irremediably sealed. Finally, the considerations of logistics and financial sustainability must be taken into account. It must be noted that sharps containers themselves may be the source of needlesticks during insertion of sharps in containers which are overfilled, unstable, or whose opening is inadequate.

One person in each health-care unit should be responsible for verifying container availability and level of fullness, and for collecting full containers. Decision makers should plan for a predictable increase in volume of medical waste, transportation, and destruction. In certain developing settings, neighboring populations must be informed of the risks associated with handling eliminated waste.

21.4.5. Safety-engineered devices

Over the past decade, the incidence of accidental exposure to blood has decreased steadily in developed countries [69]. Although blood-sampling and injection devices with safety features have contributed significantly to decreasing the rate of injury in HCW [2, 59, 65, 76], this was achieved in a setting of increased training and observance in standard precautions, which remains the cornerstone of prevention [74]. Indeed, enhanced surveillance and training yielded documented reductions in BBF rates before the advent of safety-engineered devices [1, 61, 69]. If hospital, local, or national authorities establish a policy of access to safety-engineered devices, these must also be available at all times. Personnel of all shifts must be trained in the use of a selected type of device before having to use them in a patient with suspected VHF.

21.4.6. Organizational: training, protocols, and procedures

In each unit caring for patients, the number of circulating HCW must be limited. Processes and practices must be reviewed, resorting to non- or less-invasive methods (blood gas, blood sugar, fibroscopy) whenever possible. The number of procedures per patient grouped and brought down to what is necessary. In case of injury, an investigation into the event and a "tree of causes" will help guide control efforts. Safety procedure for receiving and managing patient samples in the laboratory require careful analysis and reorganization.

At the institutional level, surveillance of BBF exposure is the starting point for improved safety and HCW awareness of risks and prevention measures. Surveillance data must be not only collected, but also analyzed and fed back to decision makers and HCW personnel. All personnel in health-care units (including cleaning personnel) on all shifts must be informed and trained. Special attention must be given to care to the deceased or to waste management in order to reduce downstream injuries in personnel not directly responsible for patient care. In many hospitals, a BBF exposure prevention day has helped bolster prevention. Occupational health departments are essential in contributing to surveillance, prevention, and immunization against hepatitis B virus. All efforts are best coordinated by a multidisciplinary BBF exposure control team knowledgeable in the principles of BBF exposure control and how they may be applied to the local setting.

At the governmental level, surveillance and primary and secondary prevention must be promoted, by legislation if necessary, as was the case in the European Union or the USA. Workers' compensation promotion in case of secondary transmission will also aid infected HCW and will improve compliance to passive surveillance efforts.

21.5. POST-EXPOSURE MANAGEMENT

Post-exposure management systems are an integral part of a well-structured effort to enhance HCW safety. Well-publicized and understood procedures will improve secondary prevention by reducing the time to first aid and a primary evaluation of risks of transmission of many existing bloodborne pathogens [70]. In many cases, including several VHF viruses, postexposure prophylaxis may be effective. This will also enhance medical follow-up and HCWs' confidence in the system.

21.5.1. What about ribavirin prophylaxis following exposure to CCHF?

Three decades after it was first used, ribavirin postexposure prophylaxis following CCHF exposure remains controversial [30, 77]. At Tygerberg hospital, ribavirin was used prophylactically in six of the nine needlestick contacts [75]. One of the HCW who received ribavirin had a mild clinical course while five others who received the drug developed neither clinical CCHF nor antibodies to the virus. Two of the three needle contacts not treated with ribavirin had a severe clinical course. One HCW who had sustained a needlestick and 42 proven blood contacts who had not received ribavirin did not become infected. The authors of this study were therefore unable to conclude regarding the prophylactic use of ribavirine [75]. In other recent studies, prophylactic ribavirin was administered to HCWs who had been exposed to the infection [54, 64]. In Pakistan, ribavirin prophylaxis was offered to nine HCWs at high risk for exposure, such as those who were directly exposed to the blood of CCHF patients through contact or needlestick injury, and five (56%) died. In another (unpublished) case an emergency room doctor in Senegal was massively exposed to the blood of a CCHF case (splash to the eyes and face), received ribavirin, and did not seroconvert. A thorough review and meta-analysis of all available experiences with ribavirine in such indications is urgently needed.

In case of exposure, we suggest rigorous daily follow-up by checking complete blood count (CBC), and biochemical tests for the exposed individuals (Fig. 21-1). Ribavirin prophylaxis is generally well tolerated, potentially useful and should therefore be recommended for HCWs who are at high risk of exposures such as percutaneous injuries. The dose and duration of the ribavirivin prophylaxis was not detailed yet. However, oral use and the same dose and duration for the treatment is practiced (Chapter 19). Although further research is needed, oral ribavirin also seems to have an improved outcome in CCHF patients in Iran and Turkey [31, 49].

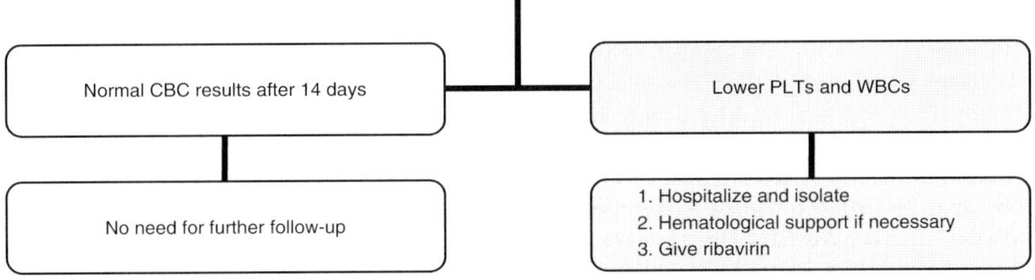

Fig. 21-1. Proposed algorithm for health-care workers who were exposed to CCHFV.

21.6. CONCLUSION

Infection control is rather simple in principle and often costs little, but may be difficult to maintain and often requires extensive reorganization for proper, continuous observance. Proper SP observance was both necessary and sufficient to prevent secondary transmission of Crimean-Congo viral hemorrhagic fever in France and Senegal, and even Lassa fever in several imported cases to industrialized countries. Excessive precautions may not only be unnecessary and costly, but counterproductive as well; impossible to observe and may actually hinder patient care. These infections are highly transmissible by contact with blood and mortality is high. Promoting adequate SP observance within a broader framework for BBF exposure control will support HCW in their efforts to continue to provide care for all patients, including those with suspected VHF.

Box 21-1. An imported case of Crimean-Congo hemorrhagic fever

In 2004, clinician and public health teams managed the first documented case of imported viral hemorrhagic fever (VHF) to France. This was also the first case of confirmed imported Crimean-Congo hemorrhagic fever (CCHF) in an industrialized, non-endemic country. Details of the case and its management are published elsewhere [45, 72].

As in most other imported cases, diagnosis was delayed and the patient's diagnosis was unknown throughout the viremic period [27]. Consequently this case was "simply" isolated in a single room and managed using standard precautions [37]. The subsequent contact tracing exercise found no clinical sign suggesting secondary transmission to any of 181 contacts traced in France, Germany, or Senegal [72]. This incident showed that complex, high-technology care delivered in an industrialized country has greatly increased the number of possible contacts. Furthermore, patient management increasingly tends to involve collaborators across specialties and across the globe. In this case, nearly 200 staff were potentially exposed in several wards of three hospitals in two countries and in a sanitary transport plane while managing a single patient.

REFERENCE

1. Abiteboul D, Antona D, Fourrier A, Brucker G, Descamps JM, Leprince A, Pernet M, Bouvet E (1992) Occupational risk of exposure to blood in nurses. Results of a one-year monitoring of the risk for nurses in 17 hospitals. Pathol Biol (Paris) 40(10):983–989
2. Adams D, Elliott TS (2006) Impact of safety needle devices on occupationally acquired needlestick injuries: a four-year prospective study. J Hosp Infect 64(1):50–55
3. Advisory Committee on Dangerous Pathogens (1996): Management and Control of Viral Haemorrhagic Fevers – December 1996, London
4. Al Tikriti SK, Al Ani F, Jurji FJ, Tantawi H, Al Moslih M, Al Janabi N, Mahmud MI, Al Bana A, Habib H, Al Munthri H, Al Janabi S, Al Jawahry K, Yonan M, Hassan F, Simpson DI (1981) Congo/Crimean haemorrhagic fever in Iraq. Bull World Health Organ 59(1):85–90
5. Altaf A, Luby S, Ahmed AJ, Zaidi N, Khan AJ, Mirza S, McCormick J, Fisher-Hoch S (1998) Outbreak of Crimean-Congo haemorrhagic fever in Quetta, Pakistan: contact tracing and risk assessment. Trop Med Int Health 3(11):878–882
6. Anonymous (1976) Ebola haemorrhagic fever in Sudan, 1976. Report of a WHO/International Study Team. Bull World Health Organ 56(2):247–270

7. Anonymous (1995) Update: management of patients with suspected viral hemorrhagic fever– United States. MMWR Morb Mortal Wkly Rep 44(25):475–479
8. Anonymous (2003) News: case of Lassa fever in a soldier returning to the United Kingdom. CDR Weekly 13(7):1
9. Anonymous (2006) Increase in cases of Crimean-Congo haemorrhagic fever, Turkey, 2006. Euro Surveill 11(7):E060720
10. Athar MN, Baqai HZ, Ahmad M, Khalid MA, Bashir N, Ahmad AM, Balouch AH, Bashir K (2002) Short report: Crimean-Congo hemorrhagic fever outbreak in Rawalpindi, Pakistan, February 2002. Am J Trop Med Hyg 69(3):284–287
11. Athar MN, Khalid MA, Ahmad AM, Bashir N, Baqai HZ, Ahmad M, Balouch AH, Bashir K (2005) Crimean-Congo hemorrhagic fever outbreak in Rawalpindi, Pakistan, February 2002: contact tracing and risk assessment. Am J Trop Med Hyg 72(4):471–473
12. Azap A, Ergonul O, Memikoglu KO, Yesilkaya A, Altunsoy A, Bozkurt GY, Tekeli E (2005) Occupational exposure to blood and body fluids among health care workers in Ankara, Turkey. Am J Infect Control 33(1):48–52
13. Baron RC, McCormick JB, Zubeir OA (1993) Ebola virus disease in southern Sudan: hospital dissemination and intrafamilial spread. Bull World Health Organ 61(6):997–1003
14. Bennett NT, Howard RJ (1994) Quantity of blood inoculated in a needlestick injury from suture needles. J Am Coll.Surg 178(2):107–110
15. Borio L, Inglesby T, Peters CJ, Schmaljohn AL, Hughes JM, Jahrling PB, Ksiazek T, Johnson KM, Meyerhoff A, O'Toole T, Ascher MS, Bartlett J, Breman JG, Eitzen EM, Jr., Hamburg M, Hauer J, Henderson DA, Johnson RT, Kwik G, Layton M, Lillibridge S, Nabel GJ, Osterholm MT, Perl TM, Russell P, Tonat K (2002) Hemorrhagic fever viruses as biological weapons: medical and public health management. JAMA 287(18):2391–2405
16. Burney MI, Ghafoor A, Saleen M, Webb PA, Casals J (1980) Nosocomial outbreak of viral hemorrhagic fever caused by Crimean Hemorrhagic Fever – Congo Virus in Pakistan, January 1976. Am J Trop Med Hyg 29:941–947
17. Burney MI, Ghafoor A, Saleen M, Webb PA, Casals J (1980) Nosocomial outbreak of viral hemorrhagic fever caused by Crimean hemorrhagic fever-Congo virus in Pakistan, January 1976. Am J Trop Med Hyg 29(5):941–947
18. Carey DE, Kemp GE, White HA, Pinneo L, Addy RF, Fom AL, Stroh G, Casals J, Henderson BE (1972) Lassa fever. Epidemiological aspects of the 1970 epidemic, Jos, Nigeria. Trans R Soc Trop Med Hyg 66(3):402–408
19. Celikbas A, Ergonul O, Dokuzoguz B, Eren S, Baykam N, Polat-Duzgun A (2005) Crimean Congo hemorrhagic fever infection simulating acute appendicitis. J Infect 50(4):363–365
20. Celikbas A, Ergonul O, Yildirim U, Zenciroglu A, Erdogan D, Ziraman I, Yilmaz N, Saracoglu F, Demirel N, Cakmak O, Dokuzoguz B (2006) Intrauterine infection of Crimean-Congo haemorrhagic fever: the courses of two episodes. Clin Microbiol Infect 12(4):1665–1665
21. Centers for Disease Control and Prevention: International notes Crimean-Congo Hemorragic Fever – Republic of South Africa. MMWR Morb Mortal Wkly Rep 34(7):94–101
22. Centers for Disease Control and Prevention (1988) Management of patients with suspected viral hemorrhagic fever. MMWR Morb Mortal Wkly Rep 37 (Suppl 3):1–16
23. Centers for Disease Control and Prevention (1995) Update: management of patients with suspected viral hemorrhagic fever– United States. JAMA 274(5):374–375
24. Centers for Disease Control and Prevention (1997) Evaluation of blunt suture needles in preventing percutaneous injuries among health care workers during gynecologic surgical procedures – New York City, March 1993–June 1994. Morb Mortal Wkly Rep 46(2):25–29
25. Centers for Disease Control and Prevention (2004) Imported Lassa fever – New Jersey, 2004. MMWR Morb Mortal Wkly Rep 53(38):894–897
26. Cooper CB, Gransden WR, Webster M, King M, O'Mahony M, Young S, Banatvala JE (1982) A case of Lassa fever: experience at St Thomas's Hospital. Br Med J (Clin Res.Ed) 285(6347):1003–1005

27. Crowcroft NS (2002) Management of Lassa fever in European countries. Euro Surveill 7(3):50–52
28. Dunster L, Dunster M, Ofula V, Beti D, Kazooba-Voskamp F, Burt F, Swanepoel R, DeCock KM (2002) First documentation of human Crimean-Congo hemorrhagic fever, Kenya. Emerg.Infect Dis 8(9):1005–1006
29. Erdem Y, Talas MS (2006) Blunt and penetrating object injuries in housekeepers working in a Turkish University Hospital. Am J Infect Control 34(4):208–214
30. Ergonul O (2006) Crimean-Congo haemorrhagic fever. Lancet Infect Dis 6(4):203–214
31. Ergonul O, Celikbas A, Dokuzoguz B, Eren S, Baykam N, Esener H (2004) Characteristics of patients with Crimean-Congo hemorrhagic fever in a recent outbreak in Turkey and impact of oral ribavirin therapy. Clin Infect Dis 39(2):284–287
32. Ergonul O, Zeller H, Celikbas A, Dokuzoguz B (2007) The lack of Crimean-Congo hemorrhagic fever virus antibodies in healthcare workers in an endemic region. Int J Infect Dis 11(1):48–51
33. Ergonul O, Zeller H, Kilic S, Kutlu S, Kutlu M, Cavusoglu S, Esen B, Dokuzoguz B (2006) Zoonotic infections among veterinarians in Turkey: Crimean-Congo hemorrhagic fever and beyond. Int J Infect Dis 10(6):465–469
34. Fisher-Hoch SP, Khan JA, Rehman S, Mirza S, Khurshid M, McCormick JB (1995) Crimean Congo-haemorrhagic fever treated with oral ribavirin. Lancet 346(8973):472–475
35. Fisher-Hoch SP, McCormick JB, Swanepoel R, Van Middlekoop A, Harvey S, Kustner HG (1992) Risk of human infections with Crimean-Congo hemorrhagic fever virus in a South African rural community. Am J Trop Med Hyg 47(3):337–345
36. Fisher-Hoch SP, Tomori O, Nasidi A, Perez-Oronoz GI, Fakile Y, Hutwagner L, McCormick JB (1995) Review of cases of nosocomial Lassa fever in Nigeria: the high price of poor medical practice. BMJ 311(7009):857–859
37. Garner JS (1996) Guideline for isolation precautions in hospitals. The Hospital Infection Control Practices Advisory Committee. Infect Control Hosp Epidemiol 17(1):53–80
38. Haas WH, Breuer T, Pfaff G, Schmitz H, Kohler P, Asper M, Emmerich P, Drosten C, Golnitz U, Fleischer K, Gunther S (2003) Imported Lassa fever in Germany: surveillance and management of contact persons. Clin Infect Dis 36(10):1254–1258
39. Harxhi A, Pilaca A, Delia Z, Pano K, Rezza G (2005) Crimean-Congo hemorrhagic fever: a case of nosocomial transmission. Infection 33(4):295–296
40. Haut Comité de la Santé Publique de France (2001) Infections virales aiguës, importées, hautement contagieuses, et leur prise en charge – Décembre 2001
41. Helmick CG, Webb PA, Scribner CL, Krebs JW, McCormick JB (1986) No evidence for increased risk of Lassa fever infection in hospital staff. Lancet 2(8517):1202–1205
42. Holmes GP, McCormick JB, Trock SC, Chase RA, Lewis SM, Mason CA, Hall PA, Brammer LS, Perez-Oronoz GI, McDonnell MK, (1990) Lassa fever in the United States. Investigation of a case and new guidelines for management. N Engl J Med 323(16):1120–1123
43. Hoogstraal H (1979) The epidemiology of tick-borne Crimean-Congo hemorrhagic fever in Asia, Europe, and Africa. J Med Entomol 15(4):307–417
44. Jaax N, Jahrling P, Geisbert T, Geisbert J, Steele K, McKee K, Nagley D, Johnson E, Jaax G, Peters C (1995) Transmission of Ebola virus (Zaire strain) to uninfected control monkeys in a biocontainment laboratory. Lancet 346(8991–8992):1669–1671
45. Jaureguiberry S, Tattevin P, Tarantola A, Legay F, Tall A, Nabeth P, Zeller H, Michelet C (2005) Imported Crimean-Congo hemorrhagic Fever. J Clin Microbiol 43(9):4905–4907
46. Johnson E, Jaax N, White J, Jahrling P (1995) Lethal experimental infections of rhesus monkeys by aerosolized Ebola virus. Int J Exp Pathol 76(4):227–236
47. Kiehl W, Haas W (2000) A case of Lassa fever imported into Wiesbaden, Germany. Eurosurveillance Wkly 4(17):000427
48. Macher AM, Wolfe MS (2006) Historical Lassa fever reports and 30-year clinical update. Emerg Infect Dis 12(5):835–837
49. Mardani M, Keshtkar Jahromi M, Holakouie Naieni K, Zeinali M (2003) The efficacy of oral ribavirin in the treatment of Crimean-Congo Hemorrhagic Fever in Iran. Clin Infect Dis 36:1613–1618

50. Mast ST, Woolwine JD, Gerberding JL (1993) Efficacy of gloves in reducing blood volumes transferred during simulated needlestick injury. J Infect Dis 168(6):1589–1592
51. Mehrabi-Tavana A, Chinikar S, Mazaheri V (2002) The seroepidemiological aspects of Crieman-Congo hemorrhagic fever in three health workers: a report from Iran. Arch Iranian Med 5(4):255–258
52. Nabeth P, Cheick DO, Lo B, Faye O, Vall IOM, Niang M, Wague B, Diop D, Diallo M, Diallo B, Diop OM, Simon F (2004) Crimean-Congo hemorrhagic fever, Mauritania. Emerg Infect Dis 10(12):2143–2149
53. Papa A, Bozovi B, Pavlidou V, Papadimitriou E, Pelemis M, Antoniadis A (2002) Genetic detection and isolation of crimean-congo hemorrhagic fever virus, Kosovo, Yugoslavia. Emerg Infect Dis 8(8):852–854
54. Papa A, Ma B, Kouidou S, Tang Q, Hang C, Antoniadis A (2002) Genetic characterization of the M RNA segment of Crimean Congo hemorrhagic fever virus strains, China. Emerg Infect Dis 8(1):50–53
55. Peters CJ, Jahrling PB, Khan AS (1996) Patients infected with high-hazard viruses: scientific basis for infection control. Arch Virol Suppl 11:141–168
56. Peters CJ, Jahrling PB, Liu CT, Kenyon RH, McKee KT, Jr., Barrera Oro JG (1987) Experimental studies of arenaviral hemorrhagic fevers. Curr Top Microbiol Immunol 1345–68
57. Pittet D, Allegranzi B, Sax H, Dharan S, Pessoa-Silva CL, Donaldson L, Boyce JM (2006) Evidence-based model for hand transmission during patient care and the role of improved practices. Lancet Infect Dis 6(10):641–652
58. Richmond JK, Baglole DJ (2003) Lassa fever: epidemiology, clinical features, and social consequences. BMJ 327(7426):1271–1275
59. Rogues AM, Verdun-Esquer C, Buisson-Valles I, Laville MF, Lasheras A, Sarrat A, Beaudelle H, Brochard P, Gachie JP (2004) Impact of safety devices for preventing percutaneous injuries related to phlebotomy procedures in health care workers. Am J Infect Control 32(8):441–444
60. Rollin PE, Sureau P (1989) Conduite à tenir devant un malade suspect de fièvre hémorragique virale. Bull Epidemiol Hebd 17:69–70
61. Saghafi L, Raselli P, Francillon C, Francioli P (1992) Exposure to blood during various procedures: results of two surveys before and after the implementation of universal precautions. Am J Infect Control 20(2):53–57
62. Scrimgeour EM, Zaki A, Mehta FR, Abraham AK, Al Busaidy S, El Khatim H, Al Rawas SF, Kamal AM, Mohammed AJ (1996) Crimean-Congo haemorrhagic fever in Oman. Trans.R.Soc.Trop Med Hyg 90(3):290–291
63. Simpson DI, Knight EM, Courtios G, Williams MC, Kibukamusoke JW (1967) Congo virus: a hitherto undescribed virus occurring in Africa. I-Human isolation-clinical notes. East Afr Med J 44:86–92
64. Smego RA, Jr., Sarwari AR, Siddiqui AR (2004) Crimean-Congo hemorrhagic fever: prevention and control limitations in a resource-poor country. Clin Infect Dis 38(12):1731–1735
65. Sohn S, Eagan J, Sepkowitz KA, Zuccotti G (2004) Effect of implementing safety-engineered devices on percutaneous injury epidemiology. Infect Control Hosp Epidemiol 25(7):536–542
66. Suleiman MNEH, Muscat-Baron JM, Harries JR, Satti AGO, Platt GS, Bowen ETW, Simpson DIH (1980) Crimean/Congo haemorrhagic fever in Dubai. An outbreak at the Rashid hospital. Lancet 2:939–940
67. Swaan CM, van den Broek PJ, Wijnands S, van Steenbergen JE (2002) Management of viral haemorrhagic fever in the Netherlands. Euro Surveill 7(3):48–50
68. Tanner J, Parkinson H (2002) Double gloving to reduce surgical cross-infection. Cochrane Review 4
69. Tarantola A, Golliot F, Astagneau P, Fleury L, Brücker G, Bouvet E, the CCLIN Paris-Nord BBF Exposure Surveillance Taskforce (2003) Occupational blood and body fluids exposures in health care workers: Four-year surveillance from the Northern France Network. Am J Infect Control 31(6):357–363
70. Tarantola A, Rachline A, Abiteboul D (2006) Infection risks following accidental exposure to blood or body fluids in health care workers: a review of pathogens transmitted in published cases. Am J Infect Control (accepted)

71. Tarantola A, Koumare A, Rachline A, Sow PS, Diallo MB, Doumbia S, Aka C, Ehui E, Brucker G, Bouvet E (2005) A descriptive, retrospective study of 567 accidental blood exposures in healthcare workers in three West African countries. J Hosp Infect 60(3):276–282
72. Tarantola A, Nabeth P, Tattevin P, Michelet C, Zeller H (2006) Lookback exercise with imported Crimean-Congo hemorrhagic fever, Senegal and France. Emerg Infect Dis 12(9):1424–1426
73. Tarantola A, Rachline A (2005) Surveillance and training, not postexposure prophylaxis, are the basis for the prevention of occupational infection by blood-borne pathogens in developing countries. J Hosp Infect 60(1):91–92
74. Tuma S, Sepkowitz KA (2006) Efficacy of safety-engineered device implementation in the prevention of percutaneous injuries: a review of published studies. Clin Infect Dis 42(8):1159–1170
75. Van de Wal BW, Joubert JR, van Eeden PJ, King JB (1985) A nosocomial outbreak of Crimean-Congo haemmorrhagic fever at Tygerberg hospital. Part IV: Preventive and prophylactic measures. S Afr Med J 68:729–732
76. Whitby RM, McLaws ML, (2002) Hollow-bore needlestick injuries in a tertiary teaching hospital: epidemiology, education and engineering. Med J Aust 177(8):418–422
77. Whitehouse CA (2004) Crimean-Congo hemorrhagic fever. Antiviral Res 64(3):145–160
78. Woodrow CJ, Eziefula AC, Agranoff D, Scott GM, Watson J, Chiodini PL, Lockwood DN, Grant AD (2007) Early risk assessment for viral haemorrhagic fever: experience at the Hospital for tropical diseases, London, UK. J Infect 54(1):6–11
79. Zweighaft RM, Fraser DW, Hattwick MA, Winkler WG, Jordan WC, Alter M, Wolfe M, Wulff H, Johnson KM (1977) Lassa fever: response to an imported case. N Engl J Med 297(15):803–807

CHAPTER 22

INTERNATIONAL SURVEILLANCE AND CONTROL OF CRIMEAN-CONGO HEMORRHAGIC FEVER OUTBREAKS

PIERRE FORMENTY, DVM, PH.D.[1]*, GLENN SCHNEPF, PH.D.[2], FERNANDO GONZALEZ-MARTIN, PH.D.[2], AND ZHENQIANG BI, PH.D.[1]

[1] *Bio-risk Reduction for Dangerous Pathogens Team (BDP), Department of Epidemic and Pandemic Alert and Response (CDS/EPR), World Health Organization, 20 Avenue Appia, CH-1211 Geneva 27, Switzerland*

[2] *World Health Organization, Department of Epidemic and Pandemic Alert and Response (CDS/EPR), 20 Avenue Appia, CH-1211 Geneva 27, Switzerland*

**Corresponding author: Tel.: +(41) 22 791 25 50; E-mail: formentyp@who.int*

22.1. INTRODUCTION

Crimean-Congo hemorrhagic fever virus (CCHFV) is a zoonotic virus transmitted by ticks that cause severe outbreaks in humans, but causes no disease in ruminants, its amplifying host [2, 5, 7, 14]. Although CCHFV is not pathogenic to animals, the disease is an important viral hemorrhagic fever (VHF) because of its high case fatality rate (CFR) (up to 50%), its potential for nosocomial transmission, and the difficulties in treatment and prevention. CCHF is endemic in all of Africa, Asia, the Balkans, and the Middle East, south of the 50° north latitude due to the geographical limitations of its primary tick vector, *Hyalomma* spp. ticks (Fig. 22-1). In these endemic areas, ecological changes, poverty, social instability, insufficient medical equipment, and absence of standard infection control practices have contributed to increased transmission of the virus. CCHF outbreaks have constituted a threat to public health services because of this increased transmission in its natural environment, in the community, and in the hospital setting. In addition, CCHF is a VHF of special regional concern, and as such, should be considered for notification to the World Health Organization (WHO) under the revised International Health Regulations (IHR 2005) adopted in May 2005 [6].

22.2. CRIMEAN-CONGO HEMORRHAGIC FEVER AND THE NEW INTERNATIONAL HEALTH REGULATIONS 2005

A revision of the IHR was unanimously adopted by the World Health Assembly on 23 May 2005 (http://www.who.int/csr/ihr/en/). The IHR 2005 significantly broadens the scope of the current Regulations (IHR 1969) from States notification

O. Ergonul and C. A. Whitehouse (eds.), Crimean-Congo Hemorrhagic Fever, 295–303.

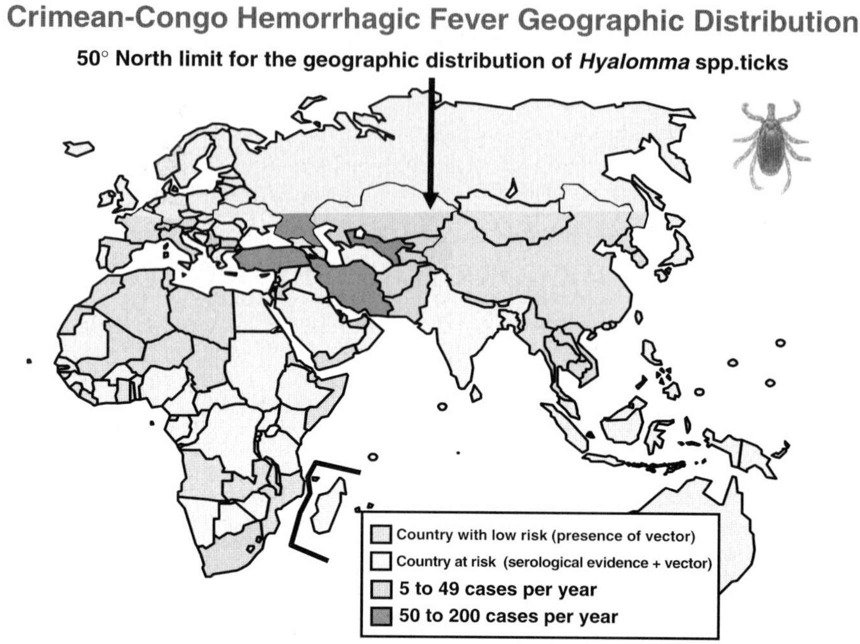

Fig. 22-1. Map showing the geographic distribution of CCHF, including the 50° north latitude limit for the geographic distribution of *Hyalomma* spp. ticks. (*See Color Plates*)

to WHO of single cases of only cholera, plague, and yellow fever, to the notification for all events that may constitute a public health emergency of international concern (PHEIC). States are also obliged to report evidence of public health risks outside their territory that may cause international disease spread. Notifications and reports should be communicated to WHO through the National IHR focal point. On receiving advice from an emergency committee, if the WHO Director-General determines that a PHEIC is occurring in a particular country, the Director-General may make temporary recommendations in order to prevent or reduce the international spread of disease and to minimize interference with international traffic and trade. In addition, the IHR 2005 updates and further develops the provisions in the current IHR with regard to routine public health measures at points of entry and matters relating to international traffic. A number of the obligations placed on WHO and States Parties under the IHR 2005 reveal a real commitment by both to adopt a new way of working. The IHR 2005 is an international legal instrument which, upon entry into force on 15 June 2007, will require States Parties and WHO to take concrete, often daily actions to prevent, protect against, control, and provide a public health response to the international spread of disease. In doing so, States Parties and WHO must also avoid unnecessary interference with international traffic and trade. To fully implement and comply with the IHR 2005, States Parties in collaboration with WHO will have to develop, maintain, and strengthen appropriate public health and administrative capacities in general; and at designated international ports, airports, and some land crossings, in particular. This will require not only close collaboration between WHO and States Parties, but also among States Parties themselves and other stakeholders. Such multilateral cooperation will better prepare the world for future public health emergencies. The spirit of the IHR 2005, however, is also being invoked by many prior to its entry into force. Indeed, the concern is that, in the absence of a clear set of rules such as the IHR 2005 to coordinate response efforts in the event of a major international public health crisis, attempts by the world community to control public health threats such as the spread of avian influenza and potential human pandemic influenza will be severely hampered. States Parties should regard CCHF as a VHF of special regional concern, and as such, may assess events involving CCHF for possible notification to WHO under the IHR 2005. Criteria for the assessment and potential notification of a CCHF outbreak and/or cases to WHO would be based on the four criteria/questions described in Annex 2 of IHR 2005:

- Is the public health impact of the CCHF outbreak and/or cases serious (e.g. high CFR)?
- Is the CCHF outbreak and/or cases unusual or unexpected?
- Is there a significant risk of international spread?
- Is there a significant risk of international travel or trade restrictions?

Following such an assessment, States Parties that answer "yes" to any two of the four questions/criteria would be required to notify WHO, in accordance with Article 6 of the IHR 2005. In this scenario, and if international support is necessary, WHO will, upon request by the affected State Party, collaborate in the response, which may include the rapid mobilization of resources and assistance to control the CCHF outbreak by mobilizing WHO's outbreak alert and response system and its network, the Global Outbreak Alert and Response Network (GOARN).

22.3. INTERNATIONAL SURVEILLANCE OF VHF AND THE GOARN

The phenomenal growth of international travel and trade has vastly increased the speed and ease with which pathogens and their vectors can cross continents and cause outbreaks and epidemics in new areas, sometimes establishing permanent residence there. No country acting alone can defend its borders against this threat. To face the threat of VHF and other emerging infectious diseases, WHO has set up a global system for gathering infectious diseases intelligence, detecting outbreaks quickly, and collaborating to contain their spread. This system has been in place at WHO since April 2000 when the GOARN was launched as a mechanism for keeping the microbial world under close surveillance and ensuring that outbreaks are quickly detected and contained.

This overarching network interlinks 110 existing networks, which together possess much of the data, expertise, and skills needed to keep the international community alert to outbreaks and be ready to respond to them. By electronically linking together existing networks in real time, the GOARN magnifies the resources of individual partners considerably. Several of these networks are specific to VHF. The international surveillance of VHF relies on the interconnectivity of national surveillance programs with coordination by the WHO regional offices and the network of WHO-collaborating centers and laboratories on VHF. International organizations working in animal health, such asUnited Nations Food and Agriculture Organization (FAO), World Organization for Animal Health (OIE), and the World Conservation Union (UICN); nongovernmental organizations working in public health (Médecins Sans Frontières) and in animal health, such as Ecosystèmes Forestiers d'Afrique Centrale (ECOFAC) and Wildlife Conservation Society (WCS) compliments the international surveillance of VHF efforts. Collaboration between animal health and human health at all levels is very important to ensure an effective international surveillance of VHF. For the surveillance of CCHF, countries should develop and intensify collaborations not only between animal and human health sectors, but also with entomologists who are tick population specialists and climate forecasting experts who would provide data for enhanced surveillance and outbreak response preparedness. Where CCHF is a seasonal disease, national active surveillance systems should mobilize resources during the tick season in the spring and summer months. National and international surveillance of CCHF and other zoonotic emerging diseases has improved in the last 10 years in many countries, but progress remains to be made through focused surveillance of the tick population infected with CCHFV, surveillance of human CCHF cases, and partnership with the animal disease sector (to include both domestic and animal wildlife specialists). These collaborations should be intensified through the Global Early Warning and Response System (GLEWS) with FAO and OIE, with more formal links with wildlife disease experts/networks, and partnership with the various laboratory networks (both animal and human).

22.4. FORECASTING MODELS IN THE OUTBREAK ALERT AND RESPONSE PROCESS

Outbreak forecasting and early detection systems based on climatic forecasting models have been developed successfully for Rift Valley fever (RVF), a VHF transmitted by mosquitoes [1], and more recently for CCHF transmitted by ticks [3] (see Chapter 14). Researchers showed that CCHF epidemics were highly correlated to *Hyalomma marginatum marginatum* distribution and closely related to altitude, temperature, and rainfall variables during the spring and summer. These observations have made it possible to develop predictive mathematical models using time-series data of vegetation index and forecasting climatic data. Collaborations with affected countries, space agencies, scientific institutions,

FAO, and WHO could make it possible to produce seasonal cartography of CCHF outbreak risk zones. This would allow, in the case of an alert, the ability to inform the affected countries so that they could intensify CCHF surveillance and preparedness for the possible epidemic. In the context of the IHR 2005 and CCHF surveillance, the forecasting and early detection of epidemics, and the analysis of their possible spread to new areas, are necessary and essential tools for the implementation of effective control measures. More progress is necessary in this field but the use of predictive climatology in the monitoring of the tick-borne viruses like CCHFV should be encouraged. Scientists should continue to develop further forecasting models for CCHF and link them with surveillance activities in order to help with CCHF outbreak readiness and preparedness.

22.5. STRATEGIES FOR CONTROLLING CCHF OUTBREAKS

For the past several years sporadic human cases of CCHF and limited outbreaks have been increasingly reported every year. Recent outbreaks in Afghanistan (2001–2006), Iran (2001), Kazakhstan (2005), Kosovo (2001), Mauritania (2002–2003), Pakistan (2001–2006), Russia (2006), Saudi Arabia (1990), Senegal (2004, with an imported case in France), South Africa (1996), Tajikistan (2002, 2004), and Turkey (2003–2006) have drawn the international community's attention to this emerging problem. Strategies for controlling CCHF outbreaks should take into consideration the nature of the disease (a VHF transmitted by ticks, possible treatment with ribavirin, absence of a vaccine) and its epidemiology (potential for nosocomial outbreaks, exposure to ticks, or direct contact with virus-infected animals or people are considered the major risk factors). Strategies to control CCHF outbreaks should be based on the following interventions (Fig. 22-2):

- Establish a coordination mechanism for response recognized by all partners involved in the outbreak response
- Develop a social mobilization and health education program that will inform the public and restrict practices that promote transmission in the community
- Organize an isolation hospital wards to treat patients with respect for their dignity, and implement standard infection control practices to ensure safe case management by health-care workers
- Implement standard infection control practices in all health-care settings and with all patients in the affected country/region in order to prevent nosocomial spread
- Ensure psychosocial support to patients, families, and health-care workers as needed
- Establish a powerful active surveillance system that allows for the identification of new cases and for the follow-up of their contacts for 12 days (the maximum estimated incubation period), as well as isolate the patient's contacts in order to stop virus transmission

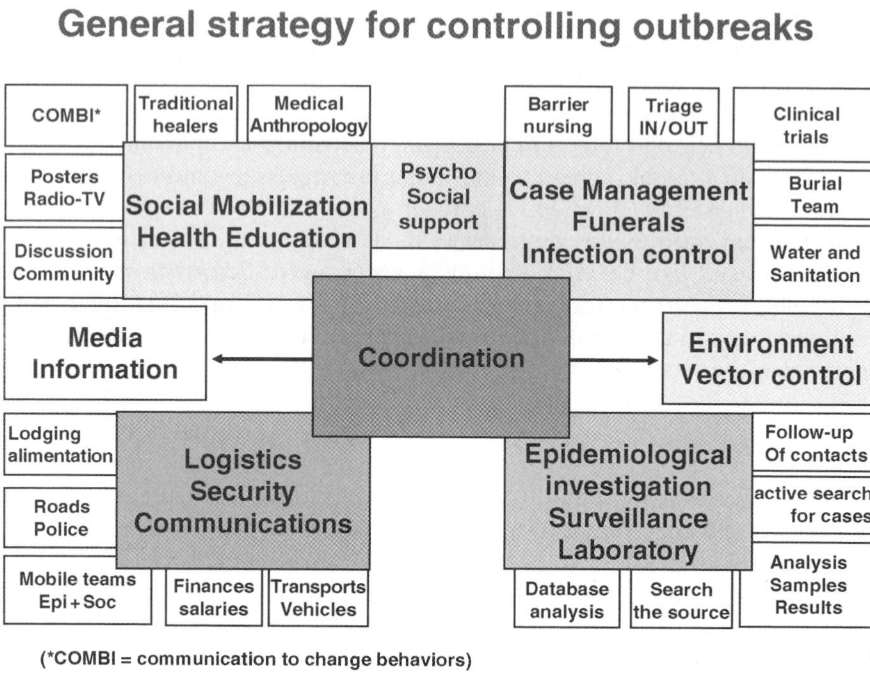

Fig. 22-2. Diagram showing the general strategy used by WHO for controlling CCHF outbreaks.

- Implement integrated vector control activities to decrease human exposure by the use of acaricides and social mobilization
- Establish an organized logistical support system, which guarantees the safety conditions needed for the correct deployment of operations

In the event of a CCHF outbreak, and with the deployment of national and international teams in the field, the first step should be to establish a coordination committee for outbreak control activities under the leadership of the Ministry of Health. This committee would be in charge of the general coordination of operations, and it should clearly identify the role and responsibilities of the different teams and the channels of information and command. It should also take care of the correct implementation of the outbreak control strategy. Medical expertise is important, but adhesion to control measures through social mobilization and health education programs are an essential component of CCHF outbreak control activities. The main objective is to inform the public and the local communities how to promote practices that decrease transmission of the disease. Such practices would include preventing contact with the blood of virus-infected animals (e.g. slaughtering activities), preventing tick bites, and preventing the transmission of disease during patient care at home or during funerals. Appropriate key behavioral interventions would include eliminating, or at least controlling, tick infestations on animals; using masks, gloves, and gowns when slaughtering and butchering animals; avoiding tick bites and removing ticks safely from the skin; reducing transmission when caring for patients at home; seeking early treatment for fever after a history of tick bites or contacts with CCHF patients; using personal protective equipment (PPE) when managing patients; reducing transmission when caring for patients in the hospital; washing hands with soap and clean water regularly; and organizing safe and respectful funerals.

Social mobilization actions should benefit from the essential contributions of medical anthropology that take into account the social and anthropological background and the local human–animal interface. Medical anthropologists assess the locals' explanatory models for misfortune, and they assist the medical team to adapt their actions according to the local culture. A social mobilization method known as "communication for behavioral impact" (COMBI), that focuses on influencing behavior at both the individual and community level, is used as the basic strategy of WHO social mobilization actions [10].

For CCHF, human-to-human transmission has been reported, and the disease does have potential for nosocomial spread. During an outbreak, it is often necessary to set up an isolation ward to prevent patients from infecting those close to them. The implementation of standard infection control practices in the isolation ward will permit proper and safe case management of patients and avoid health-care workers becoming infected. With regard to funerals, infection control procedures should be followed during the funerary ceremonies (preparation of the bodies, burials, etc.) while considering local traditions. By not taking into account these local traditions, the medical team in the community could disrupt the mourning process of the families and become isolated or even despised by the community.

The treatment of CCHF is symptomatic with replacement of the blood volume and components, rehydration, restoration of the electrolyte balance, provision of intensive care, and providing antibiotics and/or antimalarial drugs when needed. There is currently no specific antiviral therapy for CCHF. However, ribavirin has been shown to inhibit in vitro viral replication in Vero cells and reduced the mean time to death in a suckling mouse model of CCHF [12, 13]. Additionally, several case reports have been published that suggest oral or intravenous ribavirin is effective for treating CCHF infections [4, 8, 9, 11]. All published reports showed a clear benefit in patients with confirmed CCHF treated with ribavirin (intravenous and oral administration), and there were no major side effects or mortality associated with ribavirin treatment. The results of all these studies are limited by their design and sample size. Therefore, it would be advisable to promote the development of randomized clinical trials to assess the efficacy of the drug and later to provide guidelines for dosage, formulation and time of administration of the drug during the disease (see Chapter 19 for more details on the treatment of CCHF).

One of the principal objectives of CCHF outbreak control remains breaking transmission chains. The team in charge of epidemiological surveillance actively searches for new cases. Newly detected cases should be isolated if they meet the case definition adopted by the team. In addition, all the subjects who were in contact with the patients are followed for 12 days (maximum incubation period) after their last exposure, and if these contacts become ill, they would be evaluated by a clinician and eventually be isolated in the isolation ward if consider to be a suspect case by the medical team.

22.6. CONCLUSIONS

Today's technologies can help to better detect, manage, and contain the spread of CCHF, but the need for high-level governmental commitment and international collaboration remains. The new IHR 2005 supports this mechanism and will help to reinforce the countries capacities in outbreak alert and response.

We do not have an efficient vaccine or a specific treatment for CCHF. However, outbreak control measures are rather simple and appear very effective if they are fully accepted by the affected populations. The mobilization of these populations remains one of the most important missions of the outbreak response teams. The magnitude of the public health problem will continue to grow unless more effective measures are taken to reduce viral transmission. Given the worsening epidemiological trends, there is a need to renew or intensify efforts for the prevention and control of CCHF. WHO and its partners have identified five main priorities:

- Strengthening epidemiological surveillance (including laboratory diagnosis) for planning and response.
- Reducing the disease burden through WHO standard case management training for CCHF, improving emergency preparedness and response, and strengthening of national tick control programs.
- Promoting behavioral change for sustainable prevention and control of tick populations at individual, household, community, institutional, and political levels.
- Accelerating research programs, notably for CCHF disease forecasting models. Enhancing cooperation in the field of tick-borne disease research.
- Hastening the scientific agenda for new treatments and vaccines, as the lack of effective therapies or vaccines limit our control activities.

REFERENCES

1. Anyamba A, Chrétien JP, Formenty P, Small J, Tucker CJ, Malone JL, El Bushra H, Martin V, Linthicum KJ (2006) Rift Valley fever potential, Arabian Peninsula. Emerg Infect Dis 12:518–520
2. Ergonul O (2006) Crimean-Congo haemorrhagic fever. Lancet Infect Dis 6:203–214
3. Estrada-Pena A (2006) Prediction of habitat suitability for ticks. Ann N Y Acad Sci 1078:275–284
4. Fisher-Hoch SP, Khan JA, Rehman S, Mirza S, Khurshid M, McCormick JB (1995) Crimean-Congo hemorrhagic fever treated with oral ribavirin. Lancet 346:472–475

5. Flick R, Whitehouse CA (2005) Crimean-Congo hemorrhagic fever. Curr Mol Med 5:769–776
6. Formenty P, Roth CE, Gonzalez-Martin F, Grein T, Ryan M, Drury P, Kindhauser MK, et Rodier G (2005) Les pathogènes émergents, la veille internationale et le Règlement sanitaire international. Médecine et Maladies Infectieuses 36:9–15
7. Hoogstraal H (1979) The epidemiology of tick-borne Crimean-Congo hemorrhagic fever in Asia, Europe, and Africa. J Med Entomol 15:307–417
8. Mardani M, Jahromi MK, Naieni KH, Zeinali M (2003) The efficacy of oral ribavirin in the treatment of Crimean-Congo hemorrhagic fever in Iran. Clin Infect Dis 36:1613–1618
9. Papa A, Bozovi B, Pavlidou V, Papadimitriou E, Pelemis M, Antoniadis A (2002) Genetic detection and isolation of Crimean-Congo hemorrhagic fever virus, Kosovo, Yugoslavia. Emerg Infect Dis 8:852–854
10. Renganathan E, Hosein E, Parks W, Lloyd L, Suhaili MR, Odugleh A (2005) Communication-for-behavioral-impact (COMBI): a review of WHO's experiences with strategic social mobilization and communication in the prevention and control of communicable diseases. In: Haider M (ed.) Global Public Health Communications: Challenges, Perspectives, and Strategies. Jones and Bartlett Publishers, Boston, MA
11. Tang Q, Saijo M, Zhang Y, Asiguma M, Tianshu D, Han L, Shimayi B, Maeda A, Kurane I, Morikawa S (2003) A patient with Crimean-Congo hemorrhagic fever serologically diagnosed by recombinant nucleoprotein-based antibody detection systems. Clin Diagn Lab Immunol 10:489–491
12. Tignor GH, Hanham CA (1993) Ribavirin efficacy in an in vivo model of Crimean-Congo hemorrhagic fever virus (CCHF) infection. Antiviral Res 22:309–325
13. Watts DM, Ussery MA, Nash D, Peters CJ (1989) Inhibition of Crimean-Congo hemorrhagic fever viral infectivity yields in vitro by ribavirin. Am J Trop Med Hyg 41:581–585
14. Whitehouse CA (2004) Crimean-Congo hemorrhagic fever. Antiviral Res 64:145–160

SECTION VI

REMAINING QUESTIONS AND FUTURE RESEARCH

FUTURE RESEARCH

ONDER ERGONUL, M.D., M.P.H. AND CHRIS A. WHITEHOUSE, PH.D

It is our hope that this book will stimulate research, and identify new areas of study, for Crimean-Congo hemorrhagic fever (CCHF). To gain a better overall picture of this disease, a multidisciplinary approach is needed. This should include not only virologists, entomologists, clinicians, epidemiologists, and veterinarians, but also climatologists, sociologists, anthropologists, and others. The dynamics of the enzootic environment and transmission cycle of the virus need to be better understood. For example, the role, if any, of ticks other than *Hyalomma marginatum marginatum* needs to be more carefully examined. At the time of potential global warming, the role of various climatic factors on CCHF transmission will become increasingly important. Global information systems and remote sensing using satellites will be important tools to produce better spatial and temporal maps for CCHF. In addition, mathematical modeling, which can simulate *in silico* CCHF epidemics under a variety of different conditions, holds great promise. However, it must be keep in mind that all these predictive models need to be critically validated using data from real disease outbreaks whenever possible.

Laboratory research with the virus has been hampered by the need to work with it under a high level of biocontainment; in most cases, this means a biosafety level 4 (BSL-4) laboratory. With the recent increased interest and funding for biodefense and emerging infectious disease-related work, more BSL-4 laboratories are being built, especially in the USA and Europe. This increased availability of appropriate biocontainment laboratories should lead to more research on the virus itself. New antiviral drugs and improved vaccines are sorely needed for this disease. Also critical is the need for an appropriate animal model of CCHF, not only to study the pathogenesis of disease, but also to test potential new antiviral drugs and vaccines. As alluded to elsewhere in this book, studies on the pathogenesis of CCHF could also shed light on the pathogenesis of other viral hemorrhagic fevers, such as Ebola. A particularly interesting area in this regard is the study of the host response to infection. It is possible that different individuals respond differently to infection or different strains of the virus produce slightly

O. Ergonul and C. A. Whitehouse (eds.), Crimean-Congo Hemorrhagic Fever, 307–308.

different disease in their hosts. This knowledge could identify specific groups of people who are at more risk of serious disease or could provide clues for the development of new therapies for CCHF and other important viral diseases.

Of course these thoughts only scratch the surface. No one can know for sure what the future holds for CCHF or any other disease, but hopefully these ideas will provide a starting point.

FREQUENTLY ASKED QUESTIONS (FAQ) ABOUT CRIMEAN-CONGO HEMORRHAGIC FEVER

This FAQ section was structured to provide a basic knowledge for health-care workers, the media, or anyone else who is responsible for providing the public with information about Crimean-Congo hemorrhagic fever (CCHF). These questions were compiled from numerous academic and public seminars, conferences, and radio and television programs.

Q: What is a viral hemorrhagic fever?
A: A viral hemorrhagic fever is a viral disease, which has a tendency to disrupt the clotting of the blood, so that patients may develop uncontrolled bleeding. Usually fever, body aches, and other flu-like symptoms also are seen. Many common diseases can resemble viral hemorrhagic fever, but the term is reserved for a particular group of diseases associated with a high death (fatality) rate. In addition to CCHF they include Lassa fever, Rift Valley fever, Alkhumra, Omsk hemorrhagic fever, Kyasanur forest disease, Argentine, Bolivian, Brazilian, and Venezuelan hemorrhagic fevers (caused by Junin, Machupo, Sabia, and Guanarito viruses, respectively), and Marburg and Ebola hemorrhagic fevers. Although the clinical pictures are similar, the viruses are not closely related to each other, and are transmitted in a variety of ways.

Q: What is CCHF?
A: CCHF is a tick-borne viral disease of humans, which occurs in Africa, south-eastern Europe, and Asia, below the 50° parallel.

Q: Why does it have the name "Crimean-Congo hemorrhagic fever"?
A: A disease given the name Crimean hemorrhagic fever was first recognized on the Crimean Peninsula in 1944, although the virus which causes the disease was only identified in 1967. Meanwhile, in 1956 a virus given the name Congo was isolated from a child with fever in the former Belgian Congo or (now Democratic Republic of the Congo). In 1969, it was discovered that the two viruses were the

O. Ergonul and C. A. Whitehouse (eds.), Crimean-Congo Hemorrhagic Fever, 309–316.

same. Consequently, the virus and the disease were called "Crimean-Congo hemorrhagic fever."

Q: Where does the virus come from?
A: The virus is transmitted mainly by *Hyalomma* ticks, adults of which have distinctive brown and white bands on their legs. The virus can remain in the ticks for long periods, and even pass through the eggs to infect the next generation of ticks.

Immature *Hyalomma* ticks (larvae and nymphs) feed on ground-frequenting (or ground-feeding) birds (guinea fowl, partridges, rooks) and small mammals up to the size of hares. Adult *Hyalomma* ticks feed on livestock such as cattle, sheep, and goats, as well as on wild animals such as antelope, wild boar, and ostriches.

Animals bitten by infected ticks do not develop the disease, but can circulate the virus in their blood for a few days, up to 1 week, and thereafter become immune to further infection. Noninfected ticks become infected if they feed on the animals during the short period when virus is in circulation, thus ensuring that the virus is perpetuated.

Q: Can CCHFV live in vectors other than ticks, such as mosquitoes?
A: No, mosquitoes or other arthropods (other than ticks) have not been implicated as vectors of CCHFV.

Q: How do humans become infected?
A: Humans can become infected from being bitten by infected ticks, or even from squashing ticks if fluid from the ticks gets into cuts and breaks in the skin, or onto mucous membranes. Humans can also become infected if blood from infected livestock or wild animals comes into contact with broken skin (cuts and abrasions) or mucous membranes during the short period that the animals have the virus in circulation. On farms, this usually happens when young animals become infected as a result of being exposed to ticks, and humans are then exposed to blood during procedures such as the castration of calves, slaughtering of lambs, vaccination of animals, the cutting of identity notches in the ears, or the attachment of ear tags. Occasionally, animals that have been reared under tick-free conditions come into contact with ticks and the virus late in life, and so slaughtering mature animals can also result in human infection. Although the proportion of mature animals that will have virus in circulation may be extremely low, many thousands of animals are slaughtered each day at abattoirs. Hunting and butchering of wild animals can also be a source of human infection.

Similarly, humans can become infected through the contact with the blood and the body fluids of the patients. The relatives of the patients, who are in close contact with CCHF patients or the health-care workers can become infected while caring their patients. The major route is being exposed to the blood of the infected patients.

Q: What is the life cycle of the tick vector?
A: *Hyalomma marginatum* group of ticks are two-host ticks and have activity in the summer. Immature forms (larvae and nymphs) feed on small animals (hare, ground-frequenting birds) for about 2–3 weeks and drop off the ground as engorged nymphs. Here starts an inactive period of 3 weeks (can be as much as 4 months longer if interrupted by winter) when the engorged nymphs transform to adults. Adult *Hyalomma* ticks hide on the ground and actively "run" toward an animal host (large mammal) when they sense certain signals (CO_2, vibration, visual objects, body temperature). Adults feed on animals for 1–2 weeks and mate meanwhile. Engorged females drop off the host, produce up to 7,000 eggs, and die.

Q: Where do the ticks get virus from?
A: The main source of virus infection for the tick is hares. Larvae and nymph acquire the virus from infected hares and the adult stage remains infected after molting (transstadial transmission). Livestock and large wild animals, during their short viremic period, can also serve as a source of virus for ticks. Engorged infected females can transmit the virus to their eggs, and subsequently to their progeny (transovarial transmission). Transovarial transmission rate in *Hyalomma* ticks is quite low, and its role in CCHF epidemics is not known exactly. There is also the "nonviremic transmission" phenomenon, in which noninfected ticks can acquire the virus by co-feeding with infected ticks on a CCHFV refractile host (e.g. birds).

Q: Where do people get tick bites?
A: In case of CCHF, most of the human tick bites are from unfed adult *Hyalomma* ticks. With regard to the biology of *Hyalomma marginatum* group ticks, it can be supposed that the areas frequented by hares and ground-feeding birds are of potential risk.

Q: Does CCHFV transmit through eating of contaminated animals?
A: No. To our knowledge no such transmission has been reported. This would be highly due to the high acid content of our stomachs.

Q: Which people are at risk of becoming infected?
A: People who are at particular risk of becoming infected with CCHFV include those involved in agriculture and stockbreeding, such as farmers and farm

laborers, milkmaids, sheep shearers, veterinarians, abattoir workers, persons who slaughter animals, hunters, close contacts of the infected patients, and the health-care workers.

Within abattoirs those who come into contact with fresh blood are at greatest risk. Once the carcasses have been bled out and hung to mature there is a sudden increase in the acidity of the meat, and the virus cannot survive in the carcass. Ostriches appear to be the only birds in which there is similar circulation of detectable levels of virus in blood as occurs in mammals. There is no indication that meat processed and matured according to standard abattoir practices constitutes a danger to consumers. Partially fed ticks, which detach from the hides of recently slaughtered animals, may attach indiscriminately to hosts available in their environment, and thus infect slaughtermen.

Apart from people directly involved in the livestock industry, persons at risk of being bitten by ticks include those who live in the countryside and town dwellers who visit the countryside for occupational or recreational purposes, including hunting and hiking. People are not always aware of being bitten by ticks, and in patients with CCHF, ticks have been found attached in concealed areas, such as on the scalp, pubic regions, and between the toes.

Health-care workers, or close contacts of patients, can acquire the infection from contact of broken skin or mucous membranes with the blood or blood-tinged body fluids and wastes of the patient. The only time that the infection has been seen in clusters of people is when they have been exposed to a common source of the virus, for example, while slaughtering animals. In contrast, there have been several instances of secondary spread of the infection from patients to health-care personnel, and this has usually involved needlestick injuries in hospitals.

Q: How common is the disease and how often is it fatal?
A: The disease is common in Africa, southeastern countries of Europe, and Asia. The case fatality rate (the rate of dying after getting the disease) is around 5–30%.

Q: What are the signs and symptoms of the disease?
A: The disease has a short incubation period followed by a very sudden onset of illness. People usually become sick within 1–3 days of being bitten by a tick, or 5–6 days after exposure to the blood of infected livestock or humans.

People abruptly develop a severe headache with sore and reddened eyes, fever with cold chills, and intense body aches, particularly involving the muscles of the lower back and thighs, and feel extremely unwell. Body temperatures do not necessarily remain high and may fluctuate during the course of the day. There may be nausea and vomiting, and sometimes abdominal pain and diarrhea early in the course of the disease. At this stage, blood tests already show abnormal liver function, and a decrease in blood platelets, which are involved in the clotting of blood.

After about 5 days, the patients may develop a rash of pink blotches on the body, followed by various bleeding tendencies, depending on the severity of the illness. They bruise easily, often have nose bleeds, and may pass blood in their stool and/or urine. Stools seldom contain fresh blood; they usually have a dark and tarry appearance. Small or large red spots of bleeding into the skin appear, and there may be large confluent areas of bleeding into the skin around injection sites and in skin folds such as in the armpits or groin. Patients may vomit blood and bleed from the gums, and women may develop heavy uterine bleeding. Blood continues to ooze from needle puncture sites. There can also be internal bleeding, including intracerebral bleeding. Patients can go into a coma as the liver, kidney, and lung functions fail, and death can occur 5–14 days after the onset of illness, usually from heart failure.

Patients who recover show sudden improvement by day 10 of their illness. Virus remains detectable in human blood for up to 2 weeks after the onset of illness, but once the results of blood tests indicate that patients' body functions have recovered, and they feel well and are no longer bleeding; they can be discharged from hospital.

Q: What is the treatment for CCHF?
A: Treatment essentially consists of supportive therapy, which comprises intravenous feeding of the patient and replacement of blood and clotting factors. Severely ill patients may be placed on ventilators and other life support systems. The antiviral drug ribavirin has been used to treat patients, but the drug is not 100% effective; however, this is currently the only drug available to treat the disease.

Q: Why does the case fatality rate differ from country to country?
A: This remains an unanswered question, but could result from several reasons: (1) the virulence of different virus strains may be different; (2) access to health systems in different countries differ; (3) there could be more underlining "background" infections in some countries compared to others.

Q: What action should be taken if a person is suspected of having CCHF?
A: The disease may be suspected when a person suddenly becomes sick with headache, fever and chills, muscle pains, and possibly nausea, vomiting, and diarrhea, less than 1 week after being bitten by a tick, squashing ticks, or coming directly into contact with fresh blood or blood-tinged body fluids and organs of livestock, wild animals, or human CCHF patients. A doctor should be consulted immediately if the disease is suspected, and if the doctor believes that the suspicion is justified, the patient should be hospitalized and isolated immediately.

Q: What measures can be taken to prevent exposure to infection?
A: Persons in CCHF-endemic rural areas can use certain pyrethroid acaricides (permethrin, 0.5%) to treat clothing such as socks and outerwear (acaricides are insecticides used against ticks). Formulations which are generally available from

shops that sell equipment for camping and outdoor activities, include aerosol sprays and sachets of concentrated acaricide used to prepare emulsions into which clothing is dipped.

Insect repellents such as DEET can be used on the skin to preventing tick bites. But it should be kept in mind that the effective concentration of DEET to repel ticks is much higher than that used for mosquitoes. The use of DEET on small children should be avoided.

Long trousers and long-sleeved shirts should be preferred. When being in an area of high risk, personal inspection of your clothes should be made every 2 h, and total body inspection is advised at the end of the day. Virus transmission from the attached tick increases over time, so prompt tick removal is important.

Abattoir workers, veterinary staff, farm workers, and hunters should use appropriate impervious protective clothing and gloves when engaged in activities which carry a risk of exposure to animal blood. Although it is incumbent upon employers to supply protective clothing and safety instructions, employees must take responsibility for adhering to the safety regulations.

Veterinary regulations promulgated for ostrich abattoirs require that birds should be treated with an appropriate acaricide and held in tick-free facilities for 14 days before slaughter. Similar regulations would be impossible to implement for other livestock. Vast numbers of cattle, sheep, and goats are slaughtered each day, and the costs of constructing tick-free holding pens of suitable capacity would be prohibitive, as would the costs and logistics of holding and feeding the animals and supervising the operation. A potential alternative would be the development of a veterinary anti-tick and anti-CCHFV vaccine that is applied to farm animals as a public health measure, but such research would require special funding.

Q: Can the virus transmit through inhalation?
A: No, there is no report on inhalational transmission of the virus. Therefore, the universal precautions are generally considered sufficient for the protection of close contacts and health-care workers.

Q: Can a cured patient get the infection a second time?
A: After recovery, the patients are immune to further infection. It is not uncommon for recovered patients to remember little or nothing about the events of their illness. To our knowledge, there is lifelong immunity after the disease.

Q: Is there a vaccine for the infection?
A: At present there is no human vaccine, and the lack of potential demand for such a vaccine inhibits its development. A vaccine against CCHF was used in Bulgaria, but the experience is limited to one country.

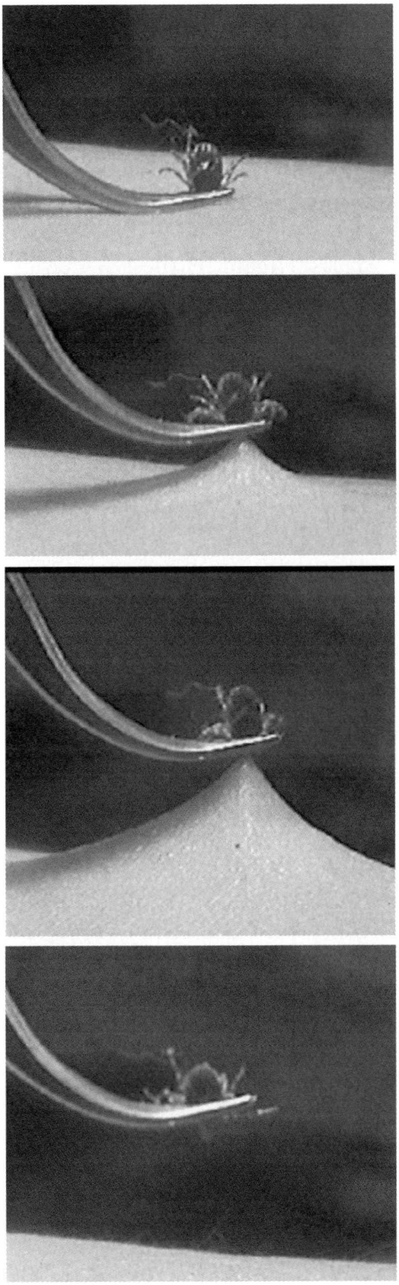

Fig. 1. Pictures showing the proper way to remove an attached tick. (Courstey of Dr. Zati Vatansever, Ankara University, Ankara, Turkey.)

Q: Is there a causal relation of global warming and the occurrence of the disease?
A: Possibly, but not demonstrated yet. Since the virus-infected ticks become more active in the warmer months, global warming will increase the number of warm days in the year, thus resulting in more active ticks and more opportunity for people to become infected.

Q: Is CCHFV a potential bioterrorism agent?
A: Yes, CCHFV is listed as a Category C bioterrorism agent by the US Centers for Disease Control and Prevention.

Q: How can I take a tick out if I see one attached to me?
A: Attached ticks should be taken out gently and cautiously (Fig. 1). Ticks' mouthparts have reverse harpoon-like barbs designed to penetrate and attach to the skin. Here is how we suggest removing an attached tick:

1. Use fine-point tweezers to grasp the tick at the place of attachment as close to the skin as possible.
2. Gently pull the tick straight out.
3. Place the tick in a small vial labeled with your name, address, and the date.
4. Wash your hands and disinfect the bite site with isopropyl alcohol.
5. Record the date, area on your body of the tick attachment, and your general health at the time.
6. Call your doctor to determine if treatment is warranted.
7. If possible, have the tick identified or tested by a laboratory, your local health department, or a veterinarian.
8. Do not attempt to prick, crush, or burn the attached tick as this may cause it to release infected fluids into your skin. Also, do not try to smother the tick (e.g. applying petroleum jelly or nail polish).

Acknowledgments

We thank Robert Swanepoel and Zati Vatansever for their contributions to this section.

INDEX

COLOR PLATES

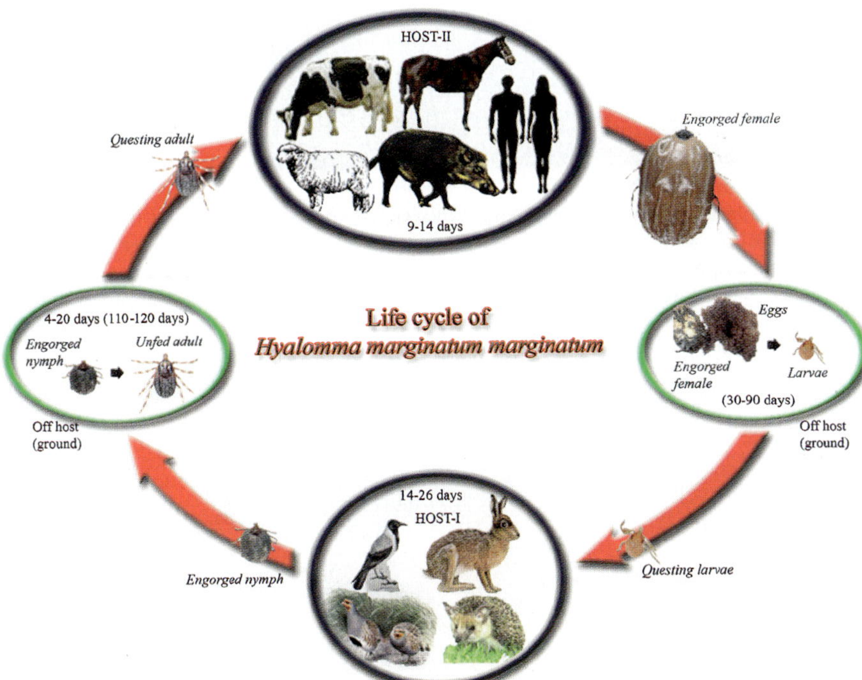

Fig. 1-1. Life cycle of *Hyalomma marginatum marginatum* ticks (Courtesy of Dr. Zati Vatansever, Ankara University, Ankara, Turkey).

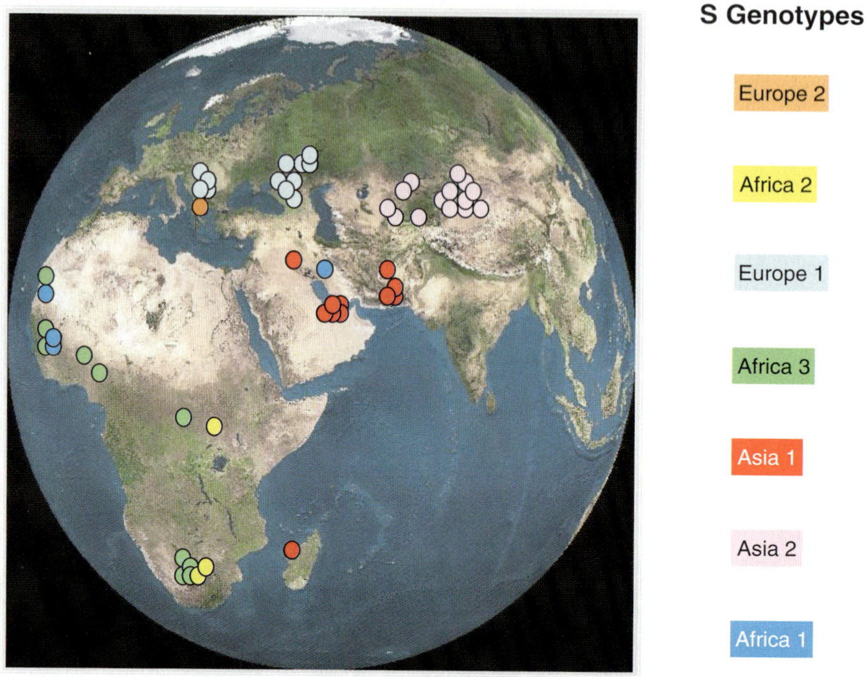

Fig. 5-2. Geographical correlation of genotypes. When superimposed onto the globe, the phylogenetic grouping of S RNA subtypes illustrates that the pattern of genetic diversity observed is largely related to the geographical distribution of the viruses. On some occasions, however, similar subtypes are sometimes found in distant geographical locations. It is possible that trade in livestock and perhaps long-distance carriage of virus or infected ticks during bird migration may have brought about links between such locations.

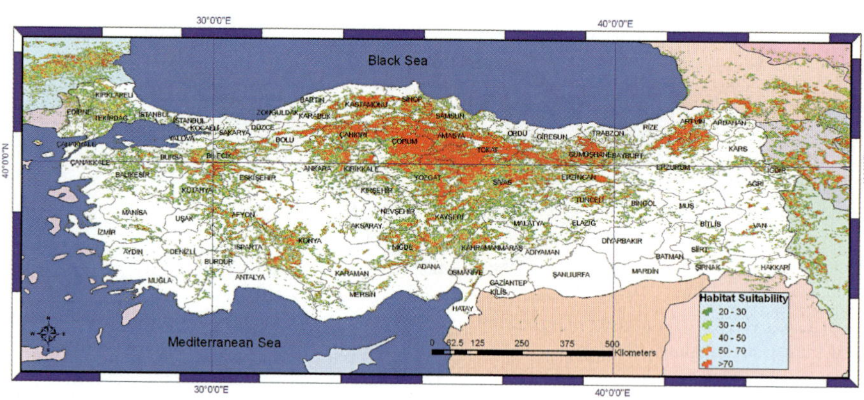

Fig. 6-8. The distribution of habitat (climate) positive suitability for the tick *Hyalomma marginatum marginatum* in Turkey, as calculated by the MaxEnt algorithm.

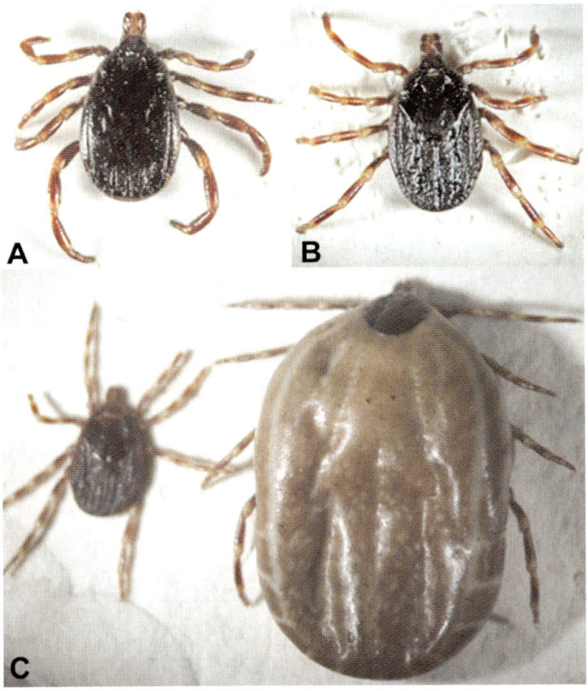

Fig. 12-1. Hyalomma spp. ticks, as illustrated here by *Hyalomma marginatum marginatum*, are the principal vectors of CCHFV. (A) Adult male, (B) adult female, (C) unfed and engorged adult females.

A

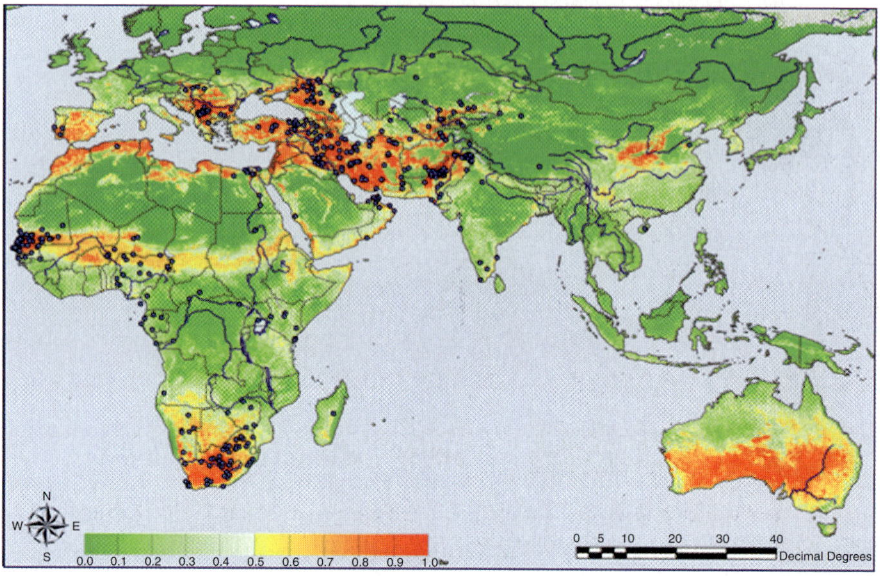

B

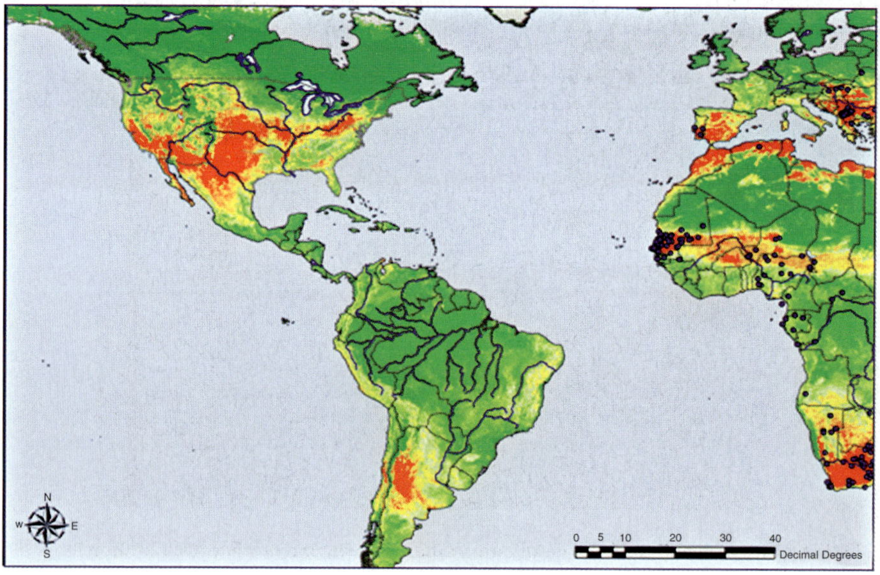

Fig. 14-4. Predictive risk map for CCHF (A) Old World and (B) New World. The data source and methods of creation are summarily described in the text (further methodological details in [12]). The mean posterior probabilities of environmental suitability for CCHF from 100 bootstrapped models are here shown on a color scale from green (low probability) to red (high probability). The original literature records of CCHF presence, on which the models were based, are shown as blue points on the map(s).

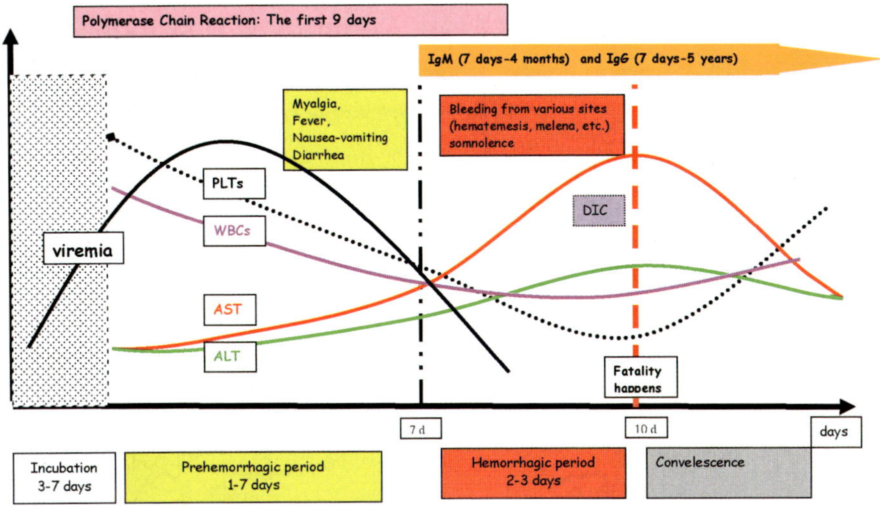

Fig. 16-1. The infection course. The starting point is the entrance of the CCHFV to the human through tick bite or a contact with infected material such as body fluids.

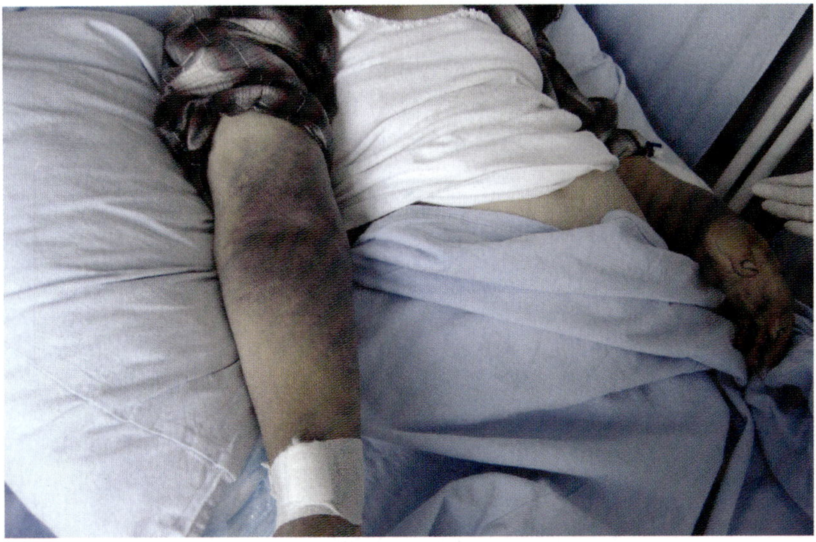

Fig. 16-2. CCHF patients, archives of Ankara Numune Education and Research Hospital.

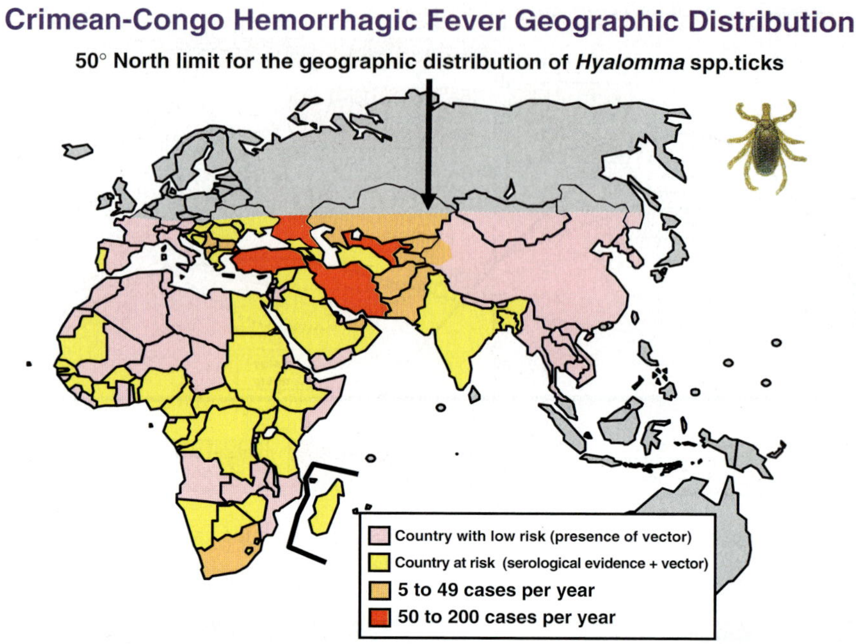

Fig. 22-1. Map showing the geographic distribution of CCHF, including the 50° north latitude limit for the geographic distribution of *Hyalomma* spp. ticks.